# SystemC and SystemC-AMS in Practice

Amal Banerjee · Balmiki Sur

# SystemC and SystemC-AMS in Practice

## SystemC 2.3, 2.2 and SystemC-AMS 1.0

Springer

Amal Banerjee
Kolkata
India

Balmiki Sur
IC Manage
Sunnyvale
CA, USA

ISBN 978-3-319-01146-2 ISBN 978-3-319-01147-9 (eBook)
DOI 10.1007/978-3-319-01147-9
Springer Cham Heidelberg New York Dordrecht London

Library of Congress Control Number: 2013942652



Printed on acid-free paper

Springer is part of Springer Science+Business Media (www.springer.com)

*To*
*My late father Sivadas Banerjee*
*My mother Meera Banerjee*
*My sister Anuradha Datta*
*And a dear friend, mentor and guide*
*Dr. Andreas Gerstlauer*

—Amal Banerjee

*To my children Debnil and Dipashreya,*
*my wife Sumana, my parents*
*And Dr. William J. B. Oldham,*
*Professor of Computer Science, Texas*
*Tech University*

—Balmiki Sur

# Contents

# Chapter 1
# Introduction to SystemC

**Abstract** We introduce SystemC [1] and provide a brief overview of its structure and main underlying concepts as transaction-level modeling (TLM).

## 1.1 Introduction

SystemC [1] is a C++ class library built on top of ANSI C ++ that allows a designer to create cycle-accurate models of system-level designs including software algorithms, hardware architectures, and their interfaces—transforming standard C/C++ to a system descriptor language (SDL). Designers create an *executable specification*, which is a C++ program that exhibits the same behavior as the system being studied, enabling its simulation, validation, optimization, and easy exploration of design space. SystemC provides three crucial constructs that are missing in standard ANSI C/C++. That is time, concurrency, and reactive behavior, essential for accurate hardware modeling and analysis. Some advantages of using an executable specification are as follows:

- Avoids inconsistencies and errors and ensures completeness of system specification.
- Allows unambiguous interpretation of the system.
- Validates system functionality and performance and enables creation of early performance models of the system under study.
- Using a test harness or test bench with a given executable specification that allows quick, accurate testing and verification of model under test.

SystemC [1] prevents manual translation of a C/C++ model into the Verilog/VHDL RTL model, thereby curtailing translation errors. A SystemC description of a system is a set of processing elements (*PE*)/modules that exchange data among each other via bi/unidirectional channels. The library provides a number of built-in communication channel classes (sc_buffer, sc_fifo, etc.,), and in addition, the user can design one's own custom channel.

A concept closely related to SystemC's [1] core execution model is the transaction-level modeling. TLM is a high-level approach to modeling digital systems that

A. Banerjee and B. Sur, *SystemC and SystemC-AMS in Practice*,
DOI: 10.1007/978-3-319-01147-9_1,

allows strict separation of details of communication among processing elements from the details of the implementation of functional units of the communication architecture. Communication mechanisms such as buses or FIFOs are modeled as channels and are presented to modules using SystemC interface classes. Transaction requests take place by calling interface functions of these channel models, which encapsulate low-level details of the information exchange. At the transaction level, the emphasis is on what data are transferred to and from which processing element—and less on the actual protocol used for data transfer. This approach makes it easier for the system-level designer to experiment, e.g., with different bus architectures (all supporting a common abstract interface) without recoding models that interact with any of the buses, provided these models interact with the bus through the common interface.

In combination, SystemC [1] and TLM are versatile and efficient toolsets to model, analyze, and validate/verify digital systems.

## Reference

1. IEEE Standards Board, IEEE Standards Association Standards Board IEEE-SA—IEEE Get Program. IEEE, 2011 *IEEE Standard 1666 Open SystemC Language Reference Manual* (*LRM*) http://standards.ieee.org/getieee/1666/download/1666-2011.pdf

# Chapter 2
# Downloading, Configuring, Installing, and Starting with SystemC

**Abstract** Having introduced the SystemC [1] library in the previous chapter, we discuss in detail how it might be downloaded, compiled, and installed on one's computer. As the SystemC [1] library was originally developed and still very widely used on computers running the Linux and Linux-like operating systems, the emphasis is on installing on Linux machines. It can also be installed and run on computers running Microsoft Windows operating system as well, though the steps are much more complicated.

## 2.1 Downloading and Installing SystemC

SystemC [1] is an ANSI C/C++ extension library originally developed and runs best on computers using Linux/Unix-like operating systems. The source file tarballs may be downloaded from the official Open SystemC Initiative (OSCI) Web site (http://www.accellera.org), decompressed, configured, compiled (with the familiar gcc C/C++ compiler or the Solaris compiler), installed, and verified for correct compilation and installation. It can be installed on a computer running the Windows operating system, following one of two methods. The straightforward method is to create a Linux/Unix-like environment by installing the popular Cygwin/MingW packages and then installing/running SystemC through it. The slightly more complicated way is to compile and run SystemC directly with Microsoft Visual C++. All code examples in this book have been compiled and tested on a machine running Red Hat Enterprise Linux 4.0 (GCC compiler version 3.2.6) and a computer running Red Hat Fedora 14 and Fedora 15 (GCC compiler version 4.5.1). All design examples in this book have been compiled and tested to run both under older stable SystemC 2.2.0 release and under the brand new stable SystemC 2.3 release. A very important distinction between SystemC 2.2.0 and 2.3.0 is that the latter runs as a shared library, and so either uses the runtime environment variable LD_LIBRARY_PATH or adds the linker flags at compile time, to the compile script. In addition, if security-enabled Linux (SELinux) is enabled,

A. Banerjee and B. Sur, *SystemC and SystemC-AMS in Practice*,
DOI: 10.1007/978-3-319-01147-9_2,

it can be disabled (temporarily using 'setenforce' or permanently using 'execstack' Linux system commands) easily. These details will be enumerated when design examples are discussed.

Compilation/installation of SystemC [1] on any computer running the Linux/Unix-like operating system is achieved by executing the familiar sequence of commands (*configure, gmake, gmake install,* and *gmake check*) from the command line. The following is an excerpt from the INSTALL file in the SystemC [1] source directory. Each design example in this book has been verified to run with the recent stable SystemC release 2.3.0 and previous stable release 2.2.0.

INSTALL NOTES FOR SystemC release 2.3.0

Contents:

1. Installation notes for Unix
2. Installation notes for Windows.

1. Installation notes for Unix
System Requirements
SystemC can be installed on the following UNIX or UNIX-like platforms:

- 32-bit Linux (x86) (Red Hat Enterprise Linux 4, 5, 6; Fedora 14, 15; Debian 5.0; Ubuntu 10.04LTS, 12.04LTS) with GNU C++ compiler versions gcc-3.3.2 through gcc-4.7.0
- 32-bit Linux (x86) (Debian, Ubuntu 12.04LTS) with Clang C++ compiler version clang-2.9 through clang-3.1
- 64-bit Linux (x86_64) (Red Hat Enterprise Linux 4, 5, 6; Fedora 17; Debian 5.0; Ubuntu 10.04LTS, 12.04LTS) with GNU C++ compiler versions gcc-3.4.5 through gcc-4.7.0
- 64-bit Linux (x86_64) (Ubuntu 12.04LTS) with Clang C++ compiler.

INSTALL NOTES FOR SystemC release 2.2.0

Contents:

1. Installation notes for Unix
2. Installation notes for Windows.

1. Installation notes for Unix
System Requirements
SystemC can be installed on the following UNIX platforms:

- Sun Solaris 2.8 with GNU C++ compiler versions gcc-2.95.3 and gcc-3.2.3.

The 'gcc/g++' is a common C/C++ compiler suite on Linux or Unix-like operating systems. All examples have been compiled/tested with both g++ 4.5.1 and 4.7.1, and SystemC versions 2.2.0 and 2.3.0, running on Fedora 14, 15 and Red Hat Enterprise Linux 4. There is a difference between how SystemC code is executed under SystemC 2.2.0 and SystemC 2.3.0. Note that any SystemC [1] code may be compiled easily with a command as:

*g++ -I. -I/<absolute or relative path to SystemC directory>/include -L. -L/<absolute or relative path to SystemC directory>/lib-linux –o <name of executable> <executable source file name>.cc -lsystemc -lm.*

Note that it is very common to name the executable file as 'sim'. There is no restriction however, and it may be named anything. A Makefile may be used.

Executing under SystemC 2.2.0 is easy from command line, just by typing in ./<executable file name> at the shell command prompt. Just like any other C/C++ program, any number of command line arguments may be passed into the executable.

Execution under SystemC 2.3.0 is different, because now SystemC is run as a shared library, and the user has to set and use the LD_LIBRARY_PATH environment variable. The following steps show the details, for a simple example named test.cc. We start with compiling:

g++ -I. -I/< absolute or relative path to SystemC directory>/include -L. -L/<absolute or relative path to SystemC directory>/lib-linux -o sim test.cc -lsystemc -lm.

Setting the LD_LIBRARY_PATH environment variable(*bash* shell):

*export LD_LIBRARY_PATH=<absolute or relative path to SystemC directory>/lib-linux.*

On the 'cshrc' or 'tcshrc' shells, use the 'setenv' command (note differences with bash shell):

*setenv LD_LIBRARY_PATH<absolute or relative path to SystemC directory>/lib-linux.*

Now execute with ./<executable file name>.

Sometimes, security-enhanced Linux (SELinux) is enabled by default. Obviously then, it has to be disabled temporarily for the SystemC 2.3.0 session. It may be permanently disabled as well, but is risky. Either way, system administrator or root access is required. Once the system administrator has logged in, SELinux may be disabled with the Linux system command '*setenforce*'—'setenforce 0' disables SELinux security temporarily, while 'setenforce 1' enables all SELinux security features. SELinux may be disabled permanently with the Linux system command '*execstack*'. All of this apply to SystemC 2.3.0 only.

## Reference

1. IEEE Standards Board, IEEE Standards Association Standards Board IEEE-SA — IEEE Get Program. IEEE, 2011 *IEEE Standard 1666 Open SystemC Language Reference Manual (LRM)* http://standards.ieee.org/getieee/1666/download/1666-2011.pdf

# Chapter 3
# SystemC Simulation Kernel, Data Types, Communication Primitives Concurrency Control, and Main Language Constructs

**Abstract** To effectively use SystemC [1], one must be familiar with some of its theoretical concepts/underpinnings and terminology. First, some general SystemC concepts are examined.

## 3.1 SystemC Language Details

We start with core SystemC [1] concepts, derived from standard ANSI C++.

| | |
|---|---|
| Module | Container class, hierarchical, can contain other modules—SystemC 'sc_module' |
| Process | The core functionality of any modules is contained in its processes, which are C++ methods. A module can have any number of three possible process types |
| Port | A module sends/receives data to/from other modules via ports—could be (uni)/(bi)directional—SystemC 'sc_port' or 'sc_export' |
| Signal | Could be resolved or unresolved—a resolved signal may have multiple drivers (bus), whereas an unresolved signal has a single driver. Single-, two-/four-valued signals are allowed. A four-valued signal is a logic signal with allowed values of 'true,' ' false,' 'don't care,' and 'high impedance' |
| Data types | Could be fixed precision (for fast simulations) or arbitrary precision. Fixed precision data types are two and four valued only, with no restriction on arbitrary precision data types |
| Clock | Special signals—multiple clocks with arbitrary phase relationships are allowed |
| Cycle-based simulation | Very lightweight cycle-based simulation kernel for very fast simulations |
| Multiple abstraction levels | Untimed models at different levels of abstraction are available—from high-level functional models to cycle-accurate RTL models. High-level models may be iteratively refined to detailed lower-level models |

A. Banerjee and B. Sur, *SystemC and SystemC-AMS in Practice*,
DOI: 10.1007/978-3-319-01147-9_3,

| | |
|---|---|
| Communication protocols | Multilevel communication semantics allow both system on chip (SoC) and system-level I/O at different abstraction levels |
| Debugging support | Runtime error checking may be enabled at compile time |
| Tracing | Common output file formats as VCD, ISDB are allowed |
| Sensitivity list | The sensitivity of a process is the set of events or timeouts which trigger that process. A process is sensitive to an event if that event has been added to its static sensitivity list or dynamic sensitivity of the process instance. *Static sensitivity* of an unspawned process is fixed during elaboration, while that of a spawned process is set when *sc_spawn* is invoked. A timeout occurs when a given time interval has elapsed |

The execution of a SystemC [1] application consists of *elaboration*, followed by *simulation*.

- Elaboration is the crucial first step, consisting of creation of the application's module hierarchy (including module primitive channels and processes), associated data structures, binding of ports and exports, execution of shell of public implementation, and the private kernel of the implementation.
- Simulation consists of execution of the scheduler part of the kernel which in turn invokes the processes of the application.

SystemC [1] provides a reactive, event-driven simulation infrastructure that supports two types of processes—*spawned* and *unspawned*. An unspawned process instance is created by invoking one of three process macros—SC_CTHREAD, SC_METHOD, and SC_THREAD—and are most widely used. Each of these is elaborated, unlike spawned processes, allowing efficient resource allocation at compile time. These three macros may be invoked from the constructor of a module, with an appropriate *sensitivity* list. A sensitivity list is tied to the core concept of an event-driven simulator—a process reacts/responds to an event, e.g., change in value of a signal. Thus, the SC_CTHREAD macro requires the rising or falling edge of any clock. The SC_THREAD macro is more flexible and robust, as it allows a general sensitivity list that may include a clock, allowing the corresponding process to respond to change in value at all ports of its container module specified in its sensitivity list. That is, if process *p* of module *M* is declared using the *SC_THREAD* macro in the constructor of *M* to be sensitized to ports *portA* and *portB* of *M*, then *p* will respond to any changes in values being read in through *portA* or *portB* or both. A SC_METHOD macro is executed each time its container module is activated and cannot be suspended with a *wait* statement (unlike SC_CTHREAD and SC_THREAD). On the contrary, both SC_CTHREAD and SC_THREAD macros are executed once. As a result, the code for any process declared as an SC_CTHREAD/SC_THREAD has an infinite loop in it, so that once invoked, the process waits for events in its sensitivity list and execution stops only when the container module is destroyed at the end of simulation. A spawned process may be created by invoking the built-in function *sc_spawn* during elaboration or simulation and may be used in *fork–join* parallel execution constructs. It must be noted that *fork–join* constructs exist strictly in the software realm and do not represent any physical hardware.

Another less-reliable technique for sensitizing processes to events is the *notify–wait* construct. Process A *notifies* on an event, and process B *waits* on that event. This scheme works for simple cases, but for complicated situations, with large number of notify–wait pairs among the processes of a module, either some events might be lost or a process might respond unexpectedly to event B, while actually waiting for event A.

Satisfying the core SystemC [1] (and TLM) execution model, ports are essential component of any SystemC module. A port may be bound to a channel, another port or export. An export may be bound to a channel, or an export, but never to a port. A port may be *bound* to a channel, but not vice versa.

A port/export may be bound by name/position, but never simultaneously by both. Relevant methods of built-in SystemC classes *sc_module*, *sc_port*, or *sc_export* may be used for the actual binding process. Port binding is flexible—port A may be bound to port B, which in turn may be bound to channel C, effectively binding port A to channel C. All port binding occurs only during elaboration, but may be deferred till the end of elaboration (e.g., port A is bound to port B, but port B is not yet bound). However, all exports have to be bound immediately. Additionally, a port may be bound to multiple channels/ports.

While all of the above-mentioned activities occur during elaboration, simulation primarily involves invoking SystemC's [1] event-driven simulator, which executes a SystemC application or module's processes in response to events (*sc_event*). Simulator time is incremented in integer steps and initialized at the start of simulation. The built-in function *sc_set_time_resolution* allows the user to set the internal simulation kernel time increment. Alternatively, a master clock may be declared, and modules and processes in each might synchronize with this. Also, a process might use a clock divider and then associated modules might synchronize with the slower clock.

The SystemC [1] scheduler can execute a process *only* if one of the following conditions is satisfied.

- Process instance has been made runnable during initialization.
- *sc_spawn* has been invoked during simulation.
- A process has been sensitized to an event and that event has occurred.
- A timeout has occurred.

The scheduler/scheduling algorithm uses four sets of runtime information.

- Set of runnable processes.
- Set of update requests.
- Set of delta notifications and timeouts.
- Set of timed notifications and timeouts.

To understand the scheduler, one must note some related constraints and definitions. The set of runnable processes can record a single instance of a process.

- An *update request* results only from a call to member function *request_update* of class *sc_prim_channel*.
- An *immediate notification* can only result from a call to member function *notify* (with no arguments) of class *sc_event*.

- A *delta notification* can only result from a call to member function *notify* (with argument 0) of class *sc_event*.
- A timed *notification* can only result from a call to member function *notify* (with non-zero argument) of class *sc_event*. The non-zero value denotes the time interval between invocation of *notify* and the actual notification message being sent.
- A timeout results only from invocation of *wait* or *next_trigger* member functions of *sc_module*, certain functions of *sc_prim_channel*, and non-member functions. A timeout with zero value results in a delta notification, and a timeout with non-zero value results in a timed notification.

Scheduler execution consists of the *initialization* phase, followed by *evaluation* phase. The initialization phase itself consists of sub-phases as *update* phase and *delta notification* phase.

The common SystemC term *delta cycle* consists of three sequential steps, exactly in that order.

- Evaluation phase.
- Update phase.
- Delta notification phase.

An application may be executed in two ways—application control and direct kernel control. If executed under direct kernel control, it cannot use *sc_main* or *sc_start*.

As an ANSI C++ compliant library, SystemC supports each of the standard C++ data types and in addition hardware-specific data types to support concurrency, the concept of time, and events. The following table lists these data types. To support hardware data representations, SystemC allows a unified string representation using C language-style strings.

*sc_string name* (*'0 base* [*sign*] *number e*[±] [*exp*]');
where *name* is the variable name, *base* is the base of the variable (b—binary, d—decimal, x—hexadecimal, and o—octal), *sign* is the sign of the variable (us—unsigned, sm—signed magnitude, and esd—canonical signed digit), and exp is the exponent (unsigned integer). Each of the standard C++ data types (int, long int, unsigned int, unsigned long int, unsigned short int, bool, char, double, float, unsigned char, and short) are supported.

The hardware-specific data types are listed in the following Table 3.1, as well as common operations on each.

Transaction-level modeling specifies that processing elements synchronize with, communicate, and exchange data with each other using channels. In low-level hardware modeling, signals allow communication between hardware components. SystemC [1], based on C++, and drawing upon the parent language constructs, achieves synchronization and communication via channels, interfaces, and ports.

- An interface declares/specifies a set of methods but do not implement any.
- A channel is an implementation of one/more interface.
- A module uses its ports to access a channel.

**Table 3.1** Common SystemC data types specific to hardware modeling

| Data type | Operations |
|---|---|
| sc_dt::sc_bit ('0', '1') | Operations—bitwise AND (&), bitwise OR (\|), bitwise XOR (^), bitwise NOT (~), AND assignment (& =), OR assignment (\| =), XOR assignment (^ =), equality (==), and inequality (! =) |
| sc_dt::sc_logic ('0', '1', 'X', and 'Z')—generalized form of sc_dt::sc_bit | Same operations as above, sc_logic and sc_bit values may be assigned to each other, but not a good practice. The AND of a '0' with an 'X' or 'Z' always gives a '0', and the AND of a '1' with an 'X' or 'Z' always gives an 'X'. The AND of an 'X' or 'Z' always gives an 'X'. OR of a '0' with an 'X' or 'Z' gives a '0', while the OR of a '1' with an 'X' or 'Z' always gives an 'X'. The NOT of an 'X' or 'Z' is always 'X' |
| sc_dt::sc_bv <n> arbitrary size bit data type, with the nth bit as the most significant bit | Supports all of the standard single bit data type operations as above, in addition to some operations specific to having a large number of bits. These are as follows: []—bit selection, concatenation (sc_dt::bv bv_1, sc_dt::bv bv_2), range (unsigned int index1, unsigned int index2), and_reduce(), or_reduce(), and xor_reduce(). Arithmetic operations may not be performed on bit vectors, but rather converted to sc_int (SystemC library method 'to_int()') or sc_uint and then reconverted to bit form once arithmetic operations are complete. *All bit vectors are indexed from the right* |
| sc_dt::sc_lv<n>—extension of sc_dt::sc_logic, just as sc_dt::sc_bv is an extension of sc_bit | All operations as applicable to sc_dt::sc_logic, as well operations similar to sc_dt::sc_bv <n> |
| sc_dt::sc_int—can be used interchangeably with C++ int data type | Bitwise AND, OR, NOT, and XOR (&, \|, ! and ^), standard decimal arithmetic addition, division, multiplication and subtraction, modulus ($) and arithmetic shift left(≪) and right (≫). Compound operations (+=, ==, /= ,* = . ^ = , ! = , ==, <=, >=, ++ and --). In addition, bit selection and range operations are available, but in such cases, a sc_dt::sc_bv is typically used |
| sc_dt::sc_uint—unsigned integer 64 bits in length | All operations as sc_dt::sc_int. May be used interchangeably with sc_int. When assigning an integer to an unsigned operand, the integer value in 2's complement form is interpreted as the unsigned number. During the reverse process, the unsigned integer is expanded to a 64-bit unsigned number and truncated to the unsigned value |
| sc_dt::sc_bigint, sc_dt::sc_ubigint are arbitrary long integer, signed, and unsigned. Maximum allowable size is 512 bits | Supports all operations as sc_dt::sc_int |

Any interface simply specifies a set of methods (and their signatures—argument list and return types) but does not provide any implementation of these methods. All SystemC [1] interfaces are derived, from the base class *sc_core::sc_interface.* A channel may implement one or more interfaces and is a container for intermodule (interprocess) communication. Different channels may implement one interface, and a channel could be primitive or hierarchical. A channel is *primitive* when it does not have any hierarchy or process and all primitive channels are derived from *sc_core::sc_prim_channel.* SystemC [1] provides four primitive channels—*sc_mutex, sc_semaphore, sc_signal*, and *sc_fifo*. A hierarchical channel is derived from *sc_core::sc_channel* and is a regular SystemC module.

A port is a SystemC [1] object that enables a module to access an interface's (correspondingly its channel's) methods. A port assumes an interface; otherwise, a channel cannot be bound to it. All ports are derived from sc_core::sc_port, and customized ports may be defined by refining sc_port or some predefined port.

Having explored the core concepts and associated ideas, we explain how these concepts and data types may be used to model real-world digital hardware. We start with simple combinational and sequential logic circuits and their combinations (combinational–combinational, combinational–sequential, sequential–sequential, and sequential–combinational) and then progress to real-world complete designs. The basis for our examples is some popular open source hardware designs as the Sun Microsystems' open source T2 microprocessor (http://www.opensparc.net) and some other open source designs from http://www.opencores.org. *It must be remembered that for the more complicated examples, with large number of input/output signals, the sample results obtained are a small subset of the possible set of results.*

While it is very important to set up the simulation correctly, it is equally (if not more) important to analyze/examine the output to ensure that the results match what is expected. SystemC [1] provides two built-in techniques to examine results—console text mode output and popular industry-standard VCD-formatted output that may be examined easily with any popular VCD trace file viewer as GTKwave. We have used both techniques, sometimes one over the other, and at other times, both together, for the same design example.

## Reference

1. IEEE Standards Board, IEEE Standards Association Standards Board IEEE-SA—IEEE Get Program. IEEE, 2011 *IEEE Standard 1666 Open SystemC Language Reference Manual (LRM)* http://standards.ieee.org/getieee/1666/download/1666-2011.pdf

# Chapter 4
# Primitive Channels: Concurrency Control

**Abstract** As SystemC [1] is based on transaction-level modeling, it must provide efficient mechanism for accurate control and execution of concurrent transactions. To this end, it offers a set of primitive channels for concurrency control. These are *sc_mutex*, *sc_semaphore*, *sc_signal*, and *sc_fifo*.

## 4.1 sc_signal and the Reader–Writer Model

The familiar reader–writer [2] or producer–consumer model [2] is examined to illustrate the use of the primitive sc_core::sc_signal channel. Figures 4.1, 4.2, and 4.3 contain the source code for the consumer classes, the producer classes, and the test harness.

```
#ifndef PRODCOM_H
#define PRODCOM_H

#include <systemc>

SC_MODULE(cons)
{
  /*Declare/define inout/output ports */
  sc_core::sc_in<bool> clk;
  sc_core::sc_in<unsigned int> in;
  unsigned int input;

  /* Consumer operation thread */
  void cons_proc0()
  {
```

**Fig. 4.1** Consumer class for producer–consumer model

A. Banerjee and B. Sur, *SystemC and SystemC-AMS in Practice*,
DOI: 10.1007/978-3-319-01147-9_4,

```
      while(1)
      { /*Synchronize with external clock */
        wait();
        input = in.read();
        std::cout<<name()<<" rec'd "<<input<<std::endl;
      }
    }

    /*Constructor */
    SC_CTOR(cons):input(0)
    {
    SC_CTHREAD(cons_proc0, clk.pos());
    /*declare/assign thread*/
    }
    /* Destructor */
    ~cons(){ }
  };
```

**Fig. 4.1** (continued)

```
SC_MODULE(prod)
{
  /*Declare/define inout/output ports */
  sc_core::sc_in<bool> clk;
  sc_core::sc_out<unsigned int> out;
  unsigned int output;

  /* Producer operation thread */
  void prod_proc0()
  {
    while(1)
    { /*Synchronize with external clock */
      wait();
      output = ((unsigned int)(10.0*drand48()));
      out.write(output);
      std::cout<<name()<<" sent "<<output<<std::endl;
    }
  }
```

**Fig. 4.2** Producer class for producer–consumer model

```
  /*Constructor */
  SC_CTOR(prod):output(0)
  {
    SC_CTHREAD(prod_proc0, clk.pos()); /*declare/assign thread*/
  }
  /* Destructor */
  ~prod(){ }
};
#endif
```

**Fig. 4.2** (continued)

```
#include "prodcom.h"

int sc_main(int argc, char **argv)
{
  /* Declare/define primitive channel --sc_signal */
  sc_core::sc_signal<unsigned int> sig;
  /* Declare/define master clock -- period 10 nanosecond, duty
 cycle 50.0% */
  sc_core::sc_clock clk("clock", 10.0, sc_core::SC_NS, 0.5);

  /* Declare consumer and producer modules */
  prod pr_od("pr_od");
  cons co_ns("co_ns");
  /* Connect module ports and channel */
  pr_od.clk(clk);
  pr_od.out(sig);
  co_ns.clk(clk);
  co_ns.in(sig);
  /*Run simulation for pre-defined time interval and stop */
  sc_core::sc_start(500.0, sc_core::SC_NS);
  sc_core::sc_stop();
  return 0;
}
```

**Fig. 4.3** Test harness for producer–consumer model

SystemC 2.2.0 output:

```
SystemC 2.2.0 --- Aug 30 2011 19:39:07
Copyright (c) 1996-2006 by all Contributors
ALL RIGHTS RESERVED
pr_od sent 0
co_ns rec'd 0
.....
.....
pr_od sent 0
co_ns rec'd 5
.....
.....
pr_od sent 2
co_ns rec'd 0
.....
.....
pr_od sent 6
co_ns rec'd 1
pr_od sent 7
.....
.....
pr_od sent 4
co_ns rec'd 1
pr_od sent 5
co_ns rec'd 4
SystemC: simulation stopped by user.
```

SystemC 2.3.0 output:

```
SystemC 2.3.0-ASI --- Jul 19 2012 18:53:11
Copyright (c) 1996-2012 by all Contributors,
ALL RIGHTS RESERVED
pr_od sent 0
co_ns rec'd 0
.....
.....
co_ns rec'd 7
pr_od sent 4
co_ns rec'd 1
pr_od sent 5
co_ns rec'd 4
Info: /OSCI/SystemC: Simulation stopped by user.
```

## 4.2 sc_mutex and Concurrency Control

The sc_mutex class is a predefined primitive channel designed to model the behavior of mutual–exclusion lock [2] used to control access to a shared resource, by concurrent processes. A mutex can be in *locked* or *unlocked* state. Only one process can lock a mutex at a time and only that process can unlock it. Once unlocked, any other process can lock the same mutex. The mutex class supplies a set of methods for efficient mutex use, which are:

*int lock(): Lock mutex if free*
*int unlock(): Unlock mutex*
*int trylock(): Check if mutex is unlocked, if so lock it or return −1*
*char *kind() : Return pointer to string 'sc_mutex'*

Figures 4.4 and 4.5 contain a sc_mutex example and its test harness.

**Fig. 4.4** Simple example showing how a SystemC mutex is used to access a shared resource

```
#ifndef MUTEXEX_H
#define MUTEXEX_H

#include <systemc>
#include <cstring>

SC_MODULE(mutexexample)
{
  private:
   /* Shared 'resource' */
   void mutextask(char *cp)
   {
     std::cout<<"mutex locked by "<<cp<<std::endl;
   }
  public:
  /* Declare/define input/output ports */
  sc_core::sc_in<bool> clk;
  /* Declare/define mutex */
  sc_core::sc_mutex mut_ex;

 void proc_A()
 {
   while(1)
   {
     wait();
     if(mut_ex.trylock() != -1)
     /* Check if mutex is locked, if not lock it */
```

Fig. 4.4 (continued)

```
        {
          mutextask("A");
          mut_ex.unlock();
        }
      }
    }
    void proc_B()
    {
      while(1)
      {
        wait();
        if(mut_ex.trylock() != −1)
        /* Check if mutex is locked, if not lock it */
        {
          mutextask("B");
          mut_ex.unlock();
        }
      }
    }

    /* Constructor */
    SC_CTOR(mutexexample)
    {
      /* Declare/assign threads */
      SC_CTHREAD(proc_A, clk.pos());
      SC_CTHREAD(proc_B, clk.pos());
    }

    /& Destructor */
    ~mutexexample(){ }
  };
  #endif
```

```
#include "mutexex.h"

int sc_main(int argc, char **argv)
{
  /* Declare/define master clock -- period 10.0 nanosecond
  dutv cvcle 50% */
```

Fig. 4.5 Test harness for simple SystemC mutex example

```
    sc_core::sc_clock clk("clk", 10.0, sc_core::SC_NS, 0.5);
    mutexexample mutex_ex("mutex_ex");
    /* Connect module port and channel */
    mutex_ex.clk(clk);
    /* Run simulation for pre-defined time interval and stop */
    sc_core::sc_start(500.0, sc_core::SC_NS);
    sc_core::sc_stop();
    return 0;
}
```

**Fig. 4.5** (continued)

SystemC 2.2.0 output:

```
        SystemC 2.2.0 --- Aug 30 2011 19:39:07
    Copyright (c) 1996-2006 by all Contributors
              ALL RIGHTS RESERVED
mutex locked by B
mutex locked by A
mutex locked by B
mutex locked by A
mutex locked by B
.....
.....
mutex locked by A
mutex locked by B
mutex locked by A
mutex locked by B
mutex locked by A
......
......
motex locked by B
mutex locked by A
SystemC: simulation stopped by user.
```

SystemC 2.3.0 output:

```
        SystemC 2.3.0-ASI --- Jul 19 2012 18:53:11
    Copyright (c) 1996-2012 by all Contributors,
    ALL RIGHTS RESERVED
mutex locked by B
mutex locked by A
.....
.....
mutex locked by B
mutex locked by A
Info: /OSCI/SystemC: Simulation stopped by user.
```

## 4.3 sc_semaphore and Concurrency Control

Like the mutex [2], the semaphore [2] is another concurrency control mechanism. SystemC's *sc_semaphore* class is designed to model the behavior of a software semaphore used to provide limited concurrency control for processes trying to access a shared resource. A semaphore has an integer value that is set to the permitted number of concurrent accesses when the semaphore is constructed. Some important library methods of this class are:

*int wait(): If the semaphore value is 0, 'wait' shall suspend until the semaphore value is incremented (by another process), when it will resume and try to decrement the semaphore value.*

*Int trywait(): If the semaphore value is 0, 'trywait' returns −1, without modifying the semaphore.*

*int post(): Increments semaphore value. If there are any suspended processes waiting for the semaphore value to be incremented, one of these processes is chosen to decrement the semaphore value, while the other processes remain suspended. The choice of the activated process is non-deterministic.*

*Int get_value(): Returns semaphore value.*

*char *kind(): Returns string. 'sc_semaphore'*

Figures 4.6 and 4.7 contain the source code and test harness for a simple semaphore.

```
#ifndef SEMAPHOREEX_H
#define SEMAPHOREEX_H
#include <systemc>
#include <cstring>

const unsigned int MAX = 5;
SC_MODULE(semaphoreexample)
{
 /*Declare/define input/output ports */
 sc_core::sc_in<bool> clk;
 /* Declare/define semaphore */
 sc_core::sc_semaphore sema_phore;
 unsigned int count;

 void sema_proc0()
 {
  while(1)
  {
   wait(); /* Wait for event - clock positive edge */
   if(sema_phore.get_value() == 0)
  {
    if(count != MAX)
    {
     sema_phore.post();
     sema_phore.post();
     std::cout<<name()<<" post to semaphore "<<std::endl;
    }
   }
  }
 }

 void sema_proc1()
 {
  while(1)
  {
   wait();
   /* Wait for event - clock positive edge */
   if(sema_phore.trywait() != -1)
   {
```

Fig. 4.6 Simple semaphore example

```
        wait(3);
        /* Wait for event - 3 clock positive edge */
        std::cout<<name()<<" process 1 wait ... "<<std::endl;
      }
    }
  }
  void sema_proc2()
  {
    while(1)
    {
      wait();
     /* Wait for event - clock positive edge */
      sema_phore.wait();
      wait(3);
     /* Wait for event - 3 clock positive edge */
      std::cout<<name()<<" process 1 wait ..."<<std::endl;
    }
  }

  /* Constructor */
  SC_CTOR(semaphoreexample):sema_phore("sema_phore",MAX)

  {
    /* Declare/assign threads */
    SC_CTHREAD(sema_proc0, clk.pos());
    SC_CTHREAD(sema_proc1, clk.pos());
    SC_CTHREAD(sema_proc2, clk.pos());
  }

  /* Destructor */
  ~semaphoreexample(){ }
};
#endif
```

**Fig. 4.6** (continued)

```
#include "semaphoreex.h"

int sc_main(int argc, char **argv)
{
  /* Declare/define master clock period 10.0 nanosecond
     duty cycle 50% */
  sc_core::sc_clock clk("clk", 10.0, sc_core::SC_NS, 0.5);
  /* declare/define semaphore */

  semaphoreexample sema_ex("sema_ex");
  /* Connect clock signal to module port */
  sema_ex.clk(clk);
  /* Run simulation for pre-defined time interval and stop */
  sc_core::sc_start(400.0, sc_core::SC_NS);
  sc_core::sc_stop();
  return 0;
}
```

**Fig. 4.7** Test harness for simple semaphore

SystemC 2.2.0 output:

```
          SystemC 2.2.0 --- Aug 30 2011 19:39:07
     Copyright (c) 1996-2006 by all Contributors
                 ALL RIGHTS RESERVED
sema_ex process 1 wait ...
sema_ex process 1 wait ...
sema_ex process 1 wait ...
sema_ex process 1 wait ...
sema_ex post to semaphore
sema_ex process 1 wait ...
sema_ex process 1 wait ...
......
......
sema_ex post to semaphore
sema_ex process 1 wait ...
sema_ex process 1 wait ...
sema_ex post to semaphore
......
......
sema_ex post to semaphore
```

```
sema_ex process 1 wait ...
sema_ex process 1 wait ...
sema_ex post to semaphore
SystemC: simulation stopped by user.
```

SystemC 2.3.0 output:

```
        SystemC 2.3.0-ASI --- Jul 19 2012 18:53:11
    Copyright (c) 1996-2012 by all Contributors,
    ALL RIGHTS RESERVED
sema_ex process 1 wait ...
sema_ex process 1 wait ...
sema_ex process 1 wait ...
sema_ex process 1 wait ...
sema_ex post to semaphore
.....
.....
sema_ex process 1 wait ...
sema_ex process 1 wait ...
sema_ex post to semaphore
Info: /OSCI/SystemC: Simulation stopped by user.
```

## References

1. IEEE Standards Board, IEEE Standards Association Standards Board IEEE-SA — IEEE Get Program. IEEE, 2011 *IEEE Standard 1666 Open SystemC Language Reference Manual (LRM)* http://standards.ieee.org/getieee/1666/download/1666-2011.pdf
2. Silberschatz A., Galvin P. B., Gagne G. (2006) *Operating Systems, Seventh edition*, John Wiley & Sons

# Chapter 5
# Modeling Combinational Logic Circuits, Implicit Events, Primitive Communication Channels and Combinations

**Abstract** Digital hardware is classified as *combinational* or *sequential*. A combinational circuit is asynchronous and responds to changes in its input, as compared to sequential circuits that can only function synchronized with an external clock signal. A number of design examples in this chapter are based on the OpenSparc T2 [1] microprocessor RTL.

## 5.1 1-Bit Input–1-Bit Output Inverter

We start with a simple inverter. The OpenSparc T2 [1] processor uses inverters with different sized input/output data bit vectors. An inverter, as the name suggests, inverts a bit—the output is logic '1' for logic '0' input and vice versa. The source code for a 1-bit input and 1-bit output inverter is shown in Fig. 5.1.

Figure 5.2 shows the corresponding test harness for this inverter and Fig. 5.3 shows typical input and output.

When executed from the standard Linux command prompt/shell, the console output is listed below, for both SystemC 2.2.0 and SystemC 2.3.0. An easier alternative, as shown in the test harness code above, is to generate VCD trace files, and use an efficient VCD trace file viewer as GTKWave (Fig. 5.3 shows VCD output for the inverter with SystemC 2.3.0). In both the console outputs below, the first column is the simulation kernel time stamp in seconds, the second column the inverter input, and the third the inverter output.

```
SC_MODULE(cl_a1_clk_inv_16x)
{
  /* Input/output ports */
  sc_core::sc_in<bool> in;
  sc_core::sc_out<bool> out;
  bool LIB;
```

Fig. 5.1 1-bit input/output inverter, LIB is an OpenSparc T2 microprocessor environment variable, set here to value 'true' at a start of simulation

A. Banerjee and B. Sur, *SystemC and SystemC-AMS in Practice*,
DOI: 10.1007/978-3-319-01147-9_5,

```
    /*Thread process - runs in infinite loop */
    void a1_clk_inv_16x_proc0()
    {
      while(1)
      {
        wait();
        if(LIB == true) out.write(!(in.read()));
      }
    }
    /* Constructor */
    SC_CTOR(cl_a1_clk_inv_16x)
    {
      SC_THREAD(a1_clk_inv_16x_proc0); /* Declare/assign thread */
      sensitive << in; /*Sensitivity list for thread*/
    }
    /* Destructor */
    ~cl_a1_clk_inv_16x(){ }
};
```

**Fig. 5.1** (continued)

```
#include "n2_cl_a1.h"
#include <cstdlib>
#include <cstring>

int sc_main(int argc, char **argv)
{
  /* Combinational logic - NO clocks */
  /* Input/output channels */
  sc_core::sc_signal<bool> sigin0;
  sc_core::sc_signal<bool> sigout0;
  /* Trace file pointer -- for VCD format output */
  sc_core::sc_trace_file *fp;

  /*Declare/define inverter and VCD trace file */
  cl_a1_clk_inv_16x inv_1_bit("inv_1_bit");
  fp = sc_create_vcd_trace_file("tr_inv_1_bit");
  fp->set_time_unit(1.0, sc_core::SC_NS);
```

**Fig. 5.2** Inverter test harness

```
 /* Connect module ports and channels */
 inv_1_bit.in(sigin0);
 inv_1_bit.out(sigout0);
 inv_1_bit.LIB = true; /*Set environmanr variable */

 sigin0.write(true);
 sc_trace(fp, sigin0, "inv_in");
 sc_trace(fp, sigout0, "inv_out");
 /* Run a single  simulation step */
 sc_core::sc_start(2.0, sc_core::SC_NS);
 /* Print time stamp and signal
   values for console display */

std::cout<<sc_core::sc_time_stamp().to_seconds()<<"¥t"<<sigin0.read()<<"¥t"<<
sigout0.read()<<std::endl;

 /* Change input and run another simulation step */
 sigin0.write(false);
 sc_core::sc_start(2.0, sc_core::SC_NS);

std::cout<<sc_core::sc_time_stamp().to_seconds()<<"¥t"<<sigin0.read()<<"¥t"<<
sigout0.read()<<std::endl;

 /* Change input and run another simulation step */
 sigin0.write(true);
 sc_core::sc_start(2.0, sc_core::SC_NS);

std::cout<<sc_core::sc_time_stamp().to_seconds()<<"¥t"<<sigin0.read()<<"¥t"<<
sigout0.read()<<std::endl;

  /* Change input and run another simulation step */
 sigin0.write(false);
 sc_core::sc_start(2.0, sc_core::SC_NS);

std::cout<<sc_core::sc_time_stamp().to_seconds()<<"¥t"<<sigin0.read()<<"¥t"<<
sigout0.read()<<std::endl;

 /* Change input and run another simulation step */
 sigin0.write(true);
 sc_core::sc_start(2.0, sc_core::SC_NS);
```

**Fig. 5.2** (continued)

```
std::cout<<sc_core::sc_time_stamp().to_seconds()<<"¥t"<<sigin0.read()<<"¥t"<<
sigout0.read()<<std::endl;

  /* Change input and run another simulation step */
  sigin0.write(false);
  sc_core::sc_start(2.0, sc_core::SC_NS);

std::cout<<sc_core::sc_time_stamp().to_seconds()<<"¥t"<<sigin0.read()<<"¥t"<<
sigout0.read()<<std::endl;

  /* Change input and run another simulation step */
  sigin0.write(true);
  sc_core::sc_start(2.0, sc_core::SC_NS);

std::cout<<sc_core::sc_time_stamp().to_seconds()<<"¥t"<<sigin0.read()<<"¥t"<<
sigout0.read()<<std::endl;

  /* Stop simulation and close VCD trace file */
  sc_core::sc_stop();
  sc_close_vcd_trace_file(fp);
  return 0;
}
```

**Fig. 5.2** (continued)

```
SystemC 2.2.0—Aug 30 2011 19:39:07
Copyright (c) 1996-2006 by all Contributors
ALL RIGHTS RESERVED
Note: VCD trace timescale unit is set by user to 1.000000e−09 s.
2e−09 1 0
4e−09 0 1
6e−09 1 0
8e−09 0 1
1e−08 1 0
1.2e−08 0 1
1.4e−08 1 0
SystemC: simulation stopped by user.
SystemC 2.3.0-ASI—Jul 19 2012 18:53:11
Copyright (c) 1996-2012 by all Contributors,
ALL RIGHTS RESERVED
Note: VCD trace timescale unit is set by user to 1.000000e−09 s.
2e−09 1 0
4e−09 0 1
```

6e−09 1 0
8e−09 0 1
1e−08 1 0
1.2e−08 0 1
1.4e−08 1 0
Info: /OSCI/SystemC: Simulation stopped by user.

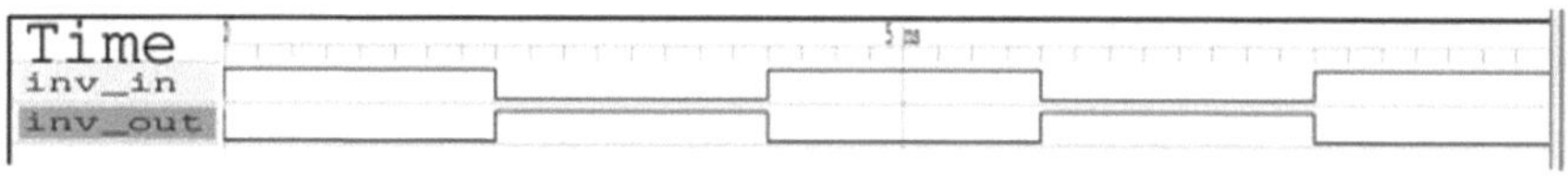


**Fig. 5.3** Inverter input output in VCD file format generated with SystemC-2.3.0

## 5.2 4 1-Bit Input NAND Gate

The operation of a 4 1-bit input NAND gate is examined next. A NAND logic gate accepts four input bits, performs the bitwise AND operation on them and inverts the result bit, to produce the final output bit. The source code is presented in Fig. 5.4 and test harness in Fig. 5.5.

In both the SystemC 2.2.0 and 2.3.0 console outputs below, the first column is the system time in seconds, followed by the 4-input bits, and the last column represents the output bit. An easier alternative, as shown in the test harness source

```
SC_MODULE(cl_u1_nand4_8x)
{
  /* Declare/define input/output ports */
  sc_core::sc_in<bool> in0;
  sc_core::sc_in<bool> in1;
  sc_core::sc_in<bool> in2;
  sc_core::sc_in<bool> in3;
  sc_core::sc_out<bool> out;
  /* Declare module variables */
  bool LIB;
  bool b0;
  bool b1;
  bool b2;
  bool b3;
  bool b4;
```

**Fig. 5.4** 4 1-bit input NAND gate. LIB is an OpenSparc T2 microprocessor environment variable, set here to value 'true' at a start of simulation

```
/* NAND operation thread process */
void u1_nand4_8x_proc0()
{
  while(1)
  {
    wait();
    if(LIB == true)
    {
       b0 = in0.read();
       b1 = in1.read();
       b2 = in2.read();
       b3 = in3.read();
       b4 = !(b0 & b1 & b2 & b3); /* NAND operation */
       out.write(b4);
    }
  }
}
/* Constructor */
SC_CTOR(cl_u1_nand4_8x)
{
  SC_THREAD(u1_nand4_8x_proc0); /*Declare/assign thread */
  sensitive << in0 << in1 << in2 << in3; /* Sensitivity list for thread */
}
/* Destructor */
~cl_u1_nand4_8x(){ }
};
```

**Fig. 5.4** (continued)

```
#include "n2_cl_u1.h"
#include <cstdlib>
#include <cstring>

int sc_main(int argc, char **argv)
{
  /* Combinational logic - NO clocks allowed */
  /* Input/output channels */
  sc_core::sc_signal<bool> sigin0;
  sc_core::sc_signal<bool> sigin1;
  sc_core::sc_signal<bool> sigin2;
```

**Fig. 5.5** Test harness for 4 1-bit input NAND gate

```
sc_core::sc_signal<bool> sigin3;
sc_core::sc_signal<bool> sigout0;

/* Declare trace file pointer - for VCD trace file */
sc_core::sc_trace_file *fp;

/* Declare/define NAND gate and VCD trace file */
cl_u1_nand4_8x nand4("nand4");
fp = sc_create_vcd_trace_file("tr_4_1_bit_nand");
fp->set_time_unit(1.0, sc_core::SC_US);

/* Connect module ports and channels */
nand4.in0(sigin0);
nand4.in1(sigin1);
nand4.in2(sigin2);
nand4.in3(sigin3);
nand4.out(sigout0);

nand4.LIB = true; /*Set environmanr variable */
/* Connect trace file and channels */
sc_trace(fp, sigin0, "nand4_in_0");
sc_trace(fp, sigin1, "nand4_in_1");
sc_trace(fp, sigin2, "nand4_in_2");
sc_trace(fp, sigin3, "nand4_in_3");
sc_trace(fp, sigout0, "nand4_out");

/* Generate data and run single simulation step */
sigin0.write(false);
sigin1.write(true);
sigin2.write(true);
sigin3.write(false);
sc_start(1.0,sc_core::SC_US);
/* Print current simulation time and signal
   values on console */
std::cout<<sc_core::sc_time_stamp().to_seconds()<<"¥t";
std::cout<<sigin0.read()<<"¥t"<<sigin1.read()<<"¥t";
std::cout<<sigin2.read()<<"¥t"<<sigin3.read()<<"¥t";
std::cout<<sigout0.read()<<std::endl;
```

**Fig. 5.5** (continued)

```
/* Change data and run simulation step */
sigin0.write(false);
sigin1.write(false);
sigin2.write(true);
sigin3.write(false);
sc_start(1.0,sc_core::SC_US);
std::cout<<sc_core::sc_time_stamp().to_seconds()<<"¥t";
std::cout<<sigin0.read()<<"¥t"<<sigin1.read()<<"¥t";
std::cout<<sigin2.read()<<"¥t"<<sigin3.read()<<"¥t";
std::cout<<sigout0.read()<<std::endl;

/* Change data and run simulation step */
sigin0.write(false);
sigin1.write(false);
sigin2.write(false);
sigin3.write(false);
sc_start(1.0,sc_core::SC_US);
std::cout<<sc_core::sc_time_stamp().to_seconds()<<"¥t";
std::cout<<sigin0.read()<<"¥t"<<sigin1.read()<<"¥t";
std::cout<<sigin2.read()<<"¥t"<<sigin3.read()<<"¥t";
std::cout<<sigout0.read()<<std::endl;

/* Change data and run simulation step */
sigin0.write(false);
sigin1.write(true);
sigin2.write(true);
sigin3.write(true);
sc_start(1.0,sc_core::SC_US);
std::cout<<sc_core::sc_time_stamp().to_seconds()<<"¥t";
std::cout<<sigin0.read()<<"¥t"<<sigin1.read()<<"¥t";
std::cout<<sigin2.read()<<"¥t"<<sigin3.read()<<"¥t";
std::cout<<sigout0.read()<<std::endl;

/* Change data and run simulation step */
sigin0.write(true);
sigin1.write(true);
sigin2.write(true);
sigin3.write(true);
sc_start(1.0,sc_core::SC_US);
```

**Fig. 5.5** (continued)

```
    std::cout<<sc_core::sc_time_stamp().to_seconds()<<"¥t";
    std::cout<<sigin0.read()<<"¥t"<<sigin1.read()<<"¥t";
    std::cout<<sigin2.read()<<"¥t"<<sigin3.read()<<"¥t";
    std::cout<<sigout0.read()<<std::endl;

    /* Change data and run simulation step */
    sigin0.write(true);
    sigin1.write(false);
    sigin2.write(false);
    sigin3.write(true);
      sc_start(1.0,sc_core::SC_US);
      std::cout<<sc_core::sc_time_stamp().to_seconds()<<"¥t";
      std::cout<<sigin0.read()<<"¥t"<<sigin1.read()<<"¥t";
      std::cout<<sigin2.read()<<"¥t"<<sigin3.read()<<"¥t";
      std::cout<<sigout0.read()<<std::endl;

      /* Change data and run simulation step */
      sigin0.write(true);
      sigin1.write(true);
      sigin2.write(false);
      sigin3.write(false);
      sc_start(1.0,sc_core::SC_US);
      std::cout<<sc_core::sc_time_stamp().to_seconds()<<"¥t";
      std::cout<<sigin0.read()<<"¥t"<<sigin1.read()<<"¥t";
      std::cout<<sigin2.read()<<"¥t"<<sigin3.read()<<"¥t";
      std::cout<<sigout0.read()<<std::endl;

      /* Stop simulation, close trace file */
      sc_core::sc_stop();
      sc_core::sc_close_vcd_trace_file(fp);
      return 0;
    }
```

**Fig. 5.5** (continued)

file, is to generate and view VCD trace files (VCD trace generated with SystemC 2.3.0 is in Fig. 5.6) with an efficient VCD trace viewer as GTKWave.

```
SystemC 2.2.0—Aug 30 2011 19:39:07
Copyright (c) 1996–2006 by all Contributors
ALL RIGHTS RESERVED
Note: VCD trace timescale unit is set by user to 1.000000e−06 s.
```

```
1e−06 0 1 1 0 0
2e−06 0 0 1 0 1
3e−06 0 0 0 0 1
4e−06 0 1 1 1 1
5e−06 1 1 1 1 0
6e−06 1 0 0 1 1
7e−06 1 1 0 0 1
SystemC: simulation stopped by user.

SystemC 2.3.0-ASI—Jul 19 2012 18:53:11
Copyright (c) 1996–2012 by all Contributors,
ALL RIGHTS RESERVED
Note: VCD trace timescale unit is set by user to 1.000000e−06 s.
1e−06 0 1 1 0 0
2e−06 0 0 1 0 1
3e−06 0 0 0 0 1
4e−06 0 1 1 1 1
5e−06 1 1 1 1 0
6e−06 1 0 0 1 1
7e−06 1 1 0 0 1
Info: /OSCI/SystemC: Simulation stopped by user.
```

Note that in this example, the simulation time steps are deliberately set different, to illustrate the point that for combinational logic, time steps do not matter, and only the behavior of the circuit is important.

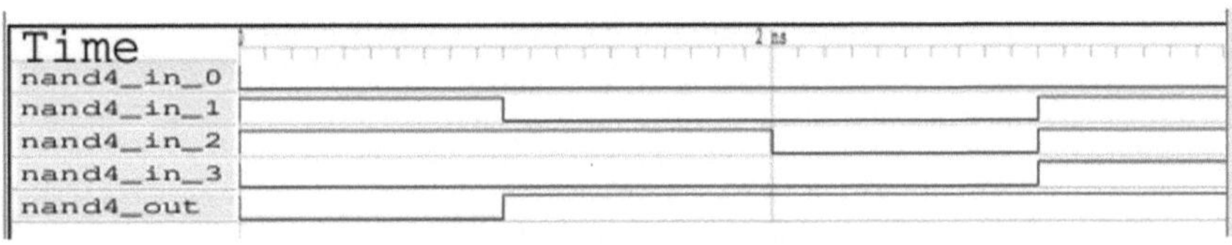


**Fig. 5.6** VCD traces obtained with SystemC 2.3.0

## 5.3 3-Bit Input Adder with 1-Bit Carry and 1-Bit Sum Output

A simple 3-bit input adder with sum and carry bit output is presented. Given three input bits $b0$, $b1$, $b2$, carry-out bit is obtained via the logical operation :$(b0 \wedge b1) \vee (b0 \wedge b2) \vee (b1 \wedge b2)$ and the sum bit is generated as: $(b0 \,\mathrm{xor}\, b1 \,\mathrm{xor}\, b2)$, where ^ is the bitwise AND operation, and v bitwise OR operation. One of the input bits is the carry-out bit from the previous stage. Figures 5.7 and 5.8 show the source code for this adder and the test harness. The truth table and logic circuit diagram are shown in Figs. 5.9, 5.10a and b, respectively. The

```
SC_MODULE(cl_u1_csa32_16x)
{
  /* Declare/define input/output ports */
sc_core::sc_in<bool> in0;
sc_core::sc_in<bool> in1;
sc_core::sc_in<bool> in2;
sc_core::sc_out<bool> carry;
sc_core::sc_out<bool> sum;
/* Declare module variables */
bool LIB;
bool b0;
bool b1;
bool b2;
bool b3;
bool b4;
/* Adder operation thread */
void u1_csa32_8x_proc0()
 {
   while(1)
   {
     wait();
     if(LIB == true)
     {
        b0 = in0.read();
        b1 = in1.read();
        b2 = in2.read();
        /* Generate carry -out bit */
        b3 = ((b0 & b1) | (b0 & b2) | (b1 & b2));
        /* Generate sum bit */
        b4 = (b0 ^ b1 ^ b2);
        carry.write(b3);
        sum .write(b4);
   }
 }
}
/* Constructor */
SC_CTOR(cl_u1_csa32_16x)
{
    /* Assign/declare thread */
    SC_THREAD(u1_csa32_16x_proc0);
```

**Fig. 5.7** 3-bit input sum, carry-out bit adder. LIB is an OpenSparc T2 microprocessor environment variable, set here to value 'true' at a start of simulation

```
        /* Sensitize thread to trigger events*/
        sensitive << in0 << in1 << in2;
    }
    /* Destructor */
      ~cl_u1_csa32_8x(){ }
    };
```

Fig. 5.7 (continued)

```
#include "n2_cl_u1.h"
#include <cstdlib>
#include <cstring>

int sc_main(int argc, char **argv)
{
   /* Combinational logic - NO clocks */
   /* Input/output channesl */
   sc_core::sc_signal<bool> sigin0;
   sc_core::sc_signal<bool> sigin1;
   sc_core::sc_signal<bool> sigin2;
   sc_core::sc_signal<bool> sigin3;
   sc_core::sc_signal<bool> sigout0;
   sc_core::sc_signal<bool> sigout1;
   /* Trace file pointer */
   sc_core::sc_trace_file *fp;
  /* Declare/define adder and trace file */
   cl_u1_csa32_16x csa32("csa32_16x");
   fp = sc_create_vcd_trace_file("tr_csa_32");
   fp->set_time_unit(1.0, sc_core::SC_NS);

  /* Connect module ports and channels */
   csa32.in0(sigin0);
   csa32.in1(sigin1);
   csa32.in2(sigin2);
   csa32.carry(sigout0);
   csa32.sum(sigout1);
   csa32.LIB = true; /*Set environmanr variable */

   /* Connect trace file with channels whose data is to be plotted */
   sc_trace(fp, sigin0, "adder_in_0");
```

Fig. 5.8 3-bit input, carry-out, sum bit output adder

```
sc_trace(fp, sigin1, "adder_in_1");
sc_trace(fp, sigin2, "adder_in_2");
sc_trace(fp, sigout0, "adder_carry");
sc_trace(fp, sigout1, "adder_sum");

/* Generate data and run a simulation step */
sigin0.write(false);
sigin1.write(false);
sigin2.write(false);
sc_start(1.0,sc_core::SC_NS);

/* Generate data and run a simulation step */
sigin0.write(true);
sigin1.write(false);
sigin2.write(false);
sc_start(1.0,sc_core::SC_NS);

/* Generate data and run a simulation step */
sigin0.write(false);
sigin1.write(true);
sigin2.write(false);
sc_start(1.0,sc_core::SC_NS);
 /* Generate data and run a simulation step */
sigin0.write(false);
sigin1.write(false);
sigin2.write(true);
sc_start(1.0,sc_core::SC_NS);

/* Generate data and run a simulation step */
sigin0.write(true);
sigin1.write(false);
sigin2.write(true);
sc_start(1.0,sc_core::SC_NS);

/* Generate data and run a simulation step */
sigin0.write(false);
sigin1.write(true);
sigin2.write(true);
sc_start(1.0,sc_core::SC_NS);
```

**Fig. 5.8** (continued)

```
    /* Generate data and run a simulation step */
    sigin0.write(true);
    sigin1.write(true);
    sigin2.write(false);
    sc_start(1.0,sc_core::SC_NS);

    /* Generate data and run a simulation step */
    sigin0.write(true);
    sigin1.write(true);
    sigin2.write(true);
    sc_start(6.0,sc_core::SC_NS);
    /* Stop simulation, close trace file */
    sc_stop();
    sc_close_vcd_trace_file(fp);
    return 0;
}
```

**Fig. 5.8** (continued)

**Fig. 5.9** 3-bit input adder truth table

| b0 | b1 | b2 | Carry | Sum | Decimal Sum |
|---|---|---|---|---|---|
| 0 | 1 | 0 | 0 | 1 | 1 |
| 1 | 0 | 0 | 0 | 1 | 1 |
| 0 | 0 | 1 | 0 | 1 | 1 |
| 1 | 0 | 1 | 1 | 0 | 2 |
| 0 | 1 | 1 | 1 | 0 | 2 |
| 1 | 1 | 0 | 1 | 0 | 2 |
| 1 | 1 | 1 | 1 | 1 | 3 |

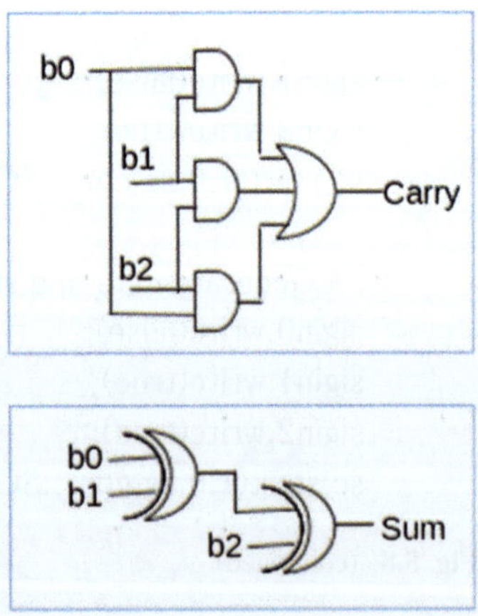


**Fig. 5.10** **a, b** 3-bit input adder with carry, sum output

reader's attention is drawn to the fact that as a 3-bit adder is almost always part of a larger adder, and the 'b2' input is often denoted as Cin (*Carry In*).

The console output is below. An easier alternative, as shown in the test harness source code, is to generate and view the VCD trace files (Fig. 5.11 has the VCD traces from SystemC 2.3.0 simulation of the adder) with an efficient viewer as GTKWave.

```
SystemC 2.2.0—Aug 30 2011 19:39:07
Copyright (c) 1996–2006 by all Contributors
ALL RIGHTS RESERVED
Note: VCD trace timescale unit is set by user to 1.000000e−09 s.
In: 1 0 0 Sum 1 Carry 0
In: 0 1 0 Sum 1 Carry 0
In: 0 0 1 Sum 1 Carry 0
In: 1 0 1 Sum 0 Carry 1
In: 0 1 1 Sum 0 Carry 1
In: 1 1 0 Sum 0 Carry 1
In: 1 1 1 Sum 1 Carry 1
SystemC: simulation stopped by user.

SystemC 2.3.0-ASI—Jul 18 2012 18:53:11
Copyright (c) 1996–2012 by all Contributors
ALL RIGHTS RESERVED
Note: VCD trace timescale unit is set by user to 1.000000e−09 s.
In: 1 0 0 Sum 1 Carry 0
In: 0 1 0 Sum 1 Carry 0
In: 0 0 1 Sum 1 Carry 0
In: 1 0 1 Sum 0 Carry 1
In: 0 1 1 Sum 0 Carry 1
In: 1 1 0 Sum 0 Carry 1
In: 1 1 1 Sum 1 Carry 1
Info: /OSCI/SystemC: Simulation stopped by user.
```

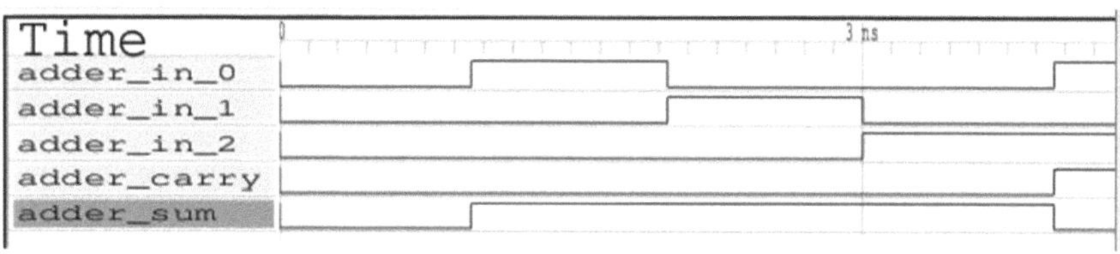


**Fig. 5.11** VCD traces from SystemC 2.3.0 simulation of bit only input/output version of adder

## 5.4 3-Bit Input Adder with Carry and Sum Output: Compound Data Types—Bit Vectors

All of the above circuits are simple with bit inputs. In real-world digital hardware, input to and outputs from most circuits are several bits long. To tackle these cases, SystemC [2] provides the bit-vector data type. The above 3 1-bit input, carry and sum bit output adder may be re-designed as follows—Figs. 5.12, 5.13 show the source code and test harness.

```
#ifndef TESTBV_H
#define TESTBV_H

#include <systemc>

SC_MODULE(adder3b)
{
  /* Input/output bit vectors */
  sc_core::sc_in< sc_dt::sc_bv<3> > input;
  sc_core::sc_out< sc_dt::sc_bv<2> > output;
  /* Declare/define module bit vectors */
  sc_dt::sc_bv<3> inp;
  sc_dt::sc_bv<2> tmp0;
  sc_dt::sc_bv<2> tmp1;
  sc_dt::sc_bv<2> tmp2;
  sc_dt::sc_bv<2> carry_sum;
  sc_dt::sc_bv<1> carry_bit;
  sc_dt::sc_bv<1> sum_bit;
  sc_dt::sc_bv<1> cb0;
  sc_dt::sc_bv<1> cb1;
  sc_dt::sc_bv<1> cb2;
  /* Adder operation thread */
  void adder3b_proc0()
  {
    while(1)
    {
      wait();
      inp = input.read();
      tmp0 = inp.range(0,1);
      tmp1 = inp.range(1,2);
      tmp2[0] = inp[0];
      tmp2[1] = inp[2];
```

**Fig. 5.12** 3-bit adder with bit-vector input/output

```
        cb0 = tmp0.and_reduce();
        cb1 = tmp1.and_reduce();
        cb2 = tmp2.and_reduce();
        carry_bit = cb0 | cb1 | cb2;  /*Carry-out bit generation */
        sum_bit = inp.xor_reduce(); /* Sum bit generation */
        carry_sum[0] = carry_bit[0];
        carry_sum[1] = sum_bit[0];
        output.write(carry_sum);
        std::cout<<"In : "<<inp<<" Carry "<<carry_bit<<" Sum
"<<sum_bit<<std::endl;
      }
    }
    /* Constructor */
    SC_CTOR(adder3b)
    {
      /* Declare/assign thread */
      SC_THREAD(adder3b_proc0);
      /* Sensitivity list for thread */
      sensitive << input;
    }
    /*Destructor */
    ~adder3b(){ }
};

#endif
```

**Fig. 5.12** (continued)

```
#include "testbv.h"
#include <cstdlib>
#include <cstring>
#include <systemc>

int sc_main(int argc, char **argv)
{
  /* Combinational logic - NO clocks */
  /* Input/output channels with bit vectors */
  sc_core::sc_signal< sc_dt::sc_bv<3> > sigin0;
  sc_core::sc_signal< sc_dt::sc_bv<2> > sigout0;
  sc_dt::sc_bv<3> input_vector;
  /*Declare/define adder, connect input/output ports with channels */
```

**Fig. 5.13** Test harness for 3-bit adder with bit-vector input/output

```
adder3b adder_3b("adder_3b");
adder_3b.input(sigin0);
adder_3b.output(sigout0);
/* Generate new input and run single simulation step */
input_vector = "000";
sigin0.write(input_vector);
sc_start(1.0, sc_core::SC_NS);
/* Generate new input and run single simulation step */
input_vector = "100";
sigin0.write(input_vector);
sc_start(1.0, sc_core::SC_NS);
/* Generate new input and run single simulation step */
input_vector = "010";
sigin0.write(input_vector);
sc_start(1.0, sc_core::SC_NS);
 /* Generate new input and run single simulation step */
input_vector = "001";
sigin0.write(input_vector);
sc_start(1.0, sc_core::SC_NS);
/* Generate new input and run single simulation step */
input_vector = "110";
sigin0.write(input_vector);
sc_start(1.0, sc_core::SC_NS);
/* Generate new input and run single simulation step */
input_vector = "011";
sigin0.write(input_vector);
sc_start(1.0, sc_core::SC_NS);
/* Generate new input and run single simulation step */
input_vector = "101";
sigin0.write(input_vector);
sc_start(1.0, sc_core::SC_NS);
/* Generate new input and run single simulation step */
input_vector = "111";
sigin0.write(input_vector);
sc_start(1.0, sc_core::SC_NS);
/* Stop simulatiom */
sc_core::sc_stop();
return 0;
}
```

**Fig. 5.13** (continued)

As seen in the source file, a number of smaller sized bit vectors are required to hold all the intermediate results. The advantage is that in this case, one can use the built-in methods of the SystemC [2] bit-vector class [*range*(), *and_reduce*(), *xor_reduce*()] to perform routine bitwise logical operations. The VCD trace file generated with SystemC 2.3.0 is in Fig. 5.14. The console output is:

SystemC 2.2.0—Aug 30 2011 19:39:07
Copyright (c) 1996–2006 by all Contributors
ALL RIGHTS RESERVED
In: 100 Carry 0 Sum 1
In: 010 Carry 0 Sum 1
In: 001 Carry 0 Sum 1
In: 110 Carry 1 Sum 0
In: 011 Carry 1 Sum 0
In: 101 Carry 1 Sum 0
In: 111 Carry 1 Sum 1
SystemC: simulation stopped by user.

SystemC 2.3.0-ASI—Jul 19 2012 18:53:11
Copyright (c) 1996–2012 by all Contributors,
ALL RIGHTS RESERVED
In: 100 Carry 0 Sum 1
In: 010 Carry 0 Sum 1
In: 001 Carry 0 Sum 1
In: 110 Carry 1 Sum 0
In: 011 Carry 1 Sum 0
In: 101 Carry 1 Sum 0
In: 111 Carry 1 Sum 1
Info: /OSCI/SystemC: Simulation stopped by user.

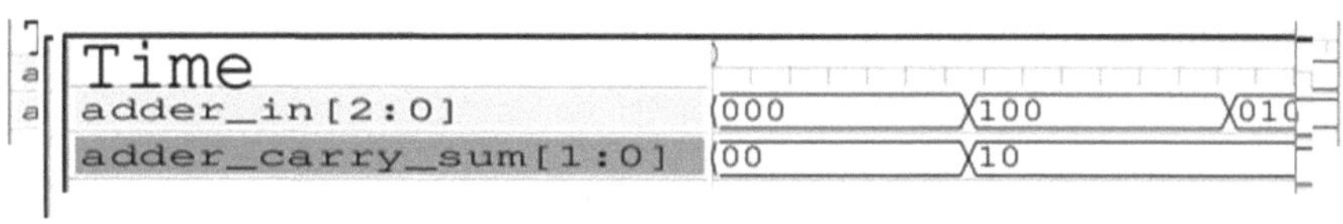


**Fig. 5.14** VCD traces for the simulation of bit vector only input/output adder with SystemC 2.3.0

## 5.5 8 × 1 Multiplexer: Bit Only Input/Output and Bit Vector Input/Output

A simple 8 × 1 multiplexer is presented next. A $n \times 1$ bit multiplexer outputs one of n input bits at any given time. Which input bit to select at a given time is determined by select bits. A multiplexer of $2^n$ inputs has n select lines, which are used to select which input line to send to the output. The source code and test bench are in Figs. 5.15 and 5.16, respectively. First, we consider the case of bit input/output

```
SC_MODULE(cl_a1_aomux8_12x)
{
/* Eight input bit ports */
sc_core::sc_in<bool> in0;
sc_core::sc_in<bool> in1;
sc_core::sc_in<bool> in2;
sc_core::sc_in<bool> in3;
sc_core::sc_in<bool> in4;
sc_core::sc_in<bool> in5;
sc_core::sc_in<bool> in6;
sc_core::sc_in<bool> in7;
/* Eight select bit ports */
sc_core::sc_in<bool> sel0;
sc_core::sc_in<bool> sel1;
sc_core::sc_in<bool> sel2;
sc_core::sc_in<bool> sel3;
sc_core::sc_in<bool> sel4;
sc_core::sc_in<bool> sel5;
sc_core::sc_in<bool> sel6;
sc_core::sc_in<bool> sel7;
/* Output port */
sc_core::sc_out<bool> out;
bool LIB;
/* Multiplexer operation thread */
void a1_aomux8_12x_proc0()
 {
  while(1)
   {
    wait();
    if(LIB == true)
     {
      out.write((sel0.read() & in0.read()) |
            (sel1.read() & in1.read()) |
            (sel2.read() & in2.read()) |
            (sel3.read() & in3.read()) |
            (sel4.read() & in4.read()) |
            (sel5.read() & in5.read()) |
            (sel6.read() & in6.read()) |
            (sel7.read() & in7.read()));
```

**Fig. 5.15** 8-bit input, 8-bit select 8 × 1 multiplexer. LIB is an OpenSparc T2 microprocessor environment variable, set here to value 'true' at a start of simulation

```
      }
    }
  }
  /* Constructor */
  SC_CTOR(cl_a1_aomux8_12x)
  {
    /*Declare/assign thread */
    SC_THREAD(a1_aomux8_12x_proc0);
    /* Sensitivity list for thread */
    sensitive << sel0 << sel1 << sel2 << sel3 << sel4 << sel5 << sel6 << sel7;
  }
  ~cl_a1_aomux8_12x(){ } /*Destructor */
};
```

**Fig. 5.15** (continued)

```
#include "n2_cl_a1.h"
#include <cstdlib>
#include <cstring>

int sc_main(int argc, char **argv)
{
  /* Eight channels for multiplexer input bit */
  sc_core::sc_signal<bool> sigin0;
  sc_core::sc_signal<bool> sigin1;
  sc_core::sc_signal<bool> sigin2;
  sc_core::sc_signal<bool> sigin3;
  sc_core::sc_signal<bool> sigin4;
  sc_core::sc_signal<bool> sigin5;
  sc_core::sc_signal<bool> sigin6;
  sc_core::sc_signal<bool> sigin7;

  /* Eight channels for multiplexer select lines */
  sc_core::sc_signal<bool> sel0;
  sc_core::sc_signal<bool> sel1;
  sc_core::sc_signal<bool> sel2;
  sc_core::sc_signal<bool> sel3;
  sc_core::sc_signal<bool> sel4;
```

**Fig. 5.16** Test harness for 8-bit input, 8-bit select 8 × 1 multiplexer

```
sc_core::sc_signal<bool> sel5;
sc_core::sc_signal<bool> sel6;
sc_core::sc_signal<bool> sel7;

/* Multiplexer output channel */
sc_core::sc_signal<bool> sigout0;

/* Trace file pointer for VCD format output */
sc_core::sc_trace_file *fp;
/* Instantiate multiplexer, trace */
cl_a1_aomux8_12x mux8("mux_8");
fp = sc_create_vcd_trace_file("tr_mux_8");
fp->set_time_unit(1.0, sc_core::SC_NS);

/* Connect multiplexer input/output ports with channels */
mux8.in0(sigin0);
mux8.in1(sigin1);
mux8.in2(sigin2);
mux8.in3(sigin3);
mux8.in4(sigin4);
mux8.in5(sigin5);
mux8.in6(sigin6);
mux8.in7(sigin7);

mux8.sel0(sel0);
mux8.sel1(sel1);
mux8.sel2(sel2);
mux8.sel3(sel3);
mux8.sel4(sel4);
mux8.sel5(sel5);
mux8.sel6(sel6);
mux8.sel7(sel7);
mux8.out(sigout0);
mux8.LIB = true;  /*Set environmanr variable */

/* Connect trace file and data input/output channels */
sc_trace(fp, sigin0, "in0");
sc_trace(fp, sigin1, "in1");
```

**Fig. 5.16** (continued)

```
sc_trace(fp, sigin2, "in2");
sc_trace(fp, sigin3, "in3");
sc_trace(fp, sigin4, "in4");
sc_trace(fp, sigin5, "in5");
sc_trace(fp, sigin6, "in6");
sc_trace(fp, sigin7, "in7");
sc_trace(fp, sel0, "sel0");
sc_trace(fp, sel1, "sel1");
sc_trace(fp, sel2, "sel2");
sc_trace(fp, sel3, "sel3");
sc_trace(fp, sel4, "sel4");
sc_trace(fp, sel5, "sel5");
sc_trace(fp, sel6, "sel6");
sc_trace(fp, sel7, "sel7");
sc_trace(fp, sigout0, "out");

/* Generate new data and run single simulation step */
sigin0.write(true);
sigin1.write(false);
sigin2.write(false);
sigin3.write(false);
sigin4.write(false);
sigin5.write(false);
sigin6.write(false);
sigin7.write(false);
sel0.write(true);
sel1.write(false);
sel2.write(false);
sel3.write(false);
sel4.write(false);
sel5.write(false);
sel6.write(false);
sel7.write(false);
sc_start(1.0, sc_core::SC_NS);
```

**Fig. 5.16** (continued)

```
/* Generate new data and run single simulation step */
 sigin0.write(false);
 sigin1.write(true);
 sigin2.write(false);
 sigin3.write(false);
 sigin4.write(false);
 sigin5.write(false);
 sigin6.write(false);
 sigin7.write(false);
 sel0.write(false);
 sel1.write(true);
 sel2.write(false);
 sel3.write(false);
 sel4.write(false);
 sel5.write(false);
 sel6.write(false);
 sel7.write(false);
 sc_start(1.0, sc_core::SC_NS);
 /* Print input signal, select signal and output
    signal bits to console */
 std::cout<<sc_core::sc_time_stamp().to_seconds()<<" "<<sigin0.read();
 std::cout<<" Data "<<sigin1.read()<<" "<<sigin2.read()<<" ";
 std::cout<<sigin3.read()<<" "<<sigin4.read()<<" ";
 std::cout<<sigin5.read()<<" "<<sigin6.read()<<" ";
 std::cout<<sigin7.read()<<" Select ";
 std::cout<<sel0.read()<<" "<<sel1.read()<<" ";
 std::cout<<sel2.read()<<" "<<sel3.read()<<" ";
 std::cout<<sel4.read()<<" "<<sel5.read()<<" ";
 std::cout<<sel6.read()<<" "<<sel7.read()<<" Output ";
 std::cout<<sigout0.read()<<std::endl;

 /* Generate new data and run single simulation step */
 sigin0.write(false);
 sigin1.write(false);
 sigin2.write(true);
 sigin3.write(false);
 sigin4.write(false);
```

**Fig. 5.16** (continued)

```
sigin5.write(false);
sigin6.write(false);
sigin7.write(false);
sel0.write(false);
sel1.write(false);
sel2.write(true);
sel3.write(false);
sel4.write(false);
sel5.write(false);
sel6.write(false);
sel7.write(false);
sc_start(1.0, sc_core::SC_NS);
std::cout<<sc_core::sc_time_stamp().to_seconds()<<" "<<sigin0.read();
std::cout<<" Data "<<sigin1.read()<<" "<<sigin2.read()<<" ";
std::cout<<sigin3.read()<<" "<<sigin4.read()<<" ";
std::cout<<sigin5.read()<<" "<<sigin6.read()<<" ";
std::cout<<sigin7.read()<<" Select ";
std::cout<<sel0.read()<<" "<<sel1.read()<<" ";
std::cout<<sel2.read()<<" "<<sel3.read()<<" ";
std::cout<<sel4.read()<<" "<<sel5.read()<<" ";
std::cout<<sel6.read()<<" "<<sel7.read()<<" Output ";
std::cout<<sigout0.read()<<std::endl;

/* Generate new data and run single simulation step */
sigin0.write(false);
sigin1.write(false);
sigin2.write(false);
sigin3.write(true);
sigin4.write(false);
sigin5.write(false);
sigin6.write(false);
sigin7.write(false);
sel0.write(false);
sel1.write(false);
sel2.write(false);
sel3.write(true);
sel4.write(false);
sel5.write(false);
```

**Fig. 5.16** (continued)

```
sel6.write(false);
sel7.write(false);
sc_start(1.0, sc_core::SC_NS);
std::cout<<sc_core::sc_time_stamp().to_seconds()<<" "<<sigin0.read();
std::cout<<" Data "<<sigin1.read()<<" "<<sigin2.read()<<" ";
std::cout<<sigin3.read()<<" "<<sigin4.read()<<" ";
std::cout<<sigin5.read()<<" "<<sigin6.read()<<" ";
std::cout<<sigin7.read()<<" Select ";
std::cout<<sel0.read()<<" "<<sel1.read()<<" ";
std::cout<<sel2.read()<<" "<<sel3.read()<<" ";
std::cout<<sel4.read()<<" "<<sel5.read()<<" ";
std::cout<<sel6.read()<<" "<<sel7.read()<<" Output ";
std::cout<<sigout0.read()<<std::endl;

/* Generate new data and run single simulation step */
sigin0.write(false);
sigin1.write(false);
sigin2.write(false);
sigin3.write(false);
sigin4.write(false);
sigin5.write(false);
sigin6.write(false);
sigin7.write(false);
sel0.write(false);
sel1.write(false);
sel2.write(false);
sel3.write(false);
sel4.write(true);
sel5.write(false);
sel6.write(false);
sel7.write(false);
sc_start(1.0, sc_core::SC_NS);
std::cout<<sc_core::sc_time_stamp().to_seconds()<<" "<<sigin0.read();
std::cout<<" Data "<<sigin1.read()<<" "<<sigin2.read()<<" ";
std::cout<<sigin3.read()<<" "<<sigin4.read()<<" ";
std::cout<<sigin5.read()<<" "<<sigin6.read()<<" ";
std::cout<<sigin7.read()<<" Select ";
std::cout<<sel0.read()<<" "<<sel1.read()<<" ";
std::cout<<sel2.read()<<" "<<sel3.read()<<" ";
```

**Fig. 5.16** (continued)

```
std::cout<<sel4.read()<<" "<<sel5.read()<<" ";
std::cout<<sel6.read()<<" "<<sel7.read()<<" Output ";
std::cout<<sigout0.read()<<std::endl;

/* Generate new data and run single simulation step */
sigin0.write(false);
sigin1.write(false);
sigin2.write(false);
sigin3.write(false);
sigin4.write(false);
sigin5.write(true);
sigin6.write(false);
sigin7.write(false);
sel0.write(false);
sel1.write(false);
sel2.write(false);
sel3.write(false);
sel4.write(false);
sel5.write(true);
sel6.write(false);
sel7.write(false);
sc_start(1.0, sc_core::SC_NS);
std::cout<<sc_core::sc_time_stamp().to_seconds()<<" "<<sigin0.read();
std::cout<<" Data "<<sigin1.read()<<" "<<sigin2.read()<<" ";
std::cout<<sigin3.read()<<" "<<sigin4.read()<<" ";
std::cout<<sigin5.read()<<" "<<sigin6.read()<<" ";
std::cout<<sigin7.read()<<" Select ";
std::cout<<sel0.read()<<" "<<sel1.read()<<" ";
std::cout<<sel2.read()<<" "<<sel3.read()<<" ";
std::cout<<sel4.read()<<" "<<sel5.read()<<" ";
std::cout<<sel6.read()<<" "<<sel7.read()<<" Output ";
std::cout<<sigout0.read()<<std::endl;

/* Generate new data and run single simulation step */
sigin0.write(false);
sigin1.write(false);
sigin2.write(false);
sigin3.write(false);
sigin4.write(false);
```

**Fig. 5.16** (continued)

```
sigin5.write(false);
sigin6.write(false);
sigin7.write(false);
sel0.write(false);
sel1.write(false);
sel2.write(false);
sel3.write(false);
sel4.write(false);
sel5.write(false);
sel6.write(true);
sel7.write(false);
sc_start(1.0, sc_core::SC_NS);
std::cout<<sc_core::sc_time_stamp().to_seconds()<<" "<<sigin0.read();
std::cout<<" Data "<<sigin1.read()<<" "<<sigin2.read()<<" ";
std::cout<<sigin3.read()<<" "<<sigin4.read()<<" ";
std::cout<<sigin5.read()<<" "<<sigin6.read()<<" ";
std::cout<<sigin7.read()<<" Select ";
std::cout<<sel0.read()<<" "<<sel1.read()<<" ";
std::cout<<sel2.read()<<" "<<sel3.read()<<" ";
std::cout<<sel4.read()<<" "<<sel5.read()<<" ";
std::cout<<sel6.read()<<" "<<sel7.read()<<" Output ";
std::cout<<sigout0.read()<<std::endl;

/* Generate new data and run single simulation step */
sigin0.write(false);
sigin1.write(false);
sigin2.write(false);
sigin3.write(false);
sigin4.write(false);
sigin5.write(false);
sigin6.write(false);
sigin7.write(true);
sel0.write(false);
sel1.write(false);
sel2.write(false);
sel3.write(false);
sel4.write(false);
sel5.write(false);
sel6.write(false);
```

**Fig. 5.16** (continued)

```
    sel7.write(true);
    sc_start(1.0, sc_core::SC_NS);
    std::cout<<sc_core::sc_time_stamp().to_seconds()<<" "<<sigin0.read();
    std::cout<<" Data "<<sigin1.read()<<" "<<sigin2.read()<<" ";
    std::cout<<sigin3.read()<<" "<<sigin4.read()<<" ";
    std::cout<<sigin5.read()<<" "<<sigin6.read()<<" ";
    std::cout<<sigin7.read()<<" Select ";
    std::cout<<sel0.read()<<" "<<sel1.read()<<" ";
    std::cout<<sel2.read()<<" "<<sel3.read()<<" ";
    std::cout<<sel4.read()<<" "<<sel5.read()<<" ";
    std::cout<<sel6.read()<<" "<<sel7.read()<<" Output ";
    std::cout<<sigout0.read()<<std::endl;

    /* Stop simulation and close trace file */
    sc_stop();
    sc_close_vcd_trace_file(fp);
    return 0;
  }
```

**Fig. 5.16** (continued)

```
SC_MODULE(mux8)
{
  /*
  Input/output ports with bit vectore for data, select and output
  */
  sc_core::sc_in< sc_dt::sc_bv<8> > in;
  sc_core::sc_in< sc_dt::sc_bv<8> > sel;
  sc_core::sc_out< sc_dt::sc_bv<1> > out;

  /* Module internal bit vectors */
  sc_dt::sc_bv<8> inbv;
  sc_dt::sc_bv<8> selbv;
  sc_dt::sc_bv<8> tmp;
  sc_dt::sc_bv<1> outbv;
```

**Fig. 5.17** Bit-vector input and bit-vector output version of 8 × 1 multiplexer

and then examine how code size may be reduced drastically by judicious use of bit vectors.

The source code and test harness for the bit vector version of the same 8 × 1 multiplexer are in Figs. 5.17 and 5.18, respectively. It is clear that in modeling real-world hardware with large number of multi-bit inputs and outputs, the

```
/*Thread for multiplexer oaeration */
void mux8_proc0()
{
  while(1)
  {
    wait();
    inbv = in.read();
    selbv = sel.read();
    tmp = inbv & selbv;
    outbv = tmp.or_reduce();
    out.write(outbv);
  }
}

/* Constructor */
SC_CTOR(mux8)
{
  /*Declare/assign thread */
  SC_THREAD(mux8_proc0);
  /Sensitivity list for thread */
  sensitive << sel;
}

/* Destructor */
~mux8(){ }
};
```

**Fig. 5.17** (continued)

```
#include "testbv.h"
#include <cstdlib>
#include <cstring>
#include "systemc.h"

int sc_main(int argc, char **argv)
{
  /* Combinational logic - NO clocks allowed */
  /* Multiplexer input/output bit-vector channels */
  sc_core::sc_signal< sc_dt::sc_bv<8> > din;
  sc_core::sc_signal< sc_dt::sc_bv<8> > sel;
  sc_core::sc_signal< sc_dt::sc_bv<1> > out;
```

**Fig. 5.18** Test bench for bit-vector input and bit-vector output version of 8 × 1 multiplexer

```
/* Local bit vectors */
sc_dt::sc_bv<8> dbv;
sc_dt::sc_bv<8> selbv;
/* Trace file pointer - for VCD output */
sc_core::sc_trace_file *fp;

/* Declare/define multiplexer and trace file */
mux8 mux_8("mux_8");
fp = sc_create_vcd_trace_file("tr_mux_8");
fp->set_time_unit(1.0, sc_core::SC_NS);

/* Connect input/output ports and channels */
mux_8.in(din);
mux_8.sel(sel);
mux_8.out(out);
/* Connect trace file and data channels */
sc_trace(fp, din, "din");
sc_trace(fp, sel, "sel");
sc_trace(fp, out, "out");

/* Generate new data and run simulation for 1 time step */
dbv="10000000";
selbv="10000000";
din.write(dbv);
sel.write(selbv);
sc_start(1.0, sc_core::SC_NS);

/* Generate new data and run simulation for 1 time step */
dbv="01000000";
selbv="01000000";
din.write(dbv);
sel.write(selbv);
sc_start(1.0, sc_core::SC_NS);
/* Print input bit vector, select bit vector and output
   bit vector to console */
std::cout<<sc_core::sc_time_stamp().to_seconds()<<" Data ";
std::cout<<din.read()<<" Select "<<sel.read()<<" Output ";
std::cout<<out.read()<<std::endl;

/* Generate new data and run simulation for 1 time step */
dbv="01000000";
```

**Fig. 5.18** (continued)

```
selbv="00100000";
din.write(dbv);
sel.write(selbv);
sc_start(1.0, sc_core::SC_NS);
std::cout<<sc_core::sc_time_stamp().to_seconds()<<" Data ";
std::cout<<din.read()<<" Select "<<sel.read()<<" Output ";
std::cout<<out.read()<<std::endl;

/* Generate new data and run simulation for 1 time step */
dbv="00010000";
selbv="00010000";
din.write(dbv);
sel.write(selbv);
sc_start(1.0, sc_core::SC_NS);
std::cout<<sc_core::sc_time_stamp().to_seconds()<<" Data ";
std::cout<<din.read()<<" Select "<<sel.read()<<" Output "
std::cout<<out.read()<<std::endl;

/* Generate new data and run simulation for 1 time step */
dbv="00010000";
selbv="00001000";
din.write(dbv);
sel.write(selbv);
sc_start(1.0, sc_core::SC_NS);
std::cout<<sc_core::sc_time_stamp().to_seconds()<<" Data ";
std::cout<<din.read()<<" Select "<<sel.read()<<" Output ";
std::cout<<out.read()<<std::endl;

/* Generate new data and run simulation for 1 time step */
dbv="00000100";
selbv="00000100";
din.write(dbv);
sel.write(selbv);
sc_start(1.0, sc_core::SC_NS);
std::cout<<sc_core::sc_time_stamp().to_seconds()<<" Data ";
std::cout<<din.read()<<" Select "<<sel.read()<<" Output ";
std::cout<<out.read()<<std::endl;

/* Generate new data and run simulation for 1 time step */
dbv="00000010";
```

**Fig. 5.18** (continued)

```
    selbv="00000010";
    din.write(dbv);
    sel.write(selbv);
    sc_start(1.0, sc_core::SC_NS);
    std::cout<<sc_core::sc_time_stamp().to_seconds()<<" Data ";
    std::cout<<din.read()<<" Select "<<sel.read()<<" Output ";
    std::cout<<out.read()<<std::endl;

    /* Generate new data and run simulation for 1 time step */
    dbv="00000001";
    selbv="00000001";
    din.write(dbv);
    sel.write(selbv);
    std::cout<<sc_core::sc_time_stamp().to_seconds()<<" Data ";
    std::cout<<din.read()<<" Select "<<sel.read()<<" Output ";
    std::cout<<out.read()<<std::endl;
    sc_start(1.0, sc_core::SC_NS);

    /* Stop simulation and close trace file */
    sc_core::sc_stop();
    sc_close_vcd_trace_file(fp);
    return 0;
}
```

**Fig. 5.18** (continued)

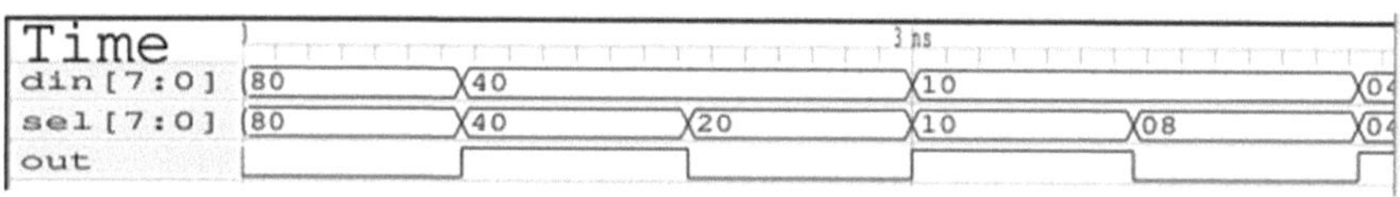


**Fig. 5.19** VCD traces generated with SystemC 2.3.0

bit-vector approach is the preferred choice to write data to or read data from a module. An easier alternative to the console output below, as shown in both the bit only input/output and bit-vector input/output versions of the 8 × 1 multiplexer, is to view the VCD trace files (VCD trace file generated with SystemC 2.3.0 is in Fig. 5.19) with an efficient trace file viewer as GTKWave.

First, the console output from the bit only input/output 8 × 1 multiplexer is presented, followed by the console output from the bit vector only input/output 9 × 1 multiplexer. The first column for each case represents the simulation kernel generated time stamp.

SystemC 2.2.0—Aug 30 2011 19:39:07
Copyright (c) 1996–2006 by all Contributors

```
ALL RIGHTS RESERVED
Note: VCD trace timescale unit is set by user to 1.000000e-09 s.
2e-09 Data 0 1 0 0 0 0 0 0 Select 0 1 0 0 0 0 0 0 Output 1
3e-09 Data 0 0 1 0 0 0 0 0 Select 0 0 1 0 0 0 0 0 Output 1
4e-09 Data 0 0 0 1 0 0 0 0 Select 0 0 0 1 0 0 0 0 Output 1
5e-09 Data 0 0 0 0 0 0 0 0 Select 0 0 0 0 1 0 0 0 Output 0
6e-09 Data 0 0 0 0 0 1 0 0 Select 0 0 0 0 0 1 0 0 Output 1
7e-09 Data 0 0 0 0 0 0 0 0 Select 0 0 0 0 0 0 1 0 Output 0
8e-09 Data 0 0 0 0 0 0 0 1 Select 0 0 0 0 0 0 0 1 Output 1
SystemC: simulation stopped by user.

SystemC 2.3.0-ASI—Jul 19 2012 18:53:11
Copyright (c) 1996–2012 by all Contributors,
ALL RIGHTS RESERVED
Note: VCD trace timescale unit is set by user to 1.000000e-09 s.
2e-09 Data 0 1 0 0 0 0 0 0 Select 0 1 0 0 0 0 0 0 Output 1
3e-09 Data 0 0 1 0 0 0 0 0 Select 0 0 1 0 0 0 0 0 Output 1
4e-09 Data 0 0 0 1 0 0 0 0 Select 0 0 0 1 0 0 0 0 Output 1
5e-09 Data 0 0 0 0 0 0 0 0 Select 0 0 0 0 1 0 0 0 Output 0
6e-09 Data 0 0 0 0 0 1 0 0 Select 0 0 0 0 0 1 0 0 Output 1
7e-09 Data 0 0 0 0 0 0 0 0 Select 0 0 0 0 0 0 1 0 Output 0
8e-09 Data 0 0 0 0 0 0 0 1 Select 0 0 0 0 0 0 0 1 Output 1
Info: /OSCI/SystemC: Simulation stopped by user.

SystemC 2.2.0—Aug 30 2011 19:39:07
Copyright (c) 1996–2006 by all Contributors
ALL RIGHTS RESERVED
Note: VCD trace timescale unit is set by user to 1.000000e-09 s.
2e-09 Data 01000000 Select 01000000 Output 1
3e-09 Data 01000000 Select 00100000 Output 0
4e-09 Data 00010000 Select 00010000 Output 1
5e-09 Data 00010000 Select 00001000 Output 0
6e-09 Data 00000100 Select 00000100 Output 1
7e-09 Data 00000010 Select 00000010 Output 1
7e-09 Data 00000010 Select 00000010 Output 1
SystemC: simulation stopped by user.

SystemC 2.3.0-ASI—Jul 19 2012 18:53:11
Copyright (c) 1996–2012 by all Contributors,
ALL RIGHTS RESERVED
Note: VCD trace timescale unit is set by user to 1.000000e-09 s.
2e-09 Data 01000000 Select 01000000 Output 1
3e-09 Data 01000000 Select 00100000 Output 0
4e-09 Data 00010000 Select 00010000 Output 1
5e-09 Data 00010000 Select 00001000 Output 0
6e-09 Data 00000100 Select 00000100 Output 1
```

**Table 5.1** 8-bit input and 4-bit output thermometer code generator

| 8-bit input and 4-bit output thermometer code generator | 8-bit input | 4-bit output |
|---|---|---|
| Note that if the 8-bit input has | 00000001 | 1000 |
| >=4 1s in it, the output is | 00000011 | 1100 |
| always 1111 | 00000111 | 1110 |
| | 00001111 | 1111 |
| | 00110111 | 1111 |
| | 00011101 | 1111 |
| | 01000001 | 1100 |
| | 10101010 | 1111 |
| | 11110011 | 1111 |
| | 00011001 | 1110 |

**Table 5.2** 4-bit input and 5-bit output 1—Hot encoder sample input/output

| 4-bit input and 5-bit output 1 hot encoder | 4-bit input | 5-bit output |
|---|---|---|
| Note that the position of the single 1 | 0000 | 00000 |
| in the 5-bit output depends on the | 0001 | 01000 |
| number of 1s in the 4-bit input | 0011 | 00100 |
| | 0111 | 00010 |
| | 1111 | 00001 |

7e−09 Data 00000010 Select 00000010 Output 1
7e−09 Data 00000010 Select 00000010 Output 1
Info: /OSCI/SystemC: Simulation stopped by user.

## 5.6 Combinational Logic Blocks Connected in Series

Having examined individual combinational logic blocks, we study the case of a number of such blocks connected in series, as in real-world hardware. We consider a simple design similar to one in [3]—a 8-bit input and 4-bit output thermometer code generator followed by a 4-bit input 1-hot encoder. A 8-bit input and 4-bit output thermometer code generator accepts a 8-bit input and emits a 4-bit output. If the 8-bit input has 4 or more 1s in it, the output is 1111, else it is the total number of ones with any extra zeros as necessary. A 4-bit input 1-hot encoder is a special encoder which accepts a 4-bit number and emits a 5-bit number at most one 1 in it, the rest being 0s. Sample inputs to and outputs from of both these circuits are summarized in the Tables 5.1 and 5.2.

Figures 5.20, 5.21 list the source code for the thermometer code generator and 1-hot encoder, while Fig. 5.22 contains the source code for the common test harness. Console output:

SystemC 2.2.0—Aug 30 2011 19:39:07
Copyright (c) 1996–2006 by all Contributors
ALL RIGHTS RESERVED
Note: VCD trace timescale unit is set by user to 1.000000e−09 s.
8 sort In: 00000001 Out: 1000

```
SC_MODULE(onecounter)
{
 /* Input/output ports for bit vector input/output */
 sc_core::sc_in< sc_dt::sc_bv<8> > in;
 sc_core::sc_out< sc_dt::sc_bv<4> > out;
 ?* Module internal bit vectors and variables */

 sc_dt::sc_bv<8> inbits;
 sc_dt::sc_bv<4> outbits;
 unsigned int count;
 unsigned int i;
 /* Thread for one counter operation */
 void process_input()
 {
  while(1)
  {
    wait();
    inbits = in.read();
    count = 0;
    outbits = "0000";
    for(i = 0; i < 8; i++)
    {
      if(count == 4) break;
      if(inbits[i] == '1') count++;
    }
    if(count < 4)
    {
      for(i = 0; i < count; i++) outbits[MAX-1-i] = "1";
    }
    else
    {
      outbits = "1111";
    }
    out.write(outbits);
    std::cout<<"8 sort In : "<<inbits<<" Out : "<<outbits<<std::endl;
  }
  count = 0;
 }
```

**Fig. 5.20** 8-bit input and 4-bit output thermometer encoder

```
    /* Constructor */
    SC_CTOR(onecounter)
    {
      SC_THREAD(process_input); /* Declare/assign thread */
      sensitive << in; /*Sensitivity list for thread */
    }
    ~onecounter(){} /*Destructor */
};
```

**Fig. 5.20** (continued)

```
SC_MODULE(onehot)
{
  /* Input/output ports for bit vector input/output */
  sc_core::sc_in< sc_dt::sc_bv<4> > in;
  sc_core::sc_out< sc_dt::sc_bv<5> > out;
   ?* Module internal bit vectors and variables */
  sc_dt::sc_bv<4> input;
  sc_dt::sc_bv<5> output;
  unsigned int count;
  unsigned int i;
   /* Thread for 1-hot encoder operation */
  void process_input()
  {
    count = 0;
    i = 0;
    input = "0000";
    while(1)
    {
      wait();
      count = 0;
      i = 0;
      input = in.read();
       for(; i < MAX; i++)
      {
        if(input[i] == '1') count++;
      }
      if(count == 0) output = "00000";
      else if(count == 1) output = "01000";
```

**Fig. 5.21** 4-bit input and 5-bit output 1-hot encoder

```
      else if(count == 2) output = "00100";
      else if(count == 3) output = "00010";
      else if(count == 4) output = "00001";
      out.write(output);
      std::cout<<"1 hot In : "<<input<<" Out : "<<output<<std::endl;
    }
  }
  /* Constructor */
  SC_CTOR(onehot)
  {
    SC_THREAD(process_input); /* Declare/assign thread */
    sensitive << in; /*Sensitivity list for thread */
  }
  ~onehot(){ } /*Destructor */
};
```

**Fig. 5.21** (continued)

```
#include "test.h"
#include <cstdlib>
#include <cstring>

#include "systemc.h"
int sc_main(int argc, char **argv)
{
  /* Bit-vector channels */
  sc_core::sc_signal< sc_dt::sc_bv<8> > thermoin;
  sc_core::sc_signal< sc_dt::sc_bv<4> > thermoout;
  sc_core::sc_signal< sc_dt::sc_bv<5> > onehotout;
  /* Local bit vector */
  sc_dt::sc_bv<8> thmin;
  /* Instantiate one counter and 1-hot encoder */
  onecounter one_count("one_count");
  onehot one_hot("one_hot");
  /* Trace file pointer */
  sc_core::sc_trace_file *fp;
  /* Create. Initialize trace file */
  fp = sc_create_vcd_trace_file("tr_thgen_onehot");
```

**Fig. 5.22** Test harness for 8-bit input and 4-bit output thermometer code generator/4-bit input and 5-bit output 1-hot encoder series combination

```
fp->set_time_unit(1.0, sc_core::SC_NS);
/* Connect module ports and channels */
one_count.in(thermoin);
one_count.out(thermoout);
one_hot.in(thermoout);
one_hot.out(onehotout);
/* Connect trace file and data channels */
sc_trace(fp, thermoin, "thermoin");
sc_trace(fp, thermoout, "thermoout");
sc_trace(fp, onehotout, "onehotout");
/* Generate data and run 1 simulation step */
thmin = "00000000";
thermoin.write(thmin);
sc_start(1.0, sc_core::SC_NS);
/* Generate data and run 1 simulation step */
thmin = "00000001";
thermoin.write(thmin);
sc_start(1.0, sc_core::SC_NS);
/* Generate data and run 1 simulation step */
thmin = "00000011";

thermoin.write(thmin);
sc_start(1.0, sc_core::SC_NS);
/* Generate data and run 1 simulation step */
thmin = "00001110";
thermoin.write(thmin);
sc_start(1.0, sc_core::SC_NS);
/* Generate data and run 1 simulation step */
thmin = "00001111";
thermoin.write(thmin);
sc_start(1.0, sc_core::SC_NS);
/* Generate data and run 1 simulation step */
thmin = "01010101";
thermoin.write(thmin);
sc_start(1.0, sc_core::SC_NS);
/* Generate data and run 1 simulation step */
thmin = "00101100";
thermoin.write(thmin);
```

**Fig. 5.22** (continued)

```
    sc_start(1.0, sc_core::SC_NS);
    /* Generate data and run 1 simulation step */
    thmin = "00111011";
    thermoin.write(thmin);
    sc_start(1.0, sc_core::SC_NS);
    /* Generate data and run 1 simulation step */  thmin = "10101111";
    thermoin.write(thmin);  sc_start(1.0, sc_core::SC_NS);
    /* Stop simulation and close trace file */
    sc_core::sc_stop();
    sc_close_vcd_trace_file(fp);

    return 0;
}
```

**Fig. 5.22** (continued)

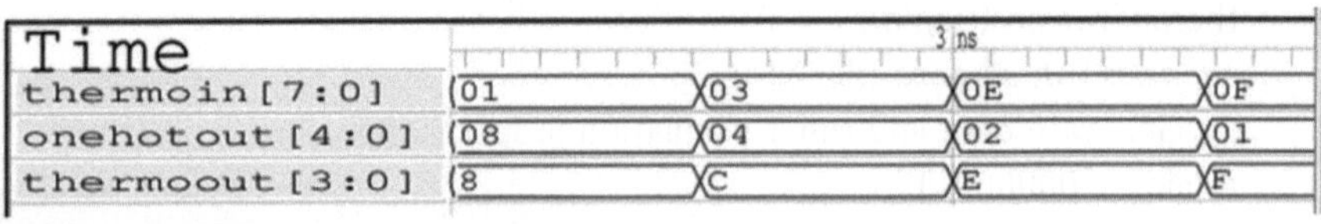


**Fig. 5.23** VCD trace generated with SystemC 2.3.0

```
1 hot In: 1000 Out: 01000
8 sort In: 00000011 Out: 1100
1 hot In: 1100 Out: 00100
8 sort In: 00001110 Out: 1110
1 hot In: 1110 Out: 00010
8 sort In: 00001111 Out: 1111
1 hot In: 1111 Out: 00001
8 sort In: 01010101 Out: 1111
8 sort In: 00101100 Out: 1110
1 hot In: 1110 Out: 00010
8 sort In: 00111011 Out: 1111
1 hot In: 1111 Out: 00001
8 sort In: 10101111 Out: 1111
SystemC: simulation stopped by user.

SystemC 2.3.0-ASI—Jul 19 2012 18:53:11
Copyright (c) 1996–2012 by all Contributors,
ALL RIGHTS RESERVED
Note: VCD trace timescale unit is set by user to 1.000000e−09 s.
8 sort In: 00000001 Out: 1000
```

```
1 hot In: 1000 Out: 01000
8 sort In: 00000011 Out: 1100
1 hot In: 1100 Out: 00100
8 sort In: 00001110 Out: 1110
1 hot In: 1110 Out: 00010
8 sort In: 00001111 Out: 1111
1 hot In: 1111 Out: 00001
8 sort In: 01010101 Out: 1111
8 sort In: 00101100 Out: 1110
1 hot In: 1110 Out: 00010
8 sort In: 00111011 Out: 1111
1 hot In: 1111 Out: 00001
8 sort In: 10101111 Out: 1111
Info: /OSCI/SystemC: Simulation stopped by user.
```

An alternative to the console output above is to view the VCD trace output [VCD trace file generated with SystemC 2.3.0 is in (Fig. 5.23)] with an efficient VCD trace fie viewer as GTKWave.

## 5.7 32-Bit Left/Right Barrel Shifter

A common sub-circuit of a modern microprocessor (especially its floating point unit) is the barrel shifter [4, 5] that implements left or right circular shift, by a specified number of bits, on a given operand. There is no loss of bits. It is often used as a sequential logic circuit synchronized with the system clock but may be analyzed as a combinational circuit in SystemC [2] whose output changes with any of its inputs—the operand bit vector whose bits need to be shifted, the shift

```
#ifndef BARRELSHIFT_H
#define BARRELSHIFT_H

#include <systemc>
const unsigned int MAX = 32;

SC_MODULE(barrelshift)
{
  /* Declare/define input/output ports */
  sc_core::sc_in< sc_dt::sc_bv<MAX> > din0;
  sc_core::sc_in< sc_dt::sc_bv<MAX> > din1;
  sc_core::sc_in<bool> shiftright;
```

**Fig. 5.24** 32-bit left/right barrel shifter

```
sc_core::sc_out< sc_dt::sc_bv<MAX> > dout;
sc_core::sc_out<bool> ready;
/*Declare/define interna data members */
sc_dt::sc_bv<MAX> tmp;
sc_dt::sc_bv<MAX> tmp_out;
unsigned int shiftlen;
bool readyvalue;

 /* Main barrel shifer thread */
 void barrelshift_proc0()
 {
  while(1)
  {
   wait();
   shiftlen = din1.read().to_int();
   /* Ensure that shift length is less than bit-vector length */
   if(shiftlen < (MAX - 1) && shiftlen > 0)
   {
    tmp = din0.read();
    if(shiftright.read() == true) /* Right shift operation */
    {
       tmp_out = (tmp(shiftlen - 1, 0),
                  tmp(MAX - 1, shiftlen));
    }
    else  if(shiftright.read() == false) /* Left shift operation */
    {
      tmp_out = (tmp(MAX -  1 - shiftlen, 0),
                 tmp(MAX - 1, MAX - shiftlen));
    }
    readyvalue = true;
   }
   else
   {
    readyvalue = false; /* No shifts allowed */
   }
   ready.write(readyvalue);
   if(readyvalue == true)
```

**Fig. 5.24** (continued)

```
      {
        dout.write(tmp_out);
      }
      else
      {
        dout.write(tmp);
      }
    }
  }

  /* Constructor with member initialization */
  SC_CTOR(barrelshift):tmp("00000000000000000000000000000000 "),
                   tmp_out("00000000000000000000000000000000 "),
                   shiftlen(0),
  {
    /* Declare/assign thread and sensitivity list */
    SC_THREAD(barrelshift_proc0);
    sensitive << din0 << din1 << shiftright;
  }

  ~barrelshift(){ } /*Destructor */
};

#endif
```

**Fig. 5.24** (continued)

```
#include "barrelshift.h"

int sc_main(int argc, char **argv)
{
  /* Declare/define channels/signals */
  sc_core::sc_signal< sc_dt::sc_bv<MAX> > sig0;
  sc_core::sc_signal< sc_dt::sc_bv<MAX> > sig1;
  sc_core::sc_signal< sc_dt::sc_bv<MAX> > sig2;
  sc_core::sc_signal<bool> sig3;

  /* Declare/define local data members */
  sc_dt::sc_bv<MAX> datain;
  sc_dt::sc_bv<MAX> shiftlen;
```

**Fig. 5.25** Test harness for 32-bit barrel shifter

```
/* Member to control shift direction */
bool shiftright;
/* Declare/define barrel shifter and trace file pointer for VCD trces */
barrelshift bs("bs");
sc_core::sc_trace_file *fp =
sc_core::sc_create_vcd_trace_file("tr_barrel_shift");
  fp->set_time_unit(1.0, sc_core::SC_NS);
  /* Connect module ports and channels */

  bs.din0(sig0);
  bs.din1(sig1);
  bs.shiftright(sig3);
  bs.dout(sig2);

  /* Connect trace file and signals */
  sc_core::sc_trace(fp, sig0, "data_in");
  sc_core::sc_trace(fp, sig1, "shift_len");
  sc_core::sc_trace(fp, sig2, "data_out");
  sc_core::sc_trace(fp, sig3, "shift_right");

  /* Generate new data and insert into channels */
  datain = "00000000000000000000000000000000";
  shiftlen = "00000000000000000000000000000010";
  shiftright = true;
  sig0.write(datain);
  sig1.write(shiftlen);
  sig3.write(shiftright);
  /* Run simulation for a pre-defined time period */
  sc_core::sc_start(5.0, sc_core::SC_NS);

  /* Print out input bit vector, output bit vector, shift length and
     shift direction left/right */
  std::cout<<"shift length "<<shiftlen.to_int()<<std::endl;
  std::cout<<"shift right "<<shiftright<<std::endl;
  std::cout<<"Data in "<<datain<<std::endl;
  std::cout<<"Data out "<<sig2.read()<<std::endl;


  /* Generate new data and insert into channels */
  datain = "10101010010101010011001111001100";
```

**Fig. 5.25** (continued)

```
shiftlen = "00000000000000000000000000000001";
shiftright = true;
sig0.write(datain);
sig1.write(shiftlen);
sig3.write(shiftright);
 /* Run simulation for a pre-defined time period */
sc_core::sc_start(5.0, sc_core::SC_NS);
std::cout<<"shift length "<<shiftlen.to_int()<<std::endl;
std::cout<<"shift right "<<shiftright<<std::endl;
std::cout<<"Data in "<<datain<<std::endl;
std::cout<<"Data out "<<sig2.read()<<std::endl;

/* Generate new data and insert into channels */
datain = "00111011010101011111001111001100";
shiftlen = "00000000000000000000000000000100";
shiftright = false;
sig0.write(datain);
sig1.write(shiftlen);
sig3.write(shiftright);
sc_core::sc_start(5.0, sc_core::SC_NS);
std::cout<<"shift length "<<shiftlen.to_int()<<std::endl;
std::cout<<"shift right "<<shiftright<<std::endl;
std::cout<<"Data in "<<datain<<std::endl;
std::cout<<"Data out "<<sig2.read()<<std::endl;

/* Generate new data and insert into channels */
datain = "10101010010101010011001111001111";
shiftlen = "00000000000000000000000000000101";
shiftright = false;
sig0.write(datain);
sig1.write(shiftlen);
sig3.write(shiftright);
sc_core::sc_start(5.0, sc_core::SC_NS);
std::cout<<"shift length "<<shiftlen.to_int()<<std::endl;
std::cout<<"shift right "<<shiftright<<std::endl;
std::cout<<"Data in "<<datain<<std::endl;
std::cout<<"Data out "<<sig2.read()<<std::endl;
```

**Fig. 5.25** (continued)

```
/* Generate new data and insert into channels */
datain = "10101010010101010011000000001111";
shiftlen = "00000000000000000000000000000011";
shiftright = true;
sig0.write(datain);
sig1.write(shiftlen);
sig3.write(shiftright);
sc_core::sc_start(5.0, sc_core::SC_NS);
std::cout<<"shift length "<<shiftlen.to_int()<<std::endl;
std::cout<<"shift right "<<shiftright<<std::endl;
std::cout<<"Data in "<<datain<<std::endl;
std::cout<<"Data out "<<sig2.read()<<std::endl;

/* Generate new data and insert into channels */
datain = "10101010010101010011000000001111";
shiftlen = "00000000000000000000000000000010";
shiftright = true;
sig0.write(datain);
sig1.write(shiftlen);
sig3.write(shiftright);
sc_core::sc_start(5.0, sc_core::SC_NS);
std::cout<<"shift length "<<shiftlen.to_int()<<std::endl;
std::cout<<"shift right "<<shiftright<<std::endl;
std::cout<<"Data in "<<datain<<std::endl;
std::cout<<"Data out "<<sig2.read()<<std::endl;

/* Stop simulation and close trace file */
sc_core::sc_stop();
sc_core::sc_close_vcd_trace_file(fp);
return 0;
}
```

**Fig. 5.25** (continued)

length and a flag variable indicating left or right shift. Figures 5.24, 5.25 contain the barrel shifter source code and its test harness.

In the console output below, 'shift right 1' indicates right shift by 'shift length' number of positions, while 'shift right 0' indicates left shift by 'shift length' number of positions. An alternative to the console output is to view the VCD trace

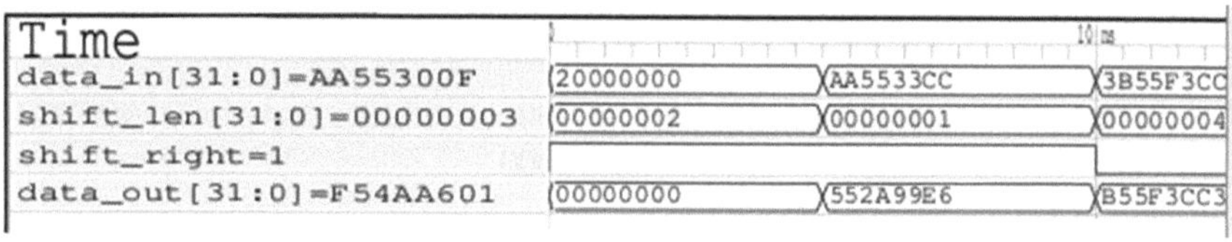


**Fig. 5.26** VCD traces generated with SystemC 2.3.0

output (Fig. 5.26 shows the VCD traces generated with SystemC 2.3.0) with an efficient VCD trace fie viewer as GTKWave.

```
SystemC 2.2.0—Aug 30 2011 19:39:07
Copyright (c) 1996–2006 by all Contributors
ALL RIGHTS RESERVED
Note: VCD trace timescale unit is set by user to 1.000000e−09 s.
shift length 2
shift right 1
Data in 00100000000000000000000000000000
Data out 00000000000000000000000000000000
shift length 1
shift right 1
Data in 10101010010101010011001111001100
Data out 01010101001010101001100111100110
shift length 4
shift right 0
Data in 00111011010101011111001111001100
Data out 10110101010111110011110011000011
shift length 5
shift right 0
Data in 10101010010101010011001111001111
Data out 01001010101001100111100111110101
shift length 3
shift right 1
Data in 10101010010101010011000000001111
Data out 11110101010010101010011000000001
shift length 2
shift right 1
Data in 10101010010101010011000000001111
Data out 11101010100101010100110000000011
SystemC: simulation stopped by user.

SystemC 2.3.0-ASI—Jul 19 2012 18:53:11
Copyright (c) 1996–2012 by all Contributors,
ALL RIGHTS RESERVED
Note: VCD trace timescale unit is set by user to 1.000000e−09 s.
```

shift length 2
shift right 1
Data in 00100000000000000000000000000000
Data out 00000000000000000000000000000000
shift length 1
shift right 1
Data in 10101010010101010011001111001100
Data out 01010101001010101001100111100110
shift length 4
shift right 0
Data in 00111011010101011111001111001100
Data out 10110101010111110011110011000011
shift length 5
shift right 0
Data in 10101010010101010011001111001111
Data out 01001010101001100111100111110101
shift length 3
shift right 1
Data in 10101010010101010011000000001111
Data out 11110101010010101010011000000001
shift length 2
shift right 1
Data in 10101010010101010011000000001111
Data out 11101010100101010100110000000011
Info: /OSCI/SystemC: Simulation stopped by user.

## 5.8 3–8 Decoder: Compound Data Types—Logic Vector

In addition to the bit and bit-vector data types, SystemC supports the logic and logic vector data types. The logic data type (*sc_logic*) can have four values '0,' '1,' 'X,' and 'Z', instead of the sc_bit's '0' and '1,' where 'X' refers to the 'don't care' and 'Z' refers to the high impedance states, respectively. A logic vector is a container for an arbitrary number of sc_logic values.

The *sc_logic* and *sc_dt*::*sc_lv* built-in data types are used to analyze a simple 3–8 decoder, a very common sub-component of a wide variety of digital circuits, and its behavior is best understood with its truth table—Fig. 5.27. In this case, A, B, and C indicate the three input bits, and $O1$, $O2$, $O3$, and $O8$ indicate the eight output bits. The source file is in Fig. 5.28 and the test harness in Fig. 5.29.

In the console traces below, the first column represents the kernel simulation time. An easier alternative, (Fig. 5.30 shows VCD traces generated with SystemC 2.3.0), is to view the VCD format trace files with an efficient viewer as GTKWave.

SystemC 2.2.0—Aug 30 2011 19:39:07

| A | B | C | 01 | 02 | 03 | 04 | 05 | 06 | 07 | 08 |
|---|---|---|---|---|---|---|---|---|---|---|
| 0 | 0 | 0 | 1 | 0 | 0 | 0 | 0 | 0 | 0 | 0 |
| 0 | 0 | 1 | 0 | 1 | 0 | 0 | 0 | 0 | 0 | 0 |
| 0 | 1 | 0 | 0 | 0 | 1 | 0 | 0 | 0 | 0 | 0 |
| 1 | 0 | 0 | 0 | 0 | 0 | 0 | 1 | 0 | 0 | 0 |
| 0 | 1 | 1 | 0 | 0 | 0 | 1 | 0 | 0 | 0 | 0 |
| 1 | 0 | 1 | 0 | 0 | 0 | 0 | 0 | 1 | 0 | 0 |
| 1 | 1 | 0 | 0 | 0 | 0 | 0 | 0 | 0 | 1 | 0 |
| 1 | 1 | 1 | 0 | 0 | 0 | 0 | 0 | 0 | 0 | 1 |

**Fig. 5.27** Truth table for 3–8 decoder

```
#ifndef THREEIGHTDEC_H
#define THREEIGHTDEC_H

#include <systemc>

const unsigned int THREE = 3;
const unsigned int EIGHT  = 8;

SC_MODULE(threeightdec)
{
  /*Declare/define input output ports */
  sc_core::sc_in< sc_dt::sc_lv<THREE> > in;
  sc_core::sc_out< sc_dt::sc_lv<EIGHT> > out;

  /* Declare/define data members */
  sc_dt::sc_lv<THREE> inlv;
  sc_dt::sc_lv<EIGHT> outlv;

  /* Main 3 - 8 decoder thread */
  void threeightdec_proc0()
  {
    while(1)
    {
      wait();
      inlv = in.read();
      if(inlv == "000") outlv = "10000000";
      else if(inlv == "001") outlv = "01000000";
```

**Fig. 5.28** 3–8 decoder

```
      else if(inlv == "010") outlv = "00100000";
      else if(inlv == "011") outlv = "00010000";
      else if(inlv == "100") outlv = "00001000";
      else if(inlv == "101") outlv = "00000100";
      else if(inlv == "110") outlv = "00000010";
      else if(inlv == "111") outlv = "00000001";
      else outlv = "ZZZZZZZZ";
      out.write(outlv);
    }
  }

  /* Constructor - initialize variables */
  SC_CTOR(threeightdec):inlv("ZZZ"), outlv("ZZZZZZZZ")
  {
    SC_THREAD(threeightdec_proc0);
    sensitive << in;
  }

  /* Destructor */
  ~threeightdec(){ }
};

#endif
```

**Fig. 5.28** (continued)

```
#include "threeightdec.h"
#include "systemc.h"

int sc_main(int argc, char **argv)
{
  /* declare/define channels and data members */
  sc_core::sc_signal< sc_dt::sc_lv<3> > insig;
  sc_core::sc_signal< sc_dt::sc_lv<8> > outsig;
  sc_dt::sc_lv<3> invect;
  sc_dt::sc_lv<8> outvect;

  /* Declare/define 3 - 8 decoder and trace generator */
  threeightdec dec_3_8("dec_3_8");
```

**Fig. 5.29** Test harness for 3–8 decoder

```
sc_core::sc_trace_file *fp;
fp = sc_create_vcd_trace_file("tr_threeightdec");
fp->set_time_unit(1.0, sc_core::SC_NS);

/* Connect module ports and channels */
dec_3_8.in(insig);
dec_3_8.out(outsig);
sc_core::sc_trace(fp, insig, "lv3in");
sc_core::sc_trace(fp, outsig, "lv8out");

/* Set data values and run simulation for pre-defind time interval */
invect = "ZZZ";
insig.write(invect);
sc_core::sc_start(5.0, sc_core::SC_NS);
 /* Print input and output signal values on console */
std::cout<<sc_core::sc_time_stamp().to_seconds()<<" Input ";

std::cout<<insig.read()<<" Output "<<outsig.read()<<std::endl;

/* Re-set data values and run simulation for pre-defind time interval */
invect = "000";
insig.write(invect);
sc_core::sc_start(5.0, sc_core::SC_NS);
std::cout<<sc_core::sc_time_stamp().to_seconds()<<" Input ";

std::cout<<insig.read()<<" Output "<<outsig.read()<<std::endl;

/* Re-set data values and run simulation for pre-defind time interval */
invect = "100";
insig.write(invect);
sc_core::sc_start(5.0, sc_core::SC_NS);
std::cout<<sc_core::sc_time_stamp().to_seconds()<<" Input ";

std::cout<<insig.read()<<" Output "<<outsig.read()<<std::endl;

/* Re-set data values and run simulation for pre-defind time interval */
invect = "010";
insig.write(invect);
sc_core::sc_start(5.0, sc_core::SC_NS);
std::cout<<sc_core::sc_time_stamp().to_seconds()<<" Input ";
```

**Fig. 5.29** (continued)

```
std::cout<<insig.read()<<" Output "<<outsig.read()<<std::endl;

/* Re-set data values and run simulation for pre-defind time interval */
invect = "001";
insig.write(invect);
sc_core::sc_start(5.0, sc_core::SC_NS);
std::cout<<sc_core::sc_time_stamp().to_seconds()<<" Input ";

std::cout<<insig.read()<<" Output "<<outsig.read()<<std::endl;

/* Re-set data values and run simulation for pre-defind time interval */
invect = "101";
insig.write(invect);
sc_core::sc_start(5.0, sc_core::SC_NS);
std::cout<<sc_core::sc_time_stamp().to_seconds()<<" Input ";

std::cout<<insig.read()<<" Output "<<outsig.read()<<std::endl;

/* Re-set data values and run simulation for pre-defind time interval */
invect = "011";
insig.write(invect);
sc_core::sc_start(5.0, sc_core::SC_NS);
std::cout<<sc_core::sc_time_stamp().to_seconds()<<" Input ";

std::cout<<insig.read()<<" Output "<<outsig.read()<<std::endl;

/* Re-set data values and run simulation for pre-defind time interval */
invect = "110";
insig.write(invect);
sc_core::sc_start(5.0, sc_core::SC_NS);
std::cout<<sc_core::sc_time_stamp().to_seconds()<<" Input ";

std::cout<<insig.read()<<" Output "<<outsig.read()<<std::endl;

/* Re-set data values and run simulation for pre-defind time interval */
invect = "111";
insig.write(invect);
sc_core::sc_start(5.0, sc_core::SC_NS);
std::cout<<sc_core::sc_time_stamp().to_seconds()<<" Input ";
```

**Fig. 5.29** (continued)

```
std::cout<<insig.read()<<" Output "<<outsig.read()<<std::endl;

/* Re-set data values and run simulation for pre-defind time interval */
invect = "XXX";
insig.write(invect);
sc_core::sc_start(5.0, sc_core::SC_NS);
std::cout<<sc_core::sc_time_stamp().to_seconds()<<" Input ";

std::cout<<insig.read()<<" Output "<<outsig.read()<<std::endl;

  /* Stop simulation and close trace file */
  sc_core::sc_stop();
  sc_core::sc_close_vcd_trace_file(fp);
  return 0;
}
```

**Fig. 5.29** (continued)

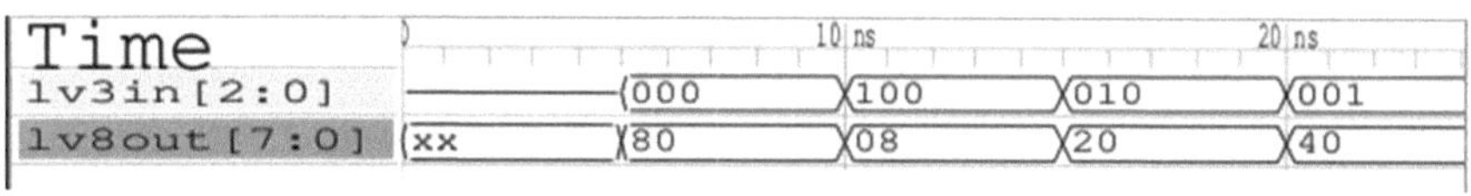


**Fig. 5.30** VCD traces generated with SystemC 2.3.0

Copyright (c) 1996–2006 by all Contributors
ALL RIGHTS RESERVED
Note: VCD trace timescale unit is set by user to 1.000000e−09 s.
5e−09 Input ZZZ Output XXXXXXXX
1e−08 Input 000 Output 10000000
1.5e−08 Input 100 Output 00001000
2e−08 Input 010 Output 00100000
2.5e−08 Input 001 Output 01000000
3e−08 Input 101 Output 00000100
3.5e−08 Input 011 Output 00010000
4e−08 Input 110 Output 00000010
4.5e−08 Input 111 Output 00000001
5e−08 Input XXX Output ZZZZZZZZ
SystemC: simulation stopped by user.

SystemC 2.3.0-ASI—Jul 19 2012 18:53:11
Copyright (c) 1996–2012 by all Contributors,
ALL RIGHTS RESERVED
Note: VCD trace timescale unit is set by user to 1.000000e−09 s.

```
5e−09 Input ZZZ Output XXXXXXXX
1e−08 Input 000 Output 10000000
1.5e−08 Input 100 Output 00001000
2e−08 Input 010 Output 00100000
2.5e−08 Input 001 Output 01000000
3e−08 Input 101 Output 00000100
3.5e−08 Input 011 Output 00010000
4e−08 Input 110 Output 00000010
4.5e−08 Input 111 Output 00000001
5e−08 Input XXX Output ZZZZZZZZ
Info: /OSCI/SystemC: Simulation stopped by user.
```

## References


1. Sun MicroSystems - Austin. IEEE Xplore. IEEE, 12 Nov. 2007 *UltraSparc T2 A highly-threaded, power-efficient SPARC SOC* http://ieeexplore.ieee.org/xpl/login.jsp?tp=&arnumber=4425786&url=http%3A%2F%2Fieeexplore.ieee.org%2Fxpls%2Fabs_all.jsp%3Farnumber%3D4425786#/xpl/articleDetails.jsp?tp=&arnumber=4425786&url=http%3A%2F
2. IEEE Standards Board, IEEE Standards Association Standards Board IEEE-SA—IEEE Get Program. IEEE, 2011 *IEEE Standard 1666 Open SystemC Language Reference Manual (LRM)* http://standards.ieee.org/getieee/1666/download/1666-2011.pdf
3. Mhambrey S. S., Clark L. T., Maurya S. K., Berezowski, K (2010) *Out_of_order Issue Logic Using Sorting Networks* – GLSVLSI - '10 Providence RI USA
4. Charles, Roth H., and Kinney I. Larry, *Fundamentals of Logic Design*. Sixth ed. Cengage Learning, 2010. Print
5. Kenneth, Martin. *Digital Integrated Circuit Design*. Oxford University Press, 2000. Print

# Chapter 6
# Modeling Sequential Logic Circuits, Implicit Events, Primitive Channels, and Their Combinations

**Abstract** As compared to a combinational logic circuit, a sequential logic circuit always runs synchronized with a master or derived clock. A number of design examples in this chapter are based on OpenSparc T2 [1] microprocessor RTL.

## 6.1 JK Master–Slave Flip-Flop

A flip-flop [2, 3] is a basic sequential logic circuit that is a crucial sub-component of a variety of more complex sequential logic circuits. A flip-flop is a classic example of a *finite state machine*, with a set of inputs, a set of control signals, and a set of outputs, and can exist in one of two stable states, often referred to as *set* and *reset*. It has built-in *memory* property, as its next state depends both on its current state and on applied inputs and control signals. Standard flip-flop output is often referred to as Q, Q′ (Q complement), its present output $Qt$, and the next output $Qt + 1$. A master–slave flip-flop is two flip-flops in series, the output of the first (master) being the input for the second (slave), and driven by complementary control signals.

Of all the flip-flops in use, the JK flip-flop is a universal one, as it can be configured to serve as an SR (set–reset), D, or T (toggle) flip-flop. An enhancement of the basic SR flip-flop ($J = Set,\ K = Reset$), with a clock (*sequential logic)* control signal, it interprets the $J = K = 1$ (or Set = Reset = 1) command, as *toggle or flip* command. The $j = 1$, $K = 0$ input is the *set* command, while the $J = 0$, $K = 1$ input is the *reset* command. The $J = 0\ K = 0$ input maintains the current state. Figure 6.1 has the truth table for the JK flip-flop, and the source code is in Fig. 6.2 and test harness in Fig. 6.3.

A. Banerjee and B. Sur, *SystemC and SystemC-AMS in Practice*,
DOI: 10.1007/978-3-319-01147-9_6,

| Q | J | K | Q(n+1) |
|---|---|---|---|
| 0 | 0 | 0 | 0 |
| 0 | 0 | 1 | 0 |
| 0 | 1 | 0 | 1 |
| 0 | 1 | 1 | 1 |
| 1 | 0 | 0 | 1 |
| 1 | 0 | 1 | 0 |
| 1 | 1 | 0 | 1 |
| 1 | 1 | 1 | 0 |

**Fig. 6.1** JK flip-flop truth table

```
#ifndef FLIPFLOP_H
#define FLIPFLOP_H

#include <systemc>

/* Inverter for complementary clock signals for master-slave */
SC_MODULE(inv)
{
  /* Declare/define input output signals */
  sc_core::sc_in<bool> din;
  sc_core::sc_out<bool> dout;

  /* Inverter operation */
  void inv_proc0()
  {
    dout.write(din.read());
  }

  SC_CTOR(inv)
  /* Constructor */
  {
     /* Register inverter operation and sensitize to input */
     SC_METHOD(inv_proc0);
     sensitive << din;
  }
```

**Fig. 6.2** JK flip-flop

```
  ~inv(){ } /* Destructor */
};

SC_MODULE(jkff)
{
  /* Declare/define input/output ports */
  sc_core::sc_in<bool> clk;
  sc_core::sc_in<bool> din0;
  sc_core::sc_in<bool> din1;
  sc_core::sc_out<bool> dout0;
  sc_core::sc_out<bool> dout1;

  /* Members */
  bool b0;
  bool b1;
  bool b2;
  bool b3;

  /* JK flip-flop operation */
  void jkff_proc0()
  {
    b1 = din0.read();
    b2 = din1.read();
/* Print out instance name, time stamp and present state */
    std::cout<<name()<<" @ "<<sc_core::sc_time_stamp().to_seconds()<<" sec
Qt "<<b0;
/* Define main transitions */
    if(b0 == false)
    {
      if(b1 == false && b2 == false) b0 = false;
      else if(b1 == false && b2 == true) b0 = false;
      else if(b1 == true && b2 == false) b0 = true;
      else if(b1 == true && b2 == true) b0 = true;
    }
    else if(b0 == true)
    {
      if(b1 == false && b2 == false) b0 = true;
      else if(b1 == false && b2 == true) b0 = false;
      else if(b1 == true && b2 == false) b0 = true;
```

**Fig. 6.2** (continued)

```
        else if(b1 == true && b2 == true) b0 = false;
      }
      /* Print out next state */
      std::cout<<" J "<<b1<<" K "<<b2<<" Q(t+1) "<<b0<<std::endl;
      /* Generate complementary output, and write both to ports */
      b3 = b0;
      dout0.write(b0);
      dout1.write(!b3);
    }

    /* Constructor */
    SC_CTOR(jkff):b0(false), b1(false), b2(false), b3(false)
    {
      /* Register flip-flop method amd sensitize it */
      SC_METHOD(jkff_proc0);
      sensitive << clk.pos() << din0 << din1;
    }

    ~jkff(){ }
   /*Destructor */
  };
  #endif
```

**Fig. 6.2** (continued)

```
  #include "flipflop.h"
  #include "syscsrcs.h"
  #include "systemc.h"

  int sc_main(int argc, char **argv)
  {
    /* Declare/define signal channels */
    sc_core::sc_signal<bool> sig0;
    sc_core::sc_signal<bool> sig1;
    sc_core::sc_signal<bool> sig2;
    sc_core::sc_signal<bool> sig3;
    sc_core::sc_signal<bool> sig4;
    sc_core::sc_signal<bool> sig5;
    sc_core::sc_signal<bool> sig6;
```

**Fig. 6.3** Test harness for master–slave JK flip-flop

```
?* Declare/define clock */
sc_core::sc_clock clk("clk", 10.0, sc_core::SC_NS, 0.5);

/* Declare/define flip-flops and inverter */
inv inv_0("inv_0");
jkff jk_ff_0("master");
jkff jk_ff_1("slave");

/* Declare/define J, K signal input sources */
syncdatasrc s_data_0("s_data_0");
syncdatasrc s_data_1("s_data_1");
/* Declare/define VCD trace file */
sc_core::sc_trace_file *fp;
fp = sc_create_vcd_trace_file("tr_jkff");
fp->set_time_unit(1.0, sc_core::SC_NS);

/* Connect module ports and channels */
s_data_0.clk(clk);
s_data_0.dout(sig0);
s_data_1.clk(clk);
s_data_1.dout(sig1);

inv_0.din(clk);
inv_0.dout(sig4);
/* Construct master-slave JK flip-flop */
jk_ff_0.clk(clk);
jk_ff_0.din0(sig0);
jk_ff_0.din1(sig1);
jk_ff_0.dout0(sig2);
jk_ff_0.dout1(sig3);
jk_ff_1.clk(sig4);
jk_ff_1.din0(sig2);
jk_ff_1.din1(sig3);
jk_ff_1.dout0(sig5);
jk_ff_1.dout1(sig6);

/* Connect signal channels and trace file */
sc_core::sc_trace(fp, clk, "clk");
sc_core::sc_trace(fp, sig0, "j");
```

**Fig. 6.3** (continued)

```
    sc_core::sc_trace(fp, sig1, "k");
    sc_core::sc_trace(fp, sig2, "qm");
    sc_core::sc_trace(fp, sig3, "qbm");
    sc_core::sc_trace(fp, sig5, "qs");
    sc_core::sc_trace(fp, sig6, "qbs");

    /* Run simulation for fixed time period, and stop, close trace file */
    sc_core::sc_start(1000.0, sc_core::SC_NS);
    sc_core::sc_stop();
    sc_core::sc_close_vcd_trace_file(fp);
    return 0;
}
```

**Fig. 6.3** (continued)

The simulation paradigm is different from the case of the combinational logic designs. In combinational logic because of the absence of clocks, simulator is executed in discrete time steps. Sequential logic requires clocks, so that the simulator may be executed for large time interval at one go, without any time stepping.

```
   SystemC 2.2.0—August 30, 2011, 19:39:07
   Copyright (c) 1996–2006 by all contributors,
   ALL RIGHTS RESERVED

Note: VCD trace timescale unit is set by user to 1.000000e-09 s.
master @ 0 s Qt 0 J 0 K 0 Q(t + 1) 0
slave @ 0 s Qt 0 J 0 K 0 Q(t + 1) 0
master @ 0 s Qt 0 J 0 K 0 Q(t + 1) 0
slave @ 0 s Qt 0 J 0 K 1 Q(t + 1) 0
master @ 1e-08 s Qt 0 J 0 K 1 Q(t + 1) 0
slave @ 1e-08 s Qt 0 J 0 K 1 Q(t + 1) 0
master @ 2e-08 s Qt 0 J 0 K 0 Q(t + 1) 0
slave @ 2e-08 s Qt 0 J 0 K 1 Q(t + 1) 0
master @ 3e-08 s Qt 0 J 0 K 0 Q(t + 1) 0
slave @ 3e-08 s Qt 0 J 0 K 1 Q(t + 1) 0
master @ 4e-08 s Qt 0 J 1 K 1 Q(t + 1) 1
slave @ 4e-08 s Qt 0 J 1 K 0 Q(t + 1) 1
……
……
master @ 9.8e-07 s Qt 1 J 1 K 1 Q(t + 1) 0
slave @ 9.8e-07 s Qt 1 J 0 K 1 Q(t + 1) 0
master @ 9.9e-07 s Qt 0 J 0 K 0 Q(t + 1) 0
slave @ 9.9e-07 s Qt 0 J 0 K 1 Q(t + 1) 0
SystemC: simulation stopped by user.
```

```
SystemC 2.3.0-ASI—July 19, 2012, 18:53:11
Copyright (c) 1996–2012 by all contributors,
ALL RIGHTS RESERVED

Note: VCD trace timescale unit is set by user to 1.000000e-09 s.
master @ 0 s Qt 0 J 0 K 0 Q(t + 1) 0
slave @ 0 s Qt 0 J 0 K 0 Q(t + 1) 0
master @ 0 s Qt 0 J 0 K 0 Q(t + 1) 0
slave @ 0 s Qt 0 J 0 K 1 Q(t + 1) 0
master @ 1e-08 s Qt 0 J 0 K 1 Q(t + 1) 0
slave @ 1e-08 s Qt 0 J 0 K 1 Q(t + 1) 0
master @ 2e-08 s Qt 0 J 0 K 0 Q(t + 1) 0
slave @ 2e-08 s Qt 0 J 0 K 1 Q(t + 1) 0
master @ 3e-08 s Qt 0 J 0 K 0 Q(t + 1) 0
slave @ 3e-08 s Qt 0 J 0 K 1 Q(t + 1) 0
.......
.......
slave @ 9.8e-07 s Qt 1 J 0 K 1 Q(t + 1) 0
master @ 9.9e-07 s Qt 0 J 0 K 0 Q(t + 1) 0
slave @ 9.9e-07 s Qt 0 J 0 K 1 Q(t + 1) 0
Info: /OSCI/SystemC: simulation stopped by user.
```

The input/output traces are in Figs. 6.4 (SystemC 2.2.0) and 6.5 (SystemC 2.3.0).

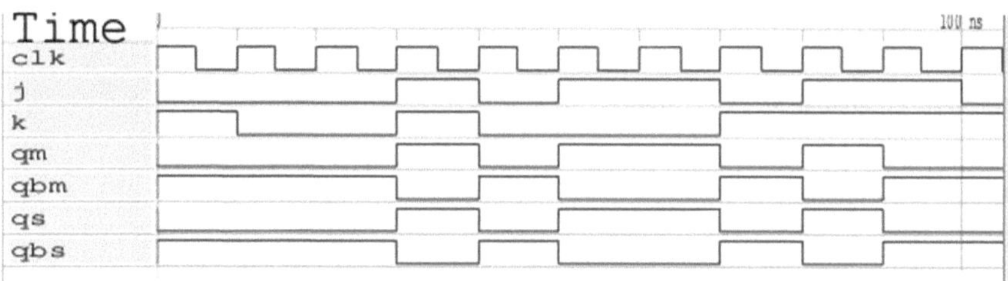


**Fig. 6.4** Input/output to/from master–slave JK flip-flop SystemC 2.2.0

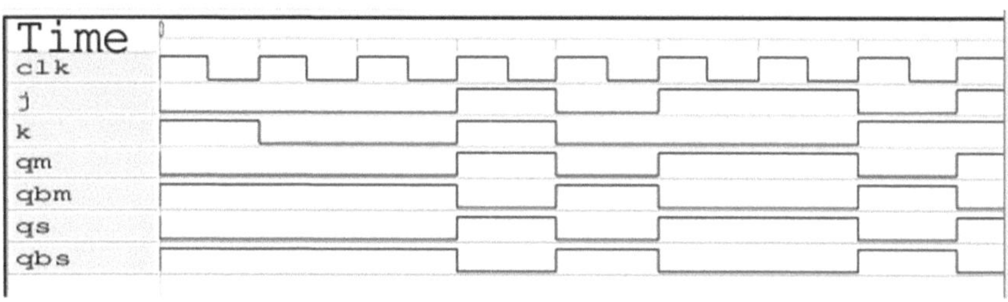


**Fig. 6.5** Input/output to/from master–slave JK flip-flop SystemC 2.3.0

## 6.2 64-Bit Serial-in Parallel-Out Shift Register

A shift register [2, 3] is a sequential logic circuit that is an important component of several complex sequential logic circuits. A shift register may be serial-in parallel-out or parallel-in serial-out. Data bits are fed in through one end of a serial-in parallel-out shift register and read out all together in parallel after a predefined number of clock ticks. This predefined number is equal to the number of register cells, each of which is a flip-flop, typically D flip-flop. A D flip-flop transfers its input (logic '1' or logic '0') at each clock pulse. For a parallel-in serial-out shift register, data bits are read in parallel at each of the register cells and then shifted out in series at one of the extremities of the register. The shifting is performed over an interval that is equal to the number of shift registers multiplied by the clock period. Here, a 64-bit serial-in parallel-out shift register is examined. The serial data input bits for the shift register are generated by a synchronized (with the same master clock to which the shift register is synchronized with) module that, with each clock tick, generates a random number—if that random number is divisible by 2, the module emits a '1', else a '0'. Figures 6.6 and 6.7 show, respectively, the 64-bit serial-in parallel-out shift register source code and the test harness.

Console output for the register is as follows. An easier alternative to examining the console output is to view the VCD trace files (as generated in the test harness) with an efficient VCD trace file viewer as GTKwave.

```
#ifndef SIPO_H
#define SIPO_H

#include <systemc>

SC_MODULE(sipo64)
{
   /* Input/output ports */
   sc_core::sc_in<bool> clk;
   sc_core::sc_in<bool> in;
   /* 64 bit output bit vector port for 64 parallel bit output */
   sc_core::sc_out< sc_dt::sc_bv<64> > out;
  /* Module internal variables */
   sc_dt::sc_bv<64> data;
   sc_dt::sc_bv<64> idlebits;
   bool b1;
   unsigned int i;
  unsigned int count;
```

**Fig. 6.6** 64-bit serial-in parallel-out shift register

```
/* Thread for shift operation */
void shift()
{
  while(1)
  {
     wait();
     if(count < 64)
     {
       /* Collect 64 serial bits in serial-in parallel-out shift register */
       b1 = in.read();
       data.range(count,count) = (b1 == true) ? "1" : "0";
       count = count + 1;
       out.write(idlebits);
     }
     else
    {
       count = 0;
       out.write(data);
       /* Print out the 64 bits collected

          over the last 64 clock ticks */

       std::cout<<sc_core::sc_time_stamp().to_seconds()<<" parallel 64 bit
output : "<<std::endl;

       std::cout<<data<<std::endl;
       /* Reset 64 bit serial-in parallel-out shift register */
       data =
"000000000000000000000000000000000000000000000000000000000000000
0";
     }
    }
 }
 /* Constructor - bit vector initialization */
 SC_CTOR(sipo64):count(0),

data("00000000000000000000000000000000000000000000000000000000000
00000"),
```

Fig. 6.6 (continued)

```
idlebits("0000000000000000000000000000000000000000000000000000000
00000000")
 {
   /* Declare/assign clocked thread */
   SC_CTHREAD(shift, clk.pos());
 }
 ~sipo64(){ } /* Destructor */
};
```

**Fig. 6.6** (continued)

```
#include "sipo.h"
#include "syscsrcs.h"

int sc_main(int argc, char **argv)
{
  /* Input/output channels */
  sc_core::sc_signal<bool> sigin;
  sc_core::sc_signal< sc_dt::sc_bv<64> > sigout;
  /* Declare/define master clock */
  sc_core::sc_clock clk("clk", 2.0,
                             sc_core::SC_NS,
                             0.5);
  /* Declare/define 64-bit serial-in parallel-out shift register */
  sipo64 sipo_64("sipo_64");
  /* Declare/define serial data bitstream source */
  syncdatasrc sync_d_src("sync_d_src");
  /* Declare/define trace VCD format file */
  sc_core::sc_trace_file *fp =  sc_core::sc_create_vcd_trace_file("tr_sipo");
  fp->set_time_unit(1.0, sc_core::SC_NS);
  /* Connect module ports and channels */
  sync_d_src.clk(clk);
  sync_d_src.dout(sigin);
  sipo_64.clk(clk);
  sipo_64.in(sigin);
  sipo_64.out(sigout);
  /* Connect trace file and data channels */
  sc_trace(fp, clk, "clock");
  sc_trace(fp, sigin, "serial_data");
```

**Fig. 6.7** 64-bit serial-in parallel-out shift register test harness

```
  sc_trace(fp, sigout, "parallel_data");
  /* Start and run simulation for pre-defined time interval */
  sc_core::sc_start(1000.0, sc_core::SC_NS);
  /* Stop simulation and close trace file */
  sc_core::sc_stop();
  sc_core::sc_close_vcd_trace_file(fp);
  return 0;
}
```

**Fig. 6.7** (continued)

```
    SystemC 2.2.0—August 30, 2011, 19:39:07
    Copyright (c) 1996–2006 by all contributors,
    ALL RIGHTS RESERVED

Note: VCD trace timescale unit is set by user to 1.000000e-09 s.
@ 1.3e-07 s parallel 64-bit output:
1000111011010101101000010100011100001110010111110010100110000100
@ 2.6e-07 s parallel 64-bit output:
0000000101000110111010001110011110111101001111101011010101000101
@ 3.9e-07 s parallel 64-bit output:
0000101001001100000101111101011100001010101011100010110010100100
@ 5.2e-07 s parallel 64-bit output:
1101100010111100110110001110110011010101011000111111100011010011
@ 6.5e-07 s parallel 64-bit output:
0010111001110101001111111110110001000011010000110011111011111011
@ 7.8e-07 s parallel 64-bit output:
0100001011011000110110001111011101001111110101111110100100100000
@ 9.1e-07 s parallel 64-bit output:
1111101110111001011011000001001110110111000101111110111011111011

    SystemC: simulation stopped by user.
    SystemC 2.3.0-ASI—July 19, 2012, 18:53:11
    Copyright (c) 1996–2012 by all contributors,
    ALL RIGHTS RESERVED

Note: VCD trace timescale unit is set by user to 1.000000e-09 s.
@ 1.3e-07 s parallel 64-bit output:
1000111011010101101000010100011100001110010111110010100110000100
@ 2.6e-07 s parallel 64-bit output:
0000000101000110111010001110011110111101001111101011010101000101
@ 3.9e-07 s parallel 64-bit output:
0000101001001100000101111101011100001010101011100010110010100100
@ 5.2e-07 s parallel 64-bit output:
1101100010111100110110001110110011010101011000111111100011010011
```

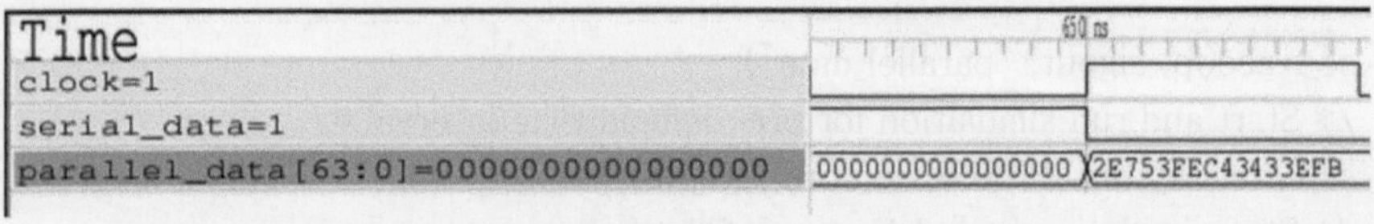


**Fig. 6.8** Input/output traces for 64-bit serial-in parallel-out shift register. Idle bytes '0000…0000' (64 0s) appear at the register output node during the time interval when 64 input bits are serially shifted through 64 register cells. Trace shows valid data appearing at the 64 parallel register cell output nodes at the end of 64 clock ticks SystemC 2.2.0

**Fig. 6.9** Input/output traces for 64-bit serial-in parallel-out shift register. Idle bytes '0000…0000' (64 0s) appear at the register output node during the time interval when 64 input bits are serially shifted through 64 register cells. Trace shows valid data appearing at the 64 parallel register cell output nodes at the end of 64 clock ticks SystemC 2.3.0

@ 6.5e-07 s parallel 64-bit output:
0010111001110101001111111110110001000011010000110011111011111011
@ 7.8e-07 s parallel 64-bit output:
010000101101100011011000111011101001111110101111110100100100000
@ 9.1e-07 s parallel 64-bit output:
1111101110111001011011000001001110110111000101111110111011111011
Info: /OSCI/SystemC: simulation stopped by user.

The input/output trace is in Figs. 6.8 (SystemC 2.2.0) and 6.9 (SystemC 2.3.0).

## 6.3 64-Bit Asynchronous Counter (Ripple Counter)

A common sequential logic circuit is the asynchronous (ripple) counter [2, 3]. It consists of a number of flip-flops (e.g., D flip-flop) connected in series. An external clock is applied *only* to the first D flip-flop's clock input. Each D flip-flop in the chain has its Q_bar output tied to its D input, with its Q output acting as the clock input for the next D flip-flop in the chain. Thus, the count value 'ripples' through the chain. A single D flip-flop, with its Q_bar output to its D input, acts as a divide-by-2 counter (frequency divider). Since a 2-bit ripple counter (or divide-by-2 frequency divider) is achieved by one D flip-flop, any $2^n$ counter may be achieved by having n D flip-flops in series, configured as above, implying that a 64-bit counter is obtained by having 6 D flip-flops in series. Our design is somewhat structural, in that we design a D flip-flop module and then connect 6 D flip-flop objects in series, as described above. Figures 6.10, 6.11, 6.12 (SystemC

```
#ifndef DFF2_H
#define DFF2_H

#include <systemc>

SC_MODULE(dff_2)
{
 /* Declare/define input/output ports */
 sc_core::sc_in<bool> din;
 sc_core::sc_in<bool> clock;
 sc_core::sc_out<bool> dout;  /* Q out */
 sc_core::sc_out<bool> doutb; * Q_bar */
 bool b0;

 /* D flip-flop operation thread */
 void dffoperation_2()
 {
   while(true)
   {
     /* Wait for positive edge of clock */
     wait();
     b0 = din.read();
     dout.write(b0);
     doutb.write(!b0);
   }
 }

 /* Constructor */
 SC_CTOR(dff_2):b0(false)
 {
   SC_CTHREAD(dffoperation_2, clock.pos()); /*Declare/assign thread */
 }
 ~dff_2(){}
};
#endif
```

**Fig. 6.10** 64-bit ripple counter

2.2.0), and 6.13 (SystemC 2.3.0) contain the source code, test harness, and input/output traces for a 64-bit ripple counter.

For both the SystemC 2.3.0 and SystemC 2.2.0 output traces, ‘q5’ indicates the final ‘divide-by-64’ output.

```
#include "dff2.h"

int sc_main(int argc, char **argv)
{
  /*Declare/define channels for Q and Q_bar outputs of
    each D flip-flop */
  sc_core::sc_signal<bool> d_qd0;
  sc_core::sc_signal<bool> d_qd1;
  sc_core::sc_signal<bool> d_qd2;
  sc_core::sc_signal<bool> d_qd3;
  sc_core::sc_signal<bool> d_qd4;
  sc_core::sc_signal<bool> d_qd5;
  sc_core::sc_signal<bool> q0;
  sc_core::sc_signal<bool> q1;
  sc_core::sc_signal<bool> q2;
  sc_core::sc_signal<bool> q3;
  sc_core::sc_signal<bool> q4;
  sc_core::sc_signal<bool> q5;

  /* Declare/define master clock */
  sc_core::sc_clock clk("clk", 2.0, sc_core::SC_NS, 0.5);
  /* Declare/define 6 D flip-flops */
  dff_2 d_f_f_2_0("d_f_f_2_0");
  dff_2 d_f_f_2_1("d_f_f_2_1");
  dff_2 d_f_f_2_2("d_f_f_2_2");
  dff_2 d_f_f_2_3("d_f_f_2_3");
  dff_2 d_f_f_2_4("d_f_f_2_4");
  dff_2 d_f_f_2_5("d_f_f_2_5");

  /* Declare/define/initialize trace file for trace data */
  sc_core::sc_trace_file *fp;
  fp = sc_core::sc_create_vcd_trace_file("tr_ripple_cntr");
  fp->set_time_unit(1.0, sc_core::SC_NS);

  /* Connect module ports and channels */
  d_f_f_2_0.clock(clk);
  d_f_f_2_0.din(d_qd0);
  d_f_f_2_0.dout(q0);
  d_f_f_2_0.doutb(d_qd0);
  d_f_f_2_1.clock(q0);
```

**Fig. 6.11** 64-bit counter test harness

```
    d_f_f_2_1.din(d_qd1);
    d_f_f_2_1.dout(q1);
    d_f_f_2_1.doutb(d_qd1);
    d_f_f_2_2.clock(q1);
    d_f_f_2_2.din(d_qd2);
    d_f_f_2_2.dout(q2);
    d_f_f_2_2.doutb(d_qd2);
    d_f_f_2_3.clock(q2);
    d_f_f_2_3.din(d_qd3);
    d_f_f_2_3.dout(q3);
    d_f_f_2_3.doutb(d_qd3);
    d_f_f_2_4.clock(q3);
    d_f_f_2_4.din(d_qd4);
    d_f_f_2_4.dout(q4);
    d_f_f_2_4.doutb(d_qd4);
    d_f_f_2_5.clock(q4);
    d_f_f_2_5.din(d_qd5);
    d_f_f_2_5.dout(q5);
    d_f_f_2_5.doutb(d_qd5);

    /* Connect trace file and data channels */
    sc_trace(fp, clk, "clk");
    sc_trace(fp, q0, "q0");
    sc_trace(fp, q1, "q1");
    sc_trace(fp, q2, "q2");
    sc_trace(fp, q3, "q3");

    /* Start and run simulation for pre-defined time interval */
    sc_core::sc_start(1000.0, sc_core::SC_NS);
    /* Stop simulation and close trace file */
    sc_core::sc_stop();
    sc_core::sc_close_vcd_trace_file(fp);
    return 0;
}
```

**Fig. 6.11** (continued)

## 6.4 Simple Finite State Machine

We examine a simple finite state machine (FSM) [2, 3]. A child is enjoying his/her holiday by riding around the rooms of his/her home on a tricycle, starting and stopping randomly. The tricycle can be in two possible states—IDLE and moving.

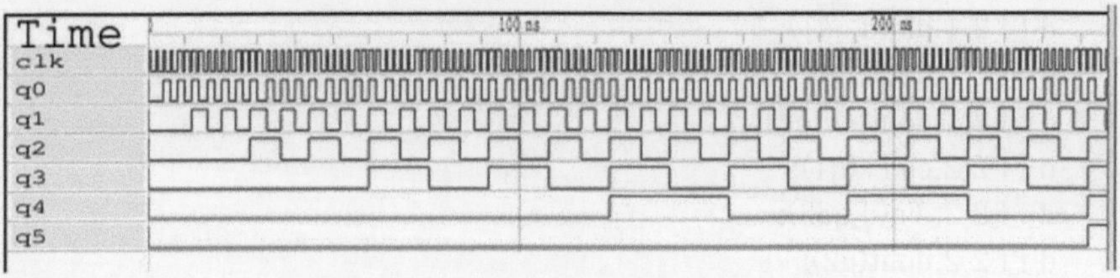


**Fig. 6.12** Input/output traces for 64-bit ripple counter SystemC 2.2.0

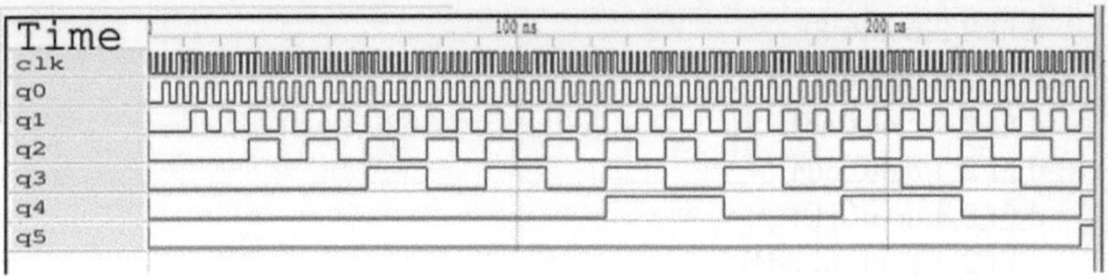


**Fig. 6.13** Input/output traces for 64-bit ripple counter SystemC 2.3.0

Figures 6.14, 6.15, and 6.16 contain the source code and test harness. It consists of a module that generates the transitions (IDLE- > MOVE and vice versa) and the FSM itself.

Console output from the FSM looks like:

SystemC 2.2.0—August 30, 2011, 19:39:07
Copyright (c) 1996–2006 by all contributors,
ALL RIGHTS RESERVED

@ 2.000000e-05 s MOVE
@ 6.000000e-05 s IDLE
@ 7.000000e-05 s MOVE
@ 1.600000e-04 s IDLE
@ 9.100000e-04 s MOVE
@ 9.500000e-04 s IDLE
SystemC: simulation stopped by user.

SystemC 2.3.0-ASI—July 19, 2012, 18:53:11
Copyright (c) 1996–2012 by all contributors,
ALL RIGHTS RESERVED

@ 2.000000e-05 s MOVE
@ 6.000000e-05 s IDLE
@ 7.000000e-05 s MOVE
@ 1.600000e-04 s IDLE
@ 9.100000e-04 s MOVE
@ 9.500000e-04 s IDLE
Info: /OSCI/SystemC: simulation stopped by user.

```
#ifndef FSM_H
#define FSM_H

#include <systemc>

const char *str0 = "IDLE";
const char *str1 = "MOVE";

SC_MODULE(fsm_input_src)
{
  public:
  /* Declare/define input/output ports */
  sc_core::sc_in<bool> clk;
  sc_core::sc_out<bool> fsm_input_out;
  /* Declare/define module variables */
  bool out_value;
  bool out_value_stored;
  unsigned int value;

  /* Main trigger generation operation thread */
  void fsm_input_src_proc0()
  {
    while(1)
      {
      wait();
      value = (unsigned int)(10.0*drand48());
      if((value % 2 == 0) && (value % 7 == 0) )
      {
        out_value_stored = out_value;
        out_value = true;
      }
      else
      {
        out_value = false;
      }
      fsm_input_out.write(out_value_stored);
    }
  }
```

**Fig. 6.14** Finite state machine transition generator

```
  /* Constructor - initialize module variables */
  SC_CTOR(fsm_input_src):out_value(false),
                         out_value_stored(false),
                         value(0)
  {
    SC_CTHREAD(fsm_input_src_proc0, clk.pos()); /*Declare/assign main
 clocked thread */
  }
  ~fsm_input_src() { }
 /* Destructor */
};
```

**Fig. 6.14** (continued)

```
SC_MODULE(fsm)
{
  private:
   /* Declare/define module private members */
   bool state_var;
   /* Enumeration for machine states */
   enum states{IDLE=1, MOVE};
   unsigned int curr_state;
  public:
   /*Declare/define input/output ports */
   sc_core::sc_in<bool> fsm_in;
   sc_core::sc_out<unsigned int> curr_state_out;
   double curr_time;

   /* Finite state machine main operation thread */
   void fsm_proc0()
   {
     while(1)
     {
       wait();
       state_var = fsm_in.read();
       curr_state = state_var == true ? MOVE : IDLE;
       curr_state_out.write(curr_state);
       curr_time = sc_core::sc_time_stamp().to_seconds();
       if(curr_state == MOVE)
        if(curr_state == MOVE)
```

**Fig. 6.15** Simple finite state machine (FSM) with two states

```
        std::cout<<"@ "<<std::scientific<<curr_time<<" seconds
"<<str1<<std::endl;
      else
        std::cout<<"@ "<<std::scientific<<curr_time<<" seconds
"<<str0<<std::endl;
    }
  }

  /* Constructor - initialize variables */
  SC_CTOR(fsm): state_var(false), curr_state(IDLE)
  {
    SC_THREAD(fsm_proc0);
 /* Declare/assign thread */
    sensitive << fsm_in;
  /* Define main thread sensitivity list */
  }
  ~fsm(){ }
 /* Destructor */
};

#endif
```

**Fig. 6.15** (continued)

```
#include "fsm.h"

int sc_main(int argc, char **argv)
{
  /* Declare/define channels */
  sc_core::sc_signal<bool> fsm_input_sig;
  sc_core::sc_signal<unsigned int> fsm_out;
  /* Declare/define master clock */
  sc_core::sc_clock clk("clk", 10.0, sc_core::SC_US, 0.5);

  /* Declare/define finite state machine state
       transition generator and finite state machine */
  fsm_input_src fsm_inp("fsm_input_src");
  fsm fs_m("fsm");
```

**Fig. 6.16** Test harness for finite state machine example

```
    /* Connect all module ports and signal channels */
    fsm_inp.clk(clk);
    fsm_inp.fsm_input_out(fsm_input_sig);

    fs_m.fsm_in(fsm_input_sig);
    fs_m.curr_state_out(fsm_out);

    /* Run simulation for finite time-interval and then stop */
    sc_core::sc_start(1500.0, sc_core::SC_US);
    sc_core::sc_stop();
    return 0;
}
```

**Fig. 6.16** (continued)

## 6.5 32-Bit Parity Generator for Forward Error Correction Module of IEEE 802.3ba Protocol

We examine in detail a 32-bit parity generator to be used as part of the forward error correction [5] module of the newly proposed IEEE 802.3ba protocol [6]. The core algorithm is simple—four 16-bit words are concatenated along with a status bit to form a 65-bit word, for a total of 32 65-bit vectors. The parity of each 65-bit word is obtained with a simple XOR operation, and the 32 1-bit XOR result bits are concatenated to a single 32-bit word, which is the output. Additional detailed source code documentation provides insight into the actual mechanics of the algorithm. Sample data for testing have been obtained from Ilango Ganga's online examples. Figures 6.17, 6.18, and 6.19 contain the parity generator source code, the input test vector generator, and the test harness.

Sample console output from the parity generator looks like

```
    SystemC 2.2.0—August 30, 2011, 19:39:07
    Copyright (c) 1996–2006 by all contributors,
    ALL RIGHTS RESERVED

0x040ea1e77eed301ec
0x1ad5a3bf86d9acf5c
0x0de55cb8fdf0f7ca0
0x1e6ccff8e8212b1c6
0x0d63bc6c309000638
0x170e3b0ce30e0497d
0x1dc8df31ec3ab4491
0x066fb9139c81cd37b
0x1b57477d4f05e3602
0x08cfd495d12947a31
```

```
#ifndef FECENC_H
#define FECENC_H

#include <systemc>
#include <cstdlib>
const unsigned int FEC_BLOCK_SIZE = 32;

SC_MODULE(fecenc)
{
 private:
  /* Declare/define internal members and data structures */
  /* 32 65-bit registers to contain the Forward Error Correction
  (FEC) code blocks, as well as 32-bit register for parity bits */
  sc_dt::sc_bv<65> arr;
  sc_dt::sc_bv<32> paritybits;
  sc_dt::sc_bv<65> fecblk0;
  sc_dt::sc_bv<65> fecblk1;
  sc_dt::sc_bv<65> fecblk2;
  sc_dt::sc_bv<65> fecblk3;
  sc_dt::sc_bv<65> fecblk4;
  sc_dt::sc_bv<65> fecblk5;
  sc_dt::sc_bv<65> fecblk6;
  sc_dt::sc_bv<65> fecblk7;
  sc_dt::sc_bv<65> fecblk8;
  sc_dt::sc_bv<65> fecblk9;
  sc_dt::sc_bv<65> fecblk10;
  sc_dt::sc_bv<65> fecblk11;
  sc_dt::sc_bv<65> fecblk12;
  sc_dt::sc_bv<65> fecblk13;
  sc_dt::sc_bv<65> fecblk14;
  sc_dt::sc_bv<65> fecblk15;
  sc_dt::sc_bv<65> fecblk16;
  sc_dt::sc_bv<65> fecblk17;
  sc_dt::sc_bv<65> fecblk18;
  sc_dt::sc_bv<65> fecblk19;
  sc_dt::sc_bv<65> fecblk20;
  sc_dt::sc_bv<65> fecblk21;
  sc_dt::sc_bv<65> fecblk22;
  sc_dt::sc_bv<65> fecblk23;
```

**Fig. 6.17** 32-bit parity bit-vector generator with 32 65-bit input bit-vectors

```
  sc_dt::sc_bv<65> fecblk24;
  sc_dt::sc_bv<65> fecblk25;
  sc_dt::sc_bv<65> fecblk26;
  sc_dt::sc_bv<65> fecblk27;
  sc_dt::sc_bv<65> fecblk28;
  sc_dt::sc_bv<65> fecblk29;
  sc_dt::sc_bv<65> fecblk30;
  sc_dt::sc_bv<65> fecblk31;

 public:
  /* Declare/define temporary data structures */
  sc_dt::sc_bv<16> segment0;
  sc_dt::sc_bv<16> segment1;
  sc_dt::sc_bv<16> segment2;
  sc_dt::sc_bv<16> segment3;
  sc_dt::sc_bv<2> datatype;
  sc_dt::sc_bv<1> tcvalue;
  sc_core::sc_in< sc_dt::sc_bv<16> > tx_bit_16;
  /* tx_clk and txc_ctrl are IEEE 802.3ba specific signals */
  sc_core::sc_in<bool> tx_clk;
  sc_core::sc_in<bool> txc_ctrl;
  unsigned int index1;
  unsigned int index2;

  void initialize_bv()
  {
    /* Initialize all data structures */

arr="0000000000000000000000000000000000000000000000000000000000000
00000";

fecblk0="0000000000000000000000000000000000000000000000000000000000
00000000";

fecblk1="0000000000000000000000000000000000000000000000000000000000
00000000";

fecblk2="0000000000000000000000000000000000000000000000000000000000
00000000";
```

Fig. 6.17 (continued)

```
fecblk3="00000000000000000000000000000000000000000000000000000000
00000000";

fecblk4="00000000000000000000000000000000000000000000000000000000
00000000";

fecblk5="00000000000000000000000000000000000000000000000000000000
00000000";

fecblk6="00000000000000000000000000000000000000000000000000000000
00000000";

fecblk7="00000000000000000000000000000000000000000000000000000000
00000000";

fecblk8="00000000000000000000000000000000000000000000000000000000
00000000";

fecblk9="00000000000000000000000000000000000000000000000000000000
00000000";

fecblk10="0000000000000000000000000000000000000000000000000000000
000000000";

fecblk11="0000000000000000000000000000000000000000000000000000000
000000000";

fecblk12="0000000000000000000000000000000000000000000000000000000
000000000";

fecblk13="0000000000000000000000000000000000000000000000000000000
000000000";

fecblk14="0000000000000000000000000000000000000000000000000000000
000000000";

fecblk15="0000000000000000000000000000000000000000000000000000000
000000000";
```

**Fig. 6.17** (continued)

```
fecblk16="0000000000000000000000000000000000000000000000000000000000000000
000000000";

fecblk17="0000000000000000000000000000000000000000000000000000000000000000
000000000";

fecblk18="0000000000000000000000000000000000000000000000000000000000000000
000000000";

fecblk19="0000000000000000000000000000000000000000000000000000000000000000
000000000";

fecblk20="0000000000000000000000000000000000000000000000000000000000000000
000000000";

fecblk21="0000000000000000000000000000000000000000000000000000000000000000
000000000";

fecblk22="0000000000000000000000000000000000000000000000000000000000000000
000000000";

fecblk23="0000000000000000000000000000000000000000000000000000000000000000
000000000";

fecblk24="0000000000000000000000000000000000000000000000000000000000000000
000000000";

fecblk25="0000000000000000000000000000000000000000000000000000000000000000
000000000";

fecblk26="0000000000000000000000000000000000000000000000000000000000000000
000000000";

fecblk27="0000000000000000000000000000000000000000000000000000000000000000
000000000";

fecblk28="0000000000000000000000000000000000000000000000000000000000000000
000000000";
```

**Fig. 6.17** (continued)

```
fecblk29="00000000000000000000000000000000000000000000000000000000
000000000";

fecblk30="00000000000000000000000000000000000000000000000000000000
000000000";

fecblk31="00000000000000000000000000000000000000000000000000000000
000000000";
   paritybits="00000000000000000000000000000000";
   segment0="0000000000000000";
   segment1="0000000000000000";
   segment2="0000000000000000";
   segment3="0000000000000000";
 }

  /* Declare/define first thread that reads in 16-bit bit-vectors,
     concatenates them with a status bit to create a 65-bit bit-vector
     for a toal of 32 65-bit bit-vectors */
  void readAggregateWord()
  {
   while(true)
   {
     wait();
     if(txc_ctrl.read() == true) datatype = "10";
     else datatype = "01";

     if(tx_bit_16.read() != "0000000000000000")
     {
      switch(index1)
      {
       case 0:
        segment0 = tx_bit_16.read();
        index1 += 1;
       break;
       case 1:
        segment1 = tx_bit_16.read();
        index1 += 1;
       break;
       case 2:
```

**Fig. 6.17** (continued)

```
     segment2 = tx_bit_16.read();
     index1 += 1;
    break;
    case 3:
     segment3 = tx_bit_16.read();
     /* Compute status bit */
     tcvalue[0] = datatype[1]^segment1[0];
    /* Concatenate four 16-bit bit-bectors and status bit to create a
    65-bit bit vector */
    arr = (tcvalue,
           segment0,
           segment1,
           segment2,
           segment3);
    index1 = 0;
    /* Populate 32 65-bit bit-vectors */
    switch(index2)
    {
      case 0:
       fecblk0 = arr;
       index2 += 1;
      break;
      case 1:
       fecblk1 = arr;
       index2 += 1;
      break;
      case 2:
       fecblk2 = arr;
       index2 += 1;
      break;
      case 3:
       fecblk3 = arr;
       index2 += 1;
      break;
      case 4:
       fecblk4 = arr;
       index2 += 1;
      break;
      case 5:
       fecblk5 = arr;
```

**Fig. 6.17** (continued)

```
        index2 += 1;
       break;
       case 6:
        fecblk6 = arr;
        index2 += 1;
       break;
       case 7:
        fecblk7 = arr;
       index2 += 1;
      break;
      case 8:
       fecblk8 = arr;
       index2 += 1;
      break;
      case 9:
       fecblk9 = arr;
       index2 += 1;
      break;
      case 10:
       fecblk10 = arr;
       index2 += 1;
      break;
      case 11:
       fecblk11 = arr;
       index2 += 1;
      break;
      case 12:
       fecblk12 = arr;
       index2 += 1;
      break;
      case 13:
       fecblk13 = arr;
       index2 += 1;
      break;
      case 14:
       fecblk14 = arr;
       index2 += 1;
      break;
      case 15:
       fecblk15 = arr;
       index2 += 1;
```

**Fig. 6.17** (continued)

```
break;
case 16:
 fecblk16 = arr;
 index2 += 1;
break;
case 17:
 fecblk17 = arr;
 index2 += 1;
break;
case 18:
 fecblk18 = arr;
 index2 += 1;
break;
case 19:
 fecblk19 = arr;
 index2 += 1;
break;
case 20:
 fecblk20 = arr;
 index2 += 1;
break;
case 21:
 fecblk1 = arr;
 index2 += 1;
break;
case 22:
 fecblk22 = arr;
 index2 += 1;
break;
case 23:
 fecblk23 = arr;
 index2 += 1;
break;
case 24:
 fecblk24 = arr;
 index2 += 1;
break;
case 25:
 fecblk25 = arr;
 index2 += 1;
```

**Fig. 6.17** (continued)

```
        break;
        case 26:
         fecblk26 = arr;
         index2 += 1;
        break;
        case 27:
         fecblk27 = arr;
         index2 += 1;
        break;
        case 28:
         fecblk28 = arr;
         index2 += 1;
        break;
        case 29:
         fecblk29 = arr;
         index2 += 1;
        break;
        case 30:
         fecblk30 = arr;
         index2 += 1;
        break;
        case 31:
         fecblk31 = arr;
         index2 += 1;
        break;
        default:
        break;
      }
     break;
     default:
     break;
    }
  }
 }
}

/* Declare/define thread to compute parity bit for each 65-bit
bit-vector and concatenate each result parity bit into a 32-bit
bit-vector */
```

**Fig. 6.17** (continued)

```
void genparity()
{
 unsigned int i = 0;
 while(true)
 {
  wait();
  if(index2 == FEC_BLOCK_SIZE)
  {
   paritybits[0] = fecblk0.xor_reduce();
   paritybits[1] = fecblk1.xor_reduce();
   paritybits[2] = fecblk2.xor_reduce();
   paritybits[3] = fecblk3.xor_reduce();
   paritybits[4] = fecblk4.xor_reduce();
   paritybits[5] = fecblk5.xor_reduce();
   paritybits[6] = fecblk6.xor_reduce();
   paritybits[7] = fecblk7.xor_reduce();
   paritybits[8] = fecblk8.xor_reduce();
   paritybits[9] = fecblk9.xor_reduce();
   paritybits[10] = fecblk10.xor_reduce();
   paritybits[11] = fecblk11.xor_reduce();
   paritybits[12] = fecblk12.xor_reduce();
   paritybits[13] = fecblk13.xor_reduce();
   paritybits[14] = fecblk14.xor_reduce();
   paritybits[15] = fecblk15.xor_reduce();
   paritybits[16] = fecblk16.xor_reduce();
   paritybits[17] = fecblk17.xor_reduce();
   paritybits[18] = fecblk18.xor_reduce();
   paritybits[19] = fecblk19.xor_reduce();
   paritybits[20] = fecblk20.xor_reduce();
   paritybits[21] = fecblk21.xor_reduce();
   paritybits[22] = fecblk22.xor_reduce();
   paritybits[23] = fecblk23.xor_reduce();
   paritybits[24] = fecblk24.xor_reduce();
   paritybits[25] = fecblk25.xor_reduce();
   paritybits[26] = fecblk26.xor_reduce();
   paritybits[27] = fecblk27.xor_reduce();
   paritybits[28] = fecblk28.xor_reduce();
   paritybits[29] = fecblk29.xor_reduce();
   paritybits[30] = fecblk30.xor_reduce();
```

**Fig. 6.17** (continued)

```
      paritybits[31] = fecblk31.xor_reduce();
      index2 = 0;
      /* Display all 32 65-bit input bit-vectors and the
      final 32-bit parity bit-vector */
      std::cout<<fecblk0.to_string(sc_dt::SC_HEX)<<std::endl;
      std::cout<<fecblk1.to_string(sc_dt::SC_HEX)<<std::endl;
      std::cout<<fecblk2.to_string(sc_dt::SC_HEX)<<std::endl;
      std::cout<<fecblk3.to_string(sc_dt::SC_HEX)<<std::endl;
      std::cout<<fecblk4.to_string(sc_dt::SC_HEX)<<std::endl;
      std::cout<<fecblk5.to_string(sc_dt::SC_HEX)<<std::endl;
      std::cout<<fecblk6.to_string(sc_dt::SC_HEX)<<std::endl;
      std::cout<<fecblk7.to_string(sc_dt::SC_HEX)<<std::endl;
      std::cout<<fecblk8.to_string(sc_dt::SC_HEX)<<std::endl;
      std::cout<<fecblk9.to_string(sc_dt::SC_HEX)<<std::endl;
      std::cout<<fecblk10.to_string(sc_dt::SC_HEX)<<std::endl;
      std::cout<<fecblk11.to_string(sc_dt::SC_HEX)<<std::endl;
      std::cout<<fecblk12.to_string(sc_dt::SC_HEX)<<std::endl;
      std::cout<<fecblk13.to_string(sc_dt::SC_HEX)<<std::endl;
      std::cout<<fecblk14.to_string(sc_dt::SC_HEX)<<std::endl;
      std::cout<<fecblk15.to_string(sc_dt::SC_HEX)<<std::endl;
      std::cout<<fecblk16.to_string(sc_dt::SC_HEX)<<std::endl;
      std::cout<<fecblk17.to_string(sc_dt::SC_HEX)<<std::endl;
      std::cout<<fecblk18.to_string(sc_dt::SC_HEX)<<std::endl;
      std::cout<<fecblk19.to_string(sc_dt::SC_HEX)<<std::endl;
      std::cout<<fecblk20.to_string(sc_dt::SC_HEX)<<std::endl;
      std::cout<<fecblk21.to_string(sc_dt::SC_HEX)<<std::endl;
      std::cout<<fecblk22.to_string(sc_dt::SC_HEX)<<std::endl;
      std::cout<<fecblk23.to_string(sc_dt::SC_HEX)<<std::endl;
      std::cout<<fecblk24.to_string(sc_dt::SC_HEX)<<std::endl;
      std::cout<<fecblk25.to_string(sc_dt::SC_HEX)<<std::endl;
      std::cout<<fecblk26.to_string(sc_dt::SC_HEX)<<std::endl;
      std::cout<<fecblk27.to_string(sc_dt::SC_HEX)<<std::endl;
      std::cout<<fecblk28.to_string(sc_dt::SC_HEX)<<std::endl;
      std::cout<<fecblk29.to_string(sc_dt::SC_HEX)<<std::endl;
      std::cout<<fecblk30.to_string(sc_dt::SC_HEX)<<std::endl;
      std::cout<<fecblk31.to_string(sc_dt::SC_HEX)<<std::endl;
      std::cout<<paritybits.to_string(sc_dt::SC_HEX)<<std::endl;
    }
  }
}
```

**Fig. 6.17** (continued)

```
    /* Constructor */
    SC_CTOR(fecenc):index1(0),index2(0)
    {
      /* Declare/assign threads */
      SC_CTHREAD(readAggregateWord, tx_clk.pos());
      SC_CTHREAD(genparity, tx_clk.pos());
      initialize_bv(); /*initialize all bit-vectors */
    }
    ~fecenc(){ } /*Destructor */
  };
  #endif
```

**Fig. 6.17** (continued)

```
#ifndef FECSRC_H
#define FECSRC_H

#include <systemc>
#include <cassert>

SC_MODULE(fecsrc)
{
  /* Declare/define internal variables and methods */
  private:
  /* Convert character string to unsigned integer - input in hexadecimal
format */
   unsigned int htoi (const char *ptr)
   {
     assert(ptr != NULL);
     unsigned int value = 0;
     char ch = *ptr;
    while (ch == ' ' || ch == '¥t') ch = *(++ptr);

    for (;;)
    {
     if (ch >= '0' && ch <= '9')
          value = (value << 4) + (ch - '0');
     else if (ch >= 'A' && ch <= 'F')
          value = (value << 4) + (ch - 'A' + 10);
```

**Fig. 6.18** Input test vector generator for parity generator

```
        else if (ch >= 'a' && ch <= 'f')
            value = (value << 4) + (ch - 'a' + 10);
        else
            return value;
        ch = *(++ptr);

    }
   }

  public:
   /* Declare/define input/output ports */
   sc_core::sc_in<bool> clk;
   sc_core::sc_out<bool> txc_ctrl;
   sc_core::sc_out< sc_dt::sc_bv<16> > tx_bit_16;

   unsigned int num;

  void output()
  {
    while(true)
    {
      wait();
      txc_ctrl.write(true);
      /* Generate 18-bit bit-vectors to be used to
         create 32 64-bit bit-vectors by concatenation */
      switch(num)
      {
        case 0:
         tx_bit_16.write("100000011101010");
         num += 1;
        break;
        case 1:
         tx_bit_16.write("0001111001110111");
         num += 1;
        break;
        case 2:
         tx_bit_16.write("1110111011010011");
         num += 1;
        break;
```

**Fig. 6.18** (continued)

```
case 3:
 tx_bit_16.write("0000000111101100");
 num += 1;
break;
case 4:
 tx_bit_16.write("1010110101011010");
 num += 1;
break;
case 5:
 tx_bit_16.write("0011101111111000");
 num += 1;
break;
case 6:
 tx_bit_16.write("0110110110011010");
 num += 1;
break;
case 7:
 tx_bit_16.write("1100111101011100");
 num += 1;
break;
case 8:
 tx_bit_16.write("1101111001010101");
 num += 1;
break;
case 9:
 tx_bit_16.write("1100101110001111");
 num += 1;
break;
case 10:
 tx_bit_16.write("1101111100001111");
 num += 1;
break;
case 11:
 tx_bit_16.write("0111110010100000");
 num += 1;
break;
case 12:
 tx_bit_16.write("1110011011001100");
 num += 1;
```

**Fig. 6.18** (continued)

```
break;
case 13:
 tx_bit_16.write("1111111110001110");
 num += 1;
break;
case 14:
 tx_bit_16.write("1000001000010010");
 num += 1;
break;
case 15:
 tx_bit_16.write("1011000111000110");
 num += 1;
break;
case 16:
 tx_bit_16.write("1101011000111011");
 num += 1;
break;
case 17:
 tx_bit_16.write("1100011011000011");
 num += 1;
break;
case 18:
 tx_bit_16.write("0000100100000000");
 num += 1;
break;
case 19:
 tx_bit_16.write("0000011000111000");
 num += 1;
break;
case 20:
 tx_bit_16.write("0111000011100011");
 num += 1;
break;
case 21:
 tx_bit_16.write("1011000011001110");
 num += 1;
break;
case 22:
 tx_bit_16.write("0011000011100000");
```

**Fig. 6.18** (continued)

```
 num += 1;
break;
case 23:
 tx_bit_16.write("0100100101111101");
 num += 1;
break;
case 24:
 tx_bit_16.write("1101110010001101");
 num += 1;
break;
case 25:
 tx_bit_16.write("1111001100011110");
 num += 1;
break;
case 26:
 tx_bit_16.write("1100001110101011"); /*C3AB */
 num += 1;
break;
case 27:
 tx_bit_16.write("0100010010010001");
 num += 1;
break;
 case 28:
 tx_bit_16.write("0110011011111011");
 num += 1;
break;
case 29:
 tx_bit_16.write("1001000100111001");
 num += 1;
break;
case 30:
 tx_bit_16.write("1100100000011100");
 num += 1;
break;
case 31:
 tx_bit_16.write("1101001101111011");
 num += 1;
break;
```

**Fig. 6.18** (continued)

```
 case 32:
 tx_bit_16.write("1011010101110100");
 num += 1;
break;
case 33:
 tx_bit_16.write("111011111010100");
 num += 1;
break;
case 34:
 tx_bit_16.write("1111000001011110");
 num += 1;
break;
case 35:
 tx_bit_16.write("0011011000000010");
 num += 1;
break;
case 36:
 tx_bit_16.write("1000110011111101");
 num += 1;
break;
 case 37:
 tx_bit_16.write("0100100101011101");
 num += 1;
break;
case 38:
 tx_bit_16.write("0001001010010100");
 num += 1;
break;
case 39:
 tx_bit_16.write("0111101000110001");
 num += 1;
break;
case 40:
 tx_bit_16.write("1110011101110111");
 num += 1;
break;
 case 41:
 tx_bit_16.write("0111110011110000");
 num += 1;
```

**Fig. 6.18** (continued)

```
break;
case 42:
 tx_bit_16.write("1100011011010000");
 num += 1;
break;
case 43:
 tx_bit_16.write("0110001010000000");
 num += 1;
break;
case 44:
 tx_bit_16.write("0100010001010010");
 num += 1;
break;
 case 45:
 tx_bit_16.write("1001110011110100");
 num += 1;
break;
case 46:
 tx_bit_16.write("1011010010010000");
 num += 1;
break;
case 47:
 tx_bit_16.write("0000010100101000");
 num += 1;
break;
case 48:
 tx_bit_16.write("1000010111001110");
 num += 1;
break;
 case 49:
 tx_bit_16.write("0001110100100111");
 num += 1;
break;
case 50:
 tx_bit_16.write("0111010100001010");
 num += 1;
break;
case 51:
 tx_bit_16.write("1101011000011011");
```

**Fig. 6.18** (continued)

```
 num += 1;
break;
case 52:
 tx_bit_16.write("0100010101101101");
 num += 1;
break;
 case 53:
 tx_bit_16.write("0101110001110001");
 num += 1;
break;
case 54:
 tx_bit_16.write("0111010000111111");
 num += 1;
break;
case 55:
 tx_bit_16.write("0101110001101001");
 num += 1;
break;
case 56:
 tx_bit_16.write("1100000110111111");
 num += 1;
break;
 case 57:
 tx_bit_16.write("0110001011100101");
 num += 1;
break;
case 58:
 tx_bit_16.write("1101110001010100");
 num += 1;
break;
case 59:
 tx_bit_16.write("0110010010110101");
 num += 1;
break;
case 60:
 tx_bit_16.write("1101110001100000");
 num += 1;
break;
```

**Fig. 6.18** (continued)

```
 case 61:
 tx_bit_16.write("0001000110111110");
 num += 1;
break;
case 62:
 tx_bit_16.write("0111111010100001");
 num += 1;
break;
case 63:
 tx_bit_16.write("1110110101010100");
 num += 1;
break;
case 64:
 tx_bit_16.write("0001110011111001");
 num += 1;
break;
 case 65:
 tx_bit_16.write("0010110001000101");
 num += 1;
break;
case 66:
 tx_bit_16.write("0000000001000010");
 num += 1;
break;
case 67:
 tx_bit_16.write("1010011101011111");
 num += 1;
break;
case 68:
 tx_bit_16.write("1100110001001011");
 num += 1;
break;
 case 69:
 tx_bit_16.write("1001010000001110");
 num += 1;
break;
case 70:
 tx_bit_16.write("1010111100110001");
 num += 1;
```

**Fig. 6.18** (continued)

```
break;
case 71:
 tx_bit_16.write("0100000011011011");
 num += 1;
break;
case 72:
 tx_bit_16.write("0111011110111011");
 num += 1;
break;
 case 73:
 tx_bit_16.write("0110000100101010");
 num += 1;
break;
case 74:
 tx_bit_16.write("0111101010111111");
 num += 1;
break;
case 75:
 tx_bit_16.write("0100000000011111"); /*401f*/
 num += 1;
break;
case 76:
 tx_bit_16.write("1100001000101101");
 num += 1;
break;
 case 77:
 tx_bit_16.write("0011010000011110");
 num += 1;
break;
case 78:
 tx_bit_16.write("1001000001010100");
 num += 1;
break;
case 79:
 tx_bit_16.write("0101110110011000");
 num += 1;
break;
 case 80:
 tx_bit_16.write("1100111001101101");
```

**Fig. 6.18** (continued)

```
 num += 1;
break;
 case 81:
 tx_bit_16.write("1010111100011111");
 num += 1;
break;
case 82:
 tx_bit_16.write("0010010010001011");
 num += 1;
break;
case 83:
 tx_bit_16.write("1011110101101101");
 num += 1;
break;
case 84:
 tx_bit_16.write("1101110100100010");
 num += 1;
break;
 case 85:
 tx_bit_16.write("1101000010110011");
 num += 1;
break;
case 86:
 tx_bit_16.write("1111100101010101");
 num += 1;
break;
case 87:
 tx_bit_16.write("0001111011010110");
 num += 1;
break;
case 88:
 tx_bit_16.write("0101011101000110");
 num += 1;
break;
 case 89:
 tx_bit_16.write("1000011011000011");
 num += 1;
break;
```

**Fig. 6.18** (continued)

```
case 90:
 tx_bit_16.write("1111100111101001");
 num += 1;
break;
case 91:
 tx_bit_16.write("0011100010011000");
 num += 1;
break;
case 92:
 tx_bit_16.write("0010111001010010");
 num += 1;
break;
 case 93:
 tx_bit_16.write("0110001010001111");
 num += 1;
break;
case 94:
 tx_bit_16.write("0100101000010010");
 num += 1;
break;
case 95:
 tx_bit_16.write("1000001011001110");
 num += 1;
break;
case 96:
 tx_bit_16.write("1111001000001100");
 num += 1;
break;
case 97:
 tx_bit_16.write("1000011011010111");
 num += 1;
break;
 case 98:
 tx_bit_16.write("0001100101000100");
 num += 1;
break;
case 99:
 tx_bit_16.write("1010101010110001");
```

**Fig. 6.18** (continued)

```
 num += 1;
break;
case 100:
 tx_bit_16.write("0101010100010011");
 num += 1;
break;
case 101:
 tx_bit_16.write("0011110010010011");
 num += 1;
break;
 case 102:
 tx_bit_16.write("0011001110000000");
 num += 1;
break;
case 103:
 tx_bit_16.write("1000101000101100");
 num += 1;
break;
case 104:
 tx_bit_16.write("0001101010101000");
 num += 1;
break;
case 105:
 tx_bit_16.write("0010010111011000");
 num += 1;
break;
 case 106:
 tx_bit_16.write("1011100000010111");
 num += 1;
break;
case 107:
 tx_bit_16.write("1101101101001101");
 num += 1;
break;
case 108:
 tx_bit_16.write("0110001101111001");
 num += 1;
break;
```

**Fig. 6.18** (continued)

```
case 109:
tx_bit_16.write("0101100110011000");
num += 1;
break;
 case 110:
 tx_bit_16.write("1001111100110000");
 num += 1;
break;
case 111:
 tx_bit_16.write("0010000111101011");/*21EB*/
 num += 1;
break;
case 112:
 tx_bit_16.write("1001011101101000");
 num += 1;
break;
 case 113:
 tx_bit_16.write("0000011001100100");
 num += 1;
break;
 case 114:
 tx_bit_16.write("0001101100100110");
 num += 1;
break;
case 115:
 tx_bit_16.write("1010101011101001"); /*AAE9*/
 num += 1;
break;
case 116:
 tx_bit_16.write("110101000110111");
 num += 1;
break;
  case 117:
 tx_bit_16.write("1101010001010011");
 num += 1;
break;
 case 118:
 tx_bit_16.write("0001101101111110");
```

**Fig. 6.18** (continued)

```
 num += 1;
break;
case 119:
 tx_bit_16.write("1101010111110010");
 num += 1;
break;
case 120:
 tx_bit_16.write("0101001111000011");
 num += 1;
break;
  case 121:
 tx_bit_16.write("1110100101101101");
 num += 1;
break;
 case 122:
 tx_bit_16.write("0011101100010010");
 num += 1;
break;
case 123:
 tx_bit_16.write("1111101101000110");
 num += 1;
break;
case 124:
 tx_bit_16.write("101001010001100");
 num += 1;
break;
case 125:
 tx_bit_16.write("111111010111000");
 num += 1;
break;
case 126:
 tx_bit_16.write("100100000011011");
 num += 1;
break;
case 127:
 tx_bit_16.write("1100100101101001");
 num += 1;
break;
default:
```

**Fig. 6.18** (continued)

```
          break;
        }
       }
    }

   /* Constructor */
    SC_CTOR(fecsrc):num(0)
    {
    SC_CTHREAD(output, clk.pos());
   /*Declare/assign clocked thread */
   }

  /* Destructor */
   ~fecsrc() { }
};

#endif
```

**Fig. 6.18** (continued)

```
#include "fecenc.h"
#include "fecsrc.h"

int sc_main(int argc, char **argv)
{
  /* Declare/define signal channels */
  sc_core::sc_signal< sc_dt::sc_bv<16> > bitchnl;
  sc_core::sc_signal<bool> ctrl_line;
  /* Declare/define master clock,
      parity generator and input test vector generator */
  sc_core::sc_clock clk("clk", 20, sc_core::SC_NS, 0.5);
  fecsrc fec_src("fec_src");
  fecenc fec_enc("fec_enc");
    /* Connect ports and channels */
  fec_src.clk(clk);
  fec_src.txc_ctrl(ctrl_line);
  fec_src.tx_bit_16(bitchnl);
  fec_enc.tx_clk(clk);
  fec_enc.txc_ctrl(ctrl_line);
  fec_enc.tx_bit_16(bitchnl);
```

**Fig. 6.19** Test harness for parity generator

```
    /* Start and run simulation for fixed time-intervaland then stop */
    sc_core::sc_start(2750.0, sc_core::SC_NS);
    sc_core::sc_stop();
    return 0;
}
```

Fig. 6.19 (continued)

```
0x1e7777cf0c6d06280
0x144529cf4b4900528
0x085ce1d27750ad61b
0x0456d5c71743f5c69
0x0c1bf62e5dc5464b5
0x1dc6011be7ea1ed54
0x01cf92c450042a75f
0x1cc4b940eaf3140db
0x177bb612a7abf401f
0x1c22d341e90545d98
0x0ce6daf1f248bbd6d
0x0dd22d0b3f9551ed6
0x0574686c3f9e93898
0x02e52628f4a1282ce
0x0f20c86d71944aab1
0x055133c9333808a2c
0x11aa825d8b817db4d
0x1637959989f3021eb
0x1976806641b26aae9
0x06a37d4531b7ed5f2
0x053c3e96d3b12fb46
0x1528c7eb8481bc969
0x0e9a60eb3
SystemC: simulation stopped by user.

    SystemC 2.3.0-ASI—July 19, 2012, 18:53:11
    Copyright (c) 1996–2012 by all contributors,
    ALL RIGHTS RESERVED

0x040ea1e77eed301ec
0x1ad5a3bf86d9acf5c
0x0de55cb8fdf0f7ca0
0x1e6ccff8e8212b1c6
0x0d63bc6c309000638
0x170e3b0ce30e0497d
0x1dc8df31ec3ab4491
0x066fb9139c81cd37b
0x1b57477d4f05e3602
```

```
0x08cfd495d12947a31
0x1e7777cf0c6d06280
0x144529cf4b4900528
0x085ce1d27750ad61b
0x0456d5c71743f5c69
0x0c1bf62e5dc5464b5
0x1dc6011be7ea1ed54
0x01cf92c450042a75f
0x1cc4b940eaf3140db
0x177bb612a7abf401f
0x1c22d341e90545d98
0x0ce6daf1f248bbd6d
0x0dd22d0b3f9551ed6
0x0574686c3f9e93898
0x02e52628f4a1282ce
0x0f20c86d71944aab1
0x055133c9333808a2c
0x11aa825d8b817db4d
0x1637959989f3021eb
0x1976806641b26aae9
0x06a37d4531b7ed5f2
0x053c3e96d3b12fb46
0x1528c7eb8481bc969
0x0e9a60eb3
Info: /OSCI/SystemC: simulation stopped by user.
```

In this case, the output on the last line is the 32-bit parity bit-vector for the 32 65-bit input bit-vectors.

## 6.6 Simple Pulse Counter for Rotary Encoder

Measuring the speed of an electric motor is a very common issue in industrial situations. This is achieved with a rotary encoder, whose output is processed digitally to extract the required information. The operation of the rotary encoder and its associated processing circuitry may be summarized as follows: given two clocks, one very fast compared to the other, how many complete cycles of the fast clock fit in one complete cycle of the slow clock. Multiplying this number with the time period of the fast clock provides the time period of the slow clock. A simple rotary encoder consists of equal-sized, alternate, transparent, and dark segments on a disc, so that light may pass through any transparent segment and be blocked by a dark segment. The light detector's output signal is very low when the light beam is interrupted (as compared to when the beam is not interrupted), and this may be encoded as logic '1' or '0'. Figure 6.20 has the pulse counter source code and test harness (Fig. 6.21), and Fig. 6.22 has the traces of the two input signals to the pulse counter.

```
#ifndef PULSECOUNTER_H
#define PULSECOUNTER_H

#include <systemc>

/* Fixed time period of fast clock */
const double time_period = 1.0E-9;

SC_MODULE(pulsecounter)
{
 /* declare/define input ports */
 sc_core::sc_in<bool> mclk;
 sc_core::sc_in<bool> sig;
 /* Declare/define internal members */
 bool start;
 double frequency;
 double h_time_interval;
 double rpm;
 unsigned int count;

 /* Pulse counter main operation thread */
 void pulsecounter_proc0()
 {
  while(1)
  {
   wait();
 /* Trigger on each rising edge of high frequency input signal */
    if(sig.read() == true)
    {
     if(start == false)
     {
      start = true;
      count += 1;
     }
     else if(start == true)
     {
      count += 1;
     }
   }
```

Fig. 6.20 Pulse counter with high- and low-frequency input signals

```
        else if(sig.read() == false)  /* Low frequency signal is low */
        {
          if(start == true)
          {
            start = false;
            /* Reset members */
            /* Compute time period and
               frequency of low frequency signal */
            h_time_interval = (double)(time_period*count);
            frequency = 1.0/(2.0*h_time_interval);
            rpm = 60*frequency;
            /* Compute RPM */
            std::cout<<std::scientific<<frequency<<" Hz "<<rpm<<"
RPM"<<std::endl;
            h_time_interval = 0.0;
            count = 0;
          }
        }
      }
    }

    /* Constructor - initialize members */
    SC_CTOR(pulsecounter):start(false),
                          count(0),
                          frequency(0.0),
                          h_time_interval(0.0),
                          rpm(0.0)
    {
      /*Declare/assign clocked thread */
      SC_CTHREAD(pulsecounter_proc0, mclk.pos());
    }
    ~pulsecounter(){ }
   /*Destructor */
  };
```

**Fig. 6.20** (continued)

```
#include "pulsecounter.h"
int sc_main(int argc, char **argv)
{
  /* Declare/define channels, high and low frequency signals */
  sc_core::sc_clock mclk("mclk",2.0,sc_core::SC_NS, 0.5);
  sc_core::sc_clock sclk("sclk",10.0, sc_core::SC_US, 0.5);
  /* Declare/define pulse counterand trace file pointer */
  pulsecounter pcntr("pcntr");
  sc_core::sc_trace_file *fp =
sc_core::sc_create_vcd_trace_file("tr_pulsecounter");
  fp->set_time_unit(1.0, sc_core::SC_NS);
  /* Connect ports and channels */
  pcntr.mclk(mclk);
  pcntr.sig(sclk);

  /* Connect trace file and channels */
  sc_trace(fp, mclk, "clock");
  sc_trace(fp, sclk, "sec_clk");

  /* Start and run simulation for pre-defind time-interval, stop and close
trace file */
  sc_core::sc_start(1500.0, sc_core::SC_US);
  sc_core::sc_stop();
  sc_core::sc_close_vcd_trace_file(fp);
  return 0;
}
```

**Fig. 6.21** Test harness for pulse counter

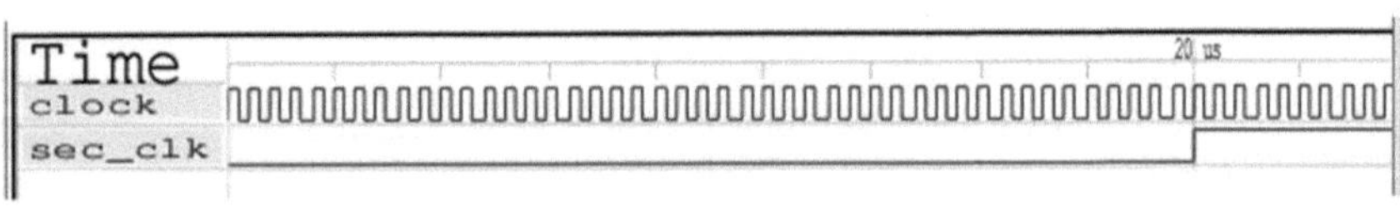


**Fig. 6.22** Traces for high- and low-frequency signals

Output at the console appears as
SystemC 2.2.0—August 30, 2011, 19:39:07
Copyright (c) 1996–2006 by all contributors,
ALL RIGHTS RESERVED

Note: VCD trace timescale unit is set by user to 1.000000e-09 s.
2.000800e + 05 Hz 1.200480e + 07 RPM
2.000000e + 05 Hz 1.200000e + 07 RPM

2.000000e + 05 Hz 1.200000e + 07 RPM
2.000000e + 05 Hz 1.200000e + 07 RPM
2.000000e + 05 Hz 1.200000e + 07 RPM
2.000000e + 05 Hz 1.200000e + 07 RPM
2.000000e + 05 Hz 1.200000e + 07 RPM
....
....
2.000000e + 05 Hz 1.200000e + 07 RPM
2.000000e + 05 Hz 1.200000e + 07 RPM
SystemC: simulation stopped by user.

SystemC 2.3.0-ASI—July 19, 2012, 18:53:11
Copyright (c) 1996–2012 by all contributors,
ALL RIGHTS RESERVED

Note: VCD trace timescale unit is set by user to 1.000000e-09 s.
2.000800e + 05 Hz 1.200480e + 07 RPM
2.000000e + 05 Hz 1.200000e + 07 RPM
2.000000e + 05 Hz 1.200000e + 07 RPM
2.000000e + 05 Hz 1.200000e + 07 RPM
2.000000e + 05 Hz 1.200000e + 07 RPM
.....
.....
2.000000e + 05 Hz 1.200000e + 07 RPM
2.000000e + 05 Hz 1.200000e + 07 RPM
Info: /OSCI/SystemC: simulation stopped by user.

## 6.7 32 Bit × 32 Bit Input 64-Bit Output Booth Multiplier

The Booth multiplication algorithm [2, 3] is a popular hardware-implementable signed-number multiplication algorithm and uses a few registers to generate the product. The algorithm is simple.

Booth's algorithm examines adjacent pairs of bits of a given n-bit multiplier Y in signed two's complement form, with an implicit bit beyond the least significant bit, $y-1 = 0$. For each bit yi, for i between 0 and N−1 inclusive, the bits yi and yi−1 are compared. If these two bits are equal, the product accumulator P is left unchanged. If $yi = 0$ and $yi-1 = 1$, the multiplicand times $2^i$ is added to P; if $yi = 1$ and $yi-1 = 0$, the multiplicand times $2^i$ is subtracted from P. The final value of P is the signed product.

Booth's algorithm can be implemented by repeatedly adding (with ordinary unsigned binary addition) one of two predetermined values A and S to a product P, then performing a rightward arithmetic shift on P. Let **m** and **r** be the multiplicand and multiplier, respectively; let x and y represent the number of bits in **m** and **r**.

1. Each of the numbers A, P, and S must be $x + y + 1$ bits long

(a) A: Fill the most significant (leftmost) bits with the value of **m**. Fill the remaining $(y + 1)$ bits with zeros.
(b) S: Fill the most significant bits with the value of $-\mathbf{m}$ in two's complement notation. Fill the remaining $(y + 1)$ bits with zeros.
(c) P: Fill the most significant $x$ bits with zeros. To the right of this, append the value of **r**. Fill the least significant (rightmost) bit with a zero.

2. Determine the two least significant (rightmost) bits of $P$
   (a) If they are 01, find the value of $P + A$. Ignore any overflow.
   (b) If they are 10, find the value of $P + S$. Ignore any overflow.
   (c) If they are 00, do nothing. Use $P$ directly in the next step.
   (d) If they are 11, do nothing. Use $P$ directly in the next step.
3. Arithmetically shift the value obtained in step 2 by a single place to the right. P now contains the new value.
4. Repeat steps 2 and 3 until they have been done $y$ times.
5. Drop the least significant (rightmost) bit from $P$. This is the product of **m** and **r**.

Figures 6.23 and 6.24 contain the algorithm source code and test harness, while Figs. 6.25 (SystemC 2.2.0) and 6.26 (SystemC 2.3.0) contain partial input and output traces. The algorithm basically uses four states IDLE, ADD, SHIFT, and OUTPUT, specified with the bit combinations 00, 01, 10, and 11.

```
#ifndef BOOTH_H
#define BOOTH_H

#include <systemc>

/* Global constants */
const unsigned int WIDTH      = 32;
const unsigned int WIDTHP1   = 33;
const unsigned int WIDTH2     = 64;
const unsigned int WIDTH2P1 = 65;

SC_MODULE(booth)
{
  /* Declare/define input/output ports */
  sc_core::sc_in<bool> clk;
  sc_core::sc_in<bool> enable;
  sc_core::sc_in< sc_dt::sc_bv<WIDTH> > multiplier;
  sc_core::sc_in< sc_dt::sc_bv<WIDTH> > multiplicand;
  sc_core::sc_out<bool> done;
```

**Fig. 6.23** 32 bit x 32 bit input 64-bit output Booth multiplier

```
    sc_core::sc_out< sc_dt::sc_bv<WIDTH2> > product;

    /* Declare/define members */
    sc_dt::sc_bv<2> current_state;
    sc_dt::sc_bv<2> next_state;
    sc_dt::sc_bv<WIDTH> iter_cnt;
    sc_dt::sc_bv<WIDTH2P1> a_reg;
    sc_dt::sc_bv<WIDTH2P1> s_reg;
    sc_dt::sc_bv<WIDTH2P1> p_reg;
    sc_dt::sc_bv<WIDTH2P1> sum_reg;
    sc_dt::sc_bv<WIDTHP1> multiplier_neg;

    /* Thread to update current state */
    void booth_proc0()
    {
      while(1)
      {
        wait();
      if(enable.read() == false) current_state = "00";
      else current_state = next_state;
    }
  }

  /* Thread to track/update current state -- next state transitions *.
  void booth_proc1()

  {
    while(1)
    {
      wait();
      if(current_state == "00")
      {
        if(enable.read() == true) next_state = "01";
        else next_state = "00";
      }
      else if(current_state == "01")
      {
         next_state = "10";
      }
```

**Fig. 6.23** (continued)

```
    else if(current_state == "10")
    {
      if(iter_cnt.to_int() == WIDTH) next_state = "11";
      else next_state = "01";
    }
    else if(current_state == "11")
    {
      next_state = "00";
    }
  }
}

/* Main Booth multiplication thread - steps 1 -> 5 of algorithm */
void booth_proc2()
{
  sc_dt::sc_bv<2> p_reg_tmp;
  sc_dt::sc_bv<32> multiplier_local;
  while(1)
  {
    wait();
if(current_state == "00")
{
 multiplier_local = multiplier.read();
 multiplier_neg = (~multiplier_local).to_int() + 1;
 a_reg = (multiplier.read(),
          "00000000000000000000000000000000");
 s_reg = (multiplier_neg,
          "00000000000000000000000000000000");
 p_reg = ("00000000000000000000000000000000",
           multiplicand.read(), "0");
 iter_cnt = "00000000000000000000000000000000";
 done.write(false);
}
else if(current_state == "01")
{
 p_reg_tmp = p_reg.range(1,0);
  if(p_reg_tmp == "01")
  {
```

Fig. 6.23 (continued)

```
      sum_reg = p_reg ^ a_reg;
     }
     else if(p_reg_tmp == "10")
     {
      sum_reg = p_reg ^ s_reg;
     }
     else if(p_reg_tmp == "11")
     {
      sum_reg = p_reg;
     }
     else if(p_reg_tmp == "00")
     {
      sum_reg = p_reg;
     }
     iter_cnt = iter_cnt.to_int() + 1;
   }
   else if(current_state == "10")
   {
     p_reg = (sum_reg.range(64,64),sum_reg.range(64,1));
   }
   else if(current_state == "11")
   {
     product.write(p_reg);

         done.write(true);
       }
     }
    }

   /* Constructor - initialize bit vectors */
    SC_CTOR(booth):current_state("00"),next_state("00"),
   iter_cnt("0000000000000000"),
   a_reg("00000000000000000000000000000000000000000000000000000000
   000000000"),
   p_reg("00000000000000000000000000000000000000000000000000000000
   000000000"),
   s_reg("00000000000000000000000000000000000000000000000000000000
   000000000"),
```

Fig. 6.23 (continued)

```
sum_reg("00000000000000000000000000000000000000000000000000000
000000000000"),
multiplier_neg("00000000000000000000000000000000")
 {
   /* Declare/assign threads and sensitivity lists */
   SC_CTHREAD(booth_proc0,clk.pos());
   SC_CTHREAD(booth_proc2,clk.pos());
   SC_THREAD(booth_proc1);
   sensitive << clk << enable << multiplier << multiplicand;
 }

 ~booth(){ }
 /* Destructor */
};

#endif
```

**Fig. 6.23** (continued)

```
#include "booth.h"

int sc_main(int argc, char **argv)
{
  /* Declare/define channels/signals */
  sc_core::sc_signal< sc_dt::sc_bv<32> > multiplier_sig;
  sc_core::sc_signal< sc_dt::sc_bv<32> > multiplicand_sig;
  sc_core::sc_signal<bool> done_sig;
  sc_core::sc_signal<bool> enable_sig;
  sc_core::sc_signal< sc_dt::sc_bv<64> > product_sig;

  /* Declare/define bit vectors for multiplier/multiplicand */
  sc_dt::sc_bv<32> multiplier_value;
  sc_dt::sc_bv<32> multiplicand_value;
  bool enable_value;
  /* Declare/define master clock, Booth multiplier and trace file */
  sc_core::sc_clock clk("clk", 2.0, sc_core::SC_NS, 0.5);
  booth booth_mult("booth_mult");
  sc_core::sc_trace_file *fp = sc_core::sc_create_vcd_trace_file("tr_booth");
```

**Fig. 6.24** Test harness for 32 bit x 32 bit input 64-bit output Booth multiplier

```
fp->set_time_unit(1.0, sc_core::SC_NS);
/* Connect Booth multiplier ports and signals */
booth_mult.clk(clk);
booth_mult.enable(enable_sig);
booth_mult.multiplier(multiplier_sig);
booth_mult.multiplicand(multiplicand_sig);
booth_mult.done(done_sig);
booth_mult.product(product_sig);

/* Connect trace file and data channels */
sc_core::sc_trace(fp, clk, "clk");
sc_core::sc_trace(fp, enable_sig, "enable");
sc_core::sc_trace(fp, multiplier_sig, "multiplier");

sc_core::sc_trace(fp, multiplicand_sig, "multiplicand");
sc_core::sc_trace(fp, done_sig, "done");
sc_core::sc_trace(fp, product_sig, "product");

/* Assign values to multiplier/multiplicand and write to multiplier
   input channels */
enable_value = true;
multiplier_value = "00111100110000110101010110101011";
multiplicand_value = "01111000100001110101110010100011";

enable_sig.write(enable_value);
multiplier_sig.write(multiplier_value);
multiplicand_sig.write(multiplicand_value);

/* Run simulation for pre-defined time-interval*/
sc_core::sc_start(400.0, sc_core::SC_NS);

/* Change input values, write to input channels */
enable_value = false;
multiplier_value = "10110100010010111110000100001111101";
multiplicand_value = "01010001100010100111010110110011";
enable_sig.write(enable_value);
multiplier_sig.write(multiplier_value);
multiplicand_sig.write(multiplicand_value);
```

**Fig. 6.24** (continued)

```
    /* Run simulation for pre-defined time-interval*/
    sc_core::sc_start(100.0, sc_core::SC_NS);

    /* Change input values, write to input channels */
    enable_value = true;
    multiplier_value = "10110100010010111110000100001111101";
    multiplicand_value = "01010001100010100111010110110011";
    enable_sig.write(enable_value);
    multiplier_sig.write(multiplier_value);
    multiplicand_sig.write(multiplicand_value);

    /* Run simulation for pre-defined time-interval*
    sc_core::sc_start(300.0, sc_core::SC_NS);

    /* Stop simulation and close trace file */
    sc_core::sc_stop();
    sc_core::sc_close_vcd_trace_file(fp);
    return 0;
}
```

**Fig. 6.24** (continued)

**Fig. 6.25** Partial input/output traces for data to/from 32 bit x 32 bit input 64-bit output Booth multiplier SystemC 2.2.0

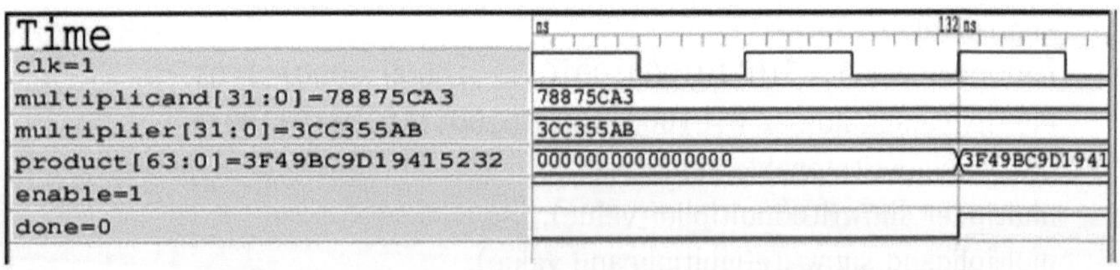


**Fig. 6.26** Partial input/output traces for data to/from 32 bit x 32 bit input 64-bit output Booth multiplier SystemC 2.3.0

## 6.8 Decimal to IEEE 754-2008 Format 32-Bit Floating-Point Converter

Decimal/real number representation in digital hardware involves judicious balancing of several competing issues, which Institute of Electrical and Electronics Engineers (IEEE) has tackled by proposing a standard (IEEE 754) [4, 7] that is used to represent real numbers on all platforms (Linux/Unix, Macintosh, and Windows). There are several ways to represent real numbers on computers. Fixed point involves a radix point somewhere in the middle of the string of digits and is equivalent to using integers that represent portions of some unit. Floating point represents numbers as a base number and exponent, a very flexible approach that can represent numbers between 1000000000000 and 0.0000000000000001.

The IEEE 754 floating-point number representation consists of the sign (1 bit), exponent (8 bits), and mantissa (23 bits) for a total of 32 bits. The binary mantissa (significand) representation is *normalized*—1. < fractional part >, i.e., a '1', followed by 23 '1s' and '0s'. However, when an IEEE 754 format decimal number is stored in memory, the '1' is not stored—'hidden bit.' The exponent must represent both positive and negative values and so has a bias of 127 added to it during storage. So, in summary, each 32-bit IEEE 754-2008 format floating-point number consists of

- 1 sign bit (1 for negative, 0 for positive).
- 8 *biased* exponent bits. A *bias* of 127 is added to the actual integer exponent value.
- 23 mantissa or significand bits, with an assumed 'hidden' bit.

Actual arithmetic involves several other issues such as *round-off (to positive/negative infinity, zero)*, *positive/negative infinity, positive/negative zero, not-a-number (NaN), overflow, underflow*, and *exponent alignment*, which do not discuss here. Figures 6.27 and 6.28 contain the source code and test harness. It is very important to note that because real number fractions in binary format are infinite series of '1s' and '0s', but physical memory can allocate a finite number of bits, when combined with the above factors (round-off, etc.,) results in real numbers slightly different than actual values—e.g., a final result of 1.0 (by hand calculation) might be 0.9999999....when outputted by the computer.

Typical console output looks like as below.

```
./sim .88 −1.25
SystemC 2.2.0—August 30, 2011, 19:39:07
Copyright (c) 1996–2006 by all contributors,
ALL RIGHTS RESERVED

0.88 −1 1.76
−1.25 0 −1.25
ieee_convt
significand 1 11000010100011110101110
```

```
#ifndef DEC2IEEE754_H
#define DEC2IEEE754_H

#include <systemc>
#include <cstdlib>
#include <cstring>
#include <cstdio>

const unsigned int MAXFRBITS32  = 23;
const unsigned int MIDPOS32     = 11;
const unsigned int EXPSIZE32    = 8;

SC_MODULE(dec2ieee754_32_conv)
{
 private:
  /* Convert decimal fraction number to binary format */
  void decfr2bin(float ff, sc_dt::sc_bv<MAXFRBITS32> &a)
  {
    unsigned int j = 0;
    float decrement = 0.0;
    float tmp0 = 0.0;
    a = "00000000000000000000000";
    ff = fabs(ff);
    for(j = 0; j < MAXFRBITS32; j++)
    {
     if(ff < 1.0)
      {
        tmp0 = EXPO*ff;
        if(tmp0 >= 1.0)
        {
          a[MAXFRBITS32 - 1 - j] = "1";
          tmp0 -= 1.0;
        }
        ff = tmp0;
      }
    }
  }
```

**Fig. 6.27** Decimal format real number to IEEE 754-2008 normalized 32-bit floating-point converter

```
/* Convert decimal exponent to binary format */
void decexp2bin(unsigned int ii, sc_dt::sc_bv<EXPSIZE32> &a)
{
   unsigned int i = 0;
   unsigned int quotient = 0;
   a = "00000000";
   while(ii > 0)
  {
     quotient = ii % EXPO;
     ii /= EXPO;
     if(quotient == 1 ) a[i] = "1";
     i += 1;
   }
}

/* Determine the largest poer of 2 that is less tahn or equal to a
    given floating point number */
int biggestpower2(float ff, sc_dt::sc_bv<1> &a)
{
  int power_2_value    = 1;
  int power_2_exponent = 0;
  float power_2_value_ltz = 1.0;
  float ffabs = fabs(ff);

  if(ff < 0.0) a = "1";
  else a = "0";
  if(ffabs > 1.0)
  {
    while((float)power_2_value < ffabs)
    {
      power_2_value *= EXPO;
      power_2_exponent += 1;
    }
  }
    else if(ffabs > 0.0 && ffabs < 1.0)
    {
      while(power_2_value_ltz > ffabs)
      {
        power_2_value_ltz /= EXPO;
```

**Fig. 6.27** (continued)

```
      power_2_exponent -= 1;
    }
  }

  if(power_2_exponent > 0) return (power_2_exponent - 1);
  else if(power_2_exponent < 0) return power_2_exponent;
}

float power2(int ii)
{
  float power_2_value = 1;
  unsigned int i = 0;
  if(ii > 0)
  {
    for(i = 0; i < ii; i++) power_2_value *= EXPO;
  }
  else
  {
    for(i = 0; i < abs(ii); i++) power_2_value /= EXPO;
  }
  return power_2_value;
}

public:
 sc_core::sc_in<float> floatin0;
 sc_core::sc_in<float> floatin1;
 sc_core::sc_out< sc_dt::sc_bv<EXPSIZE32> > exponentout0;
 sc_core::sc_out< sc_dt::sc_bv<MAXFRBITS32> > significandout0;
 sc_core::sc_out< sc_dt::sc_bv<EXPSIZE32> > exponentout1;
 sc_core::sc_out< sc_dt::sc_bv<MAXFRBITS32> > significandout1;
 sc_core::sc_out< sc_dt::sc_bv<1> > signout0;
 sc_core::sc_out< sc_dt::sc_bv<1> > signout1;

 sc_dt::sc_bv<MAXFRBITS32> significandvalue0;
 sc_dt::sc_bv<EXPSIZE32> exponentbinvalue0;
 sc_dt::sc_bv<MAXFRBITS32> significandvalue1;
 sc_dt::sc_bv<EXPSIZE32> exponentbinvalue1;
 sc_dt::sc_bv<1> signbit0;
 sc_dt::sc_bv<1> signbit1;
```

**Fig. 6.27** (continued)

```
        int exponentvalue0;
        int exponentvalue1;
        float floatvalue0;
        float convfloat0;
        float floatvalue1;
        float convfloat1;

        void dec2ieee754_32_conv_proc0()
        {
          while(1)
          {
            wait();
            floatvalue0 = floatin0.read();
            floatvalue1 = floatin1.read();
            exponentvalue0 = biggestpower2(floatvalue0, signbit0);
            exponentvalue1 = biggestpower2(floatvalue1, signbit1);
            convfloat0 = floatvalue0/power2((float)exponentvalue0);
            convfloat1 = floatvalue1/power2((float)exponentvalue1);
            std::cout<<floatvalue0<<" "<<exponentvalue0<<"
"<<convfloat0<<std::endl;
            std::cout<<floatvalue1<<" "<<exponentvalue1<<"
"<<convfloat1<<std::endl;
            if(convfloat0 > 1.0) convfloat0 -= 1.0;
            else if(convfloat0 < 0.0) convfloat0 += 1.0;
            if(convfloat1 > 1.0) convfloat1 -= 1.0;
            else if(convfloat1 < 0.0) convfloat1 += 1.0;
            exponentvalue0 += 127;
            exponentvalue1 += 127;
            decfr2bin(convfloat0, significandvalue0);
           decfr2bin(convfloat1, significandvalue1);
            decexp2bin(exponentvalue0, exponentbinvalue0);
            decexp2bin(exponentvalue1, exponentbinvalue1);
            exponentout0.write(exponentbinvalue0);
            significandout0.write(significandvalue0);
            signout0.write(signbit0);
             signout1.write(signbit1);
            exponentout1.write(exponentbinvalue1);
            significandout1.write(significandvalue1);
            std::cout<<name()<<std::endl;
```

**Fig. 6.27** (continued)

```
        std::cout<<" significand 1 "<<significandvalue0<<std::endl;
        std::cout<<" exponent 1  "<<exponentbinvalue0<<std::endl;
        std::cout<<" sign 1 "<<signbit0<<std::endl;
        std::cout<<" significand 2 "<<significandvalue1<<std::endl;
        std::cout<<" exponent 2 "<<exponentbinvalue1<<std::endl;
        std::cout<<" sign 2 "<<signbit1<<std::endl;
      }
    }

    /*Constructor - initialize members */
    SC_CTOR(dec2ieee754_32_conv):exponentbinvalue0("00000000"),
                          exponentbinvalue1("00000000"),
                          significandvalue0("00000000000000000000000"),
                          significandvalue1("00000000000000000000000"),
                          signbit0("0"), signbit1("0")
    {
      /* Declare/define main execution thread and sensitize it */
      SC_THREAD(dec2ieee754_32_conv_proc0);
      sensitive << floatin0 << floatin1;
    }
      ~dec2ieee754_32_conv(){ }
  };
  #endif
```

**Fig. 6.27** (continued)

```
#include "dec2ieee754.h"
const unsigned int SIZE = 10;

int sc_main(int argc, char **argv)
{
  if(argc < 3)
  {
    std::cout<<"Insufficient parameters ... "<<std::endl;
    std::cout<<"usage ./sim <floating point number>
                <floating point number>"<<std::endl;
    exit(0);
  }
```

**Fig. 6.28** Test harness for decimal format real number to IEEE 754-2008 normalized 32-bit floating-point converter

```
sc_core::sc_signal<float> numsig0;
sc_core::sc_signal<float> numsig1;
sc_core::sc_signal< sc_dt::sc_bv<1> > signoutsig0;
sc_core::sc_signal< sc_dt::sc_bv<1> > signoutsig1;
sc_core::sc_signal< sc_dt::sc_bv<EXPSIZE32> > exponentsig0;
sc_core::sc_signal< sc_dt::sc_bv<MAXFRBITS32> > significandsig0;
sc_core::sc_signal< sc_dt::sc_bv<EXPSIZE32> > exponentsig1;
sc_core::sc_signal< sc_dt::sc_bv<MAXFRBITS32> > significandsig1;

dec2ieee754_32_conv ieee_convt("ieee_convt");
char number0[SIZE];
char number1[SIZE];
float num0;
float num1;
unsigned int len0;
unsigned int len1;
ieee_convt.floatin0(numsig0);
ieee_convt.floatin1(numsig1);
ieee_convt.exponentout0(exponentsig0);
ieee_convt.significandout0(significandsig0);
ieee_convt.exponentout1(exponentsig1);
ieee_convt.significandout1(significandsig1);
ieee_convt.signout0(signoutsig0);
ieee_convt.signout1(signoutsig1);

len0 = strlen(argv[1]);
len1 = strlen(argv[2]);
if(len0 > 0 && len1 > 0)
{
  strcpy(number0, argv[1]);
  strcpy(number1, argv[2]);
}

else
{
  std::cout<<"invalid input ... exiting ..."<<std::endl;
  exit(0);
}
```

**Fig. 6.28** (continued)

```
    num0 = strtof(number0, NULL);
    num1 = strtof(number1, NULL);
    numsig0.write(0.0);
    numsig1.write(0.0);
    sc_core::sc_start(10.0, sc_core::SC_NS);
    numsig0.write(num0);
    numsig1.write(num1);
    sc_core::sc_start(10.0, sc_core::SC_NS);
    sc_core::sc_stop();
    return 0;
}
```

**Fig. 6.28** (continued)

```
exponent 1 01111110
sign 1 0
significand 2 01000000000000000000000
exponent 2 01111111
sign 2 1
SystemC: simulation stopped by user.

   SystemC 2.3.0-ASI—July 19, 2012, 18:53:11
   Copyright (c) 1996–2012 by all contributors,
   ALL RIGHTS RESERVED

0.88 −1 1.76
−1.25 0 −1.25
ieee_convt
significand 1 11000010100011110101110
exponent 1 01111110
sign 1 0
significand 2 01000000000000000000000
exponent 2 01111111
sign 2 1
Info: /OSCI/SystemC: simulation stopped by user.
```

## 6.9 IEEE 754-2008 Format 32-Bit Floating-Point Number Addition

Addition of two IEEE 754-2008 [4, 7] format 32-bit floating-point numbers is analyzed, utilizing the knowledge gained from the previous example. The key concept underlying IEEE 754-2008 format 32-bit floating-point number addition is *radix alignment,* as it is for subtraction, but not for either multiplication or division. This example reuses the code developed in example 6.8. Figure 6.29 illustrates the

```
0.25  =  0 01111101 00000000000000000000000
100.0 =  0 10000101 10010000000000000000000
choose to shift the .25, since we want to increase its exponent.
shift by  10000101
         -01111101
         ---------
          00001000   (8) places.

   0 01111101 00000000000000000000000 (original value)
   0 01111110 10000000000000000000000 (shifted 1 place)
              (note that hidden bit is shifted into MSB of mantissa)
   0 01111111 01000000000000000000000 (shifted 2 places)
   0 10000000 00100000000000000000000 (shifted 3 places)
   0 10000001 00010000000000000000000 (shifted 4 places)
   0 10000010 00001000000000000000000 (shifted 5 places)
   0 10000011 00000100000000000000000 (shifted 6 places)
   0 10000100 00000010000000000000000 (shifted 7 places)
   0 10000101 00000001000000000000000 (shifted 8 places)

add (hidden bit for 100.0 must be kept in mind)

   0 10000101 1.10010000000000000000000  (100.0)
+  0 10000101 0.00000001000000000000000  (0.25)
   ---------------------------------------
   0 10000101 1.10010001000000000000000
normalize the result ("hidden bit" must be 1). It already is for this example.

Final result: 0 10000101 10010001000000000000000
```

**Fig. 6.29** Addition of two IEEE 754-2008 format 32-bit floating-point numbers

concept of radix alignment, as applied to two floating-point numbers 100.0 and 0.25. Two key points must be kept in mind with respect to any IEEE 754-2008 format number:

- Shifting the mantissa *left* by 1 bit *decreases the exponent* by 1.
- Shifting the mantissa *right* by 1 bit *increases the exponent* by 1.

A mantissa must ideally be shifted to the right, so that the bits that fall off are the least significant bits.

For each of the IEEE 754-2008 format 32-bit floating-point numbers in Fig. 6.29, the leftmost bit is the sign bit, followed by 8 exponent bits and then by 23 mantissa or significand bits. Figures 6.30 and 6.31 contain the source code and

```
#ifndef DEC2IEEE754_H
#define DEC2IEEE754_H
#include <systemc>
#include <cstdlib>
#include <cstring>
#include <cstdio>

const unsigned int MAXFRBITS32    = 23;
const unsigned int MIDPOS32       = 11;
const unsigned int EXPSIZE32      = 8;
const unsigned int EXPO           = 2;

SC_MODULE(ieee754_32_arithmetic_all_bin)
{
  private:
    /* Method to convert unsigned int exponent
       to 8 bit binary value */
    void decexp2bin(unsigned int ii, sc_dt::sc_bv<EXPSIZE32> &a)
    {
      unsigned int i = 0;
      unsigned int quotient = 0;
      a = "00000000";
      while(ii > 0)
      {
        quotient = ii % EXPO;
        ii /= EXPO;
        if(quotient == 1 ) a[i] = "1";
        i += 1;
      }
    }
   /* Method to generate unsigned integer power of 2 value */
    unsigned int intpower2(unsigned int pwr)
    {
      unsigned int powervalue = 1;
      unsigned int i = 0;

      while(i < pwr)
```

**Fig. 6.30** IEEE 754-2008 format 32-bit floating-point number adder

```
  {
    powervalue *= EXPO;
    i++;
  }
  return powervalue;
}

/* Convert 8 bit exponent value to integer */
int bv8_2int(sc_dt::sc_bv<EXPSIZE32> &a)
{
  unsigned int i = 0;
  unsigned int pwr2 = 0;

  for(i = 0; i < EXPSIZE32; i++)
  {
    if(a(i,i) == "1") pwr2 += intpower2(i);
  }
  return pwr2;
}

/* Read input from decimal to IEEE-754-2008 converter */
void readinput()
{
  exponent0 = exponentin0.read();
  exponent1 = exponentin1.read();
  significand0 = significandin0.read();
  significand1 = significandin1.read();
  sign0 = signin0.read();
  sign1 = signin1.read();
}

/* Method to generate two's complement
   value for 32 bit number */
void twocomplement(sc_dt::sc_bv<MAXFRBITS32> &a)
{
  sc_dt::sc_bv<MAXFRBITS32> tmp0;
  sc_dt::sc_bv<1> tmp1;
  sc_dt::sc_bv<1> carrybit;
  unsigned int i;
```

**Fig. 6.30** (continued)

```
  tmp0(MAXFRBITS32 - 1, 0) = a(MAXFRBITS32 - 1, 0);
  tmp0 = ~tmp0;
  for(i = 0; i < MAXFRBITS32; i++)
  {
    tmp1 = tmp0(i,i);
    if(i == 0)
    {
      if(tmp1(0,0) == "1") { carrybit = "1"; tmp0(0,0) = "0"; }
      else if(tmp1(0,0) == "0"){ carrybit = "0"; tmp0(0,0) = "1"; }
    }
    else
    {
      if(tmp1(0,0) == "1" && carrybit == "1")
        { carrybit = "1"; tmp0(i,i) = "0"; }
      else if(tmp1(0,0) == "0" && carrybit == "1")
        { carrybit = "0"; tmp0(i,i) = "1"; }
      else if(tmp1(0,0) == "1" && carrybit == "0")
        { carrybit = "0"; tmp0(i,i) = "1"; }
      else if(tmp1(0,0) == "0" && carrybit == "0")
        { carrybit = "0"; tmp0(i,i) = "0"; }
    }
  }
  a(MAXFRBITS32 - 1, 0) = tmp0(MAXFRBITS32 - 1, 0);
}

/* Method for normalization - all IEEE-754-2008
   mantissa/significand  have the format 1.<fraction>  */
void normalize(sc_dt::sc_bv<MAXFRBITS32> &a,
               sc_dt::sc_bv<EXPSIZE32> &ee,
               sc_dt::sc_bv<2> &c)
{
  sc_dt::sc_bv<MAXFRBITS32> tmp0;
  sc_dt::sc_bv<MAXFRBITS32> tmp00 = "000000000000000000000000";
  sc_dt::sc_bv<2> tmp1;
  int tmpexp = 0;
  tmp0 = a;
  tmp1 = c;
  if(tmp1 == "11")
```

**Fig. 6.30** (continued)

```
  {
    tmpexp = bv8_2int(ee) - 127;
    tmpexp += 128;
    tmp1(1, 1) = "0";
    tmp1(0, 0) = "1";
    tmp00(MAXFRBITS32 - 1, 0) = tmp0(MAXFRBITS32 - 1, 0);
    tmp0(MAXFRBITS32 - 2, 0) = tmp00(MAXFRBITS32 - 1, 1);
    tmp0(MAXFRBITS32 - 1, MAXFRBITS32 - 1) = "0";
    decexp2bin((unsigned int)tmpexp, ee);
  }
  else if(tmp1 == "10")
  {
    tmp1(1, 1) = "0";
    tmp1(0, 0) = "1";
  }
  a(MAXFRBITS32 - 1, 0) = tmp0(MAXFRBITS32 - 1, 0);
  c(1, 0) = tmp1(1, 0);
}

/* For each input number, examine exponent, sign and perform
   appropriate radix alignment */
void checkexponentandshift(sc_dt::sc_bv<MAXFRBITS32> &a,
                           sc_dt::sc_bv<MAXFRBITS32> &b,
                           sc_dt::sc_bv<EXPSIZE32> &ae,
                           sc_dt::sc_bv<EXPSIZE32> &be,
                           sc_dt::sc_bv<EXPSIZE32> &ee,
                           sc_dt::sc_bv<1> &c,
                           sc_dt::sc_bv<1> &d,
                           unsigned int ui)
{
  if(ui == 1 || ui == 2)
   /* Radix alignment is only for addition and subtraction */
  {
sc_dt::sc_bv<MAXFRBITS32> tmpsignificand;
int pwr0 = bv8_2int(ae) - 127;
int pwr1 = bv8_2int(be) - 127;
int shiftlen = abs(pwr0 - pwr1);
if(sign0(0, 0) == "0" && sign1(0, 0) == "0")
```

**Fig. 6.30** (continued)

```
{
 if(pwr1 < pwr0)
 {
  tmpsignificand = "000000000000000000000000";
  tmpsignificand(MAXFRBITS32 - 1,
                         MAXFRBITS32 - 1) = "1";
  tmpsignificand(MAXFRBITS32 - 2,
                         0) = b(MAXFRBITS32 - 1, 1);
  shiftlen -= 1;
  tmpsignificand = tmpsignificand >> shiftlen;
  b(MAXFRBITS32 - 1, 0) =
   tmpsignificand(MAXFRBITS32 - 1, 0);
  ee(EXPSIZE32 - 1, 0) = ae(EXPSIZE32 - 1, 0);
  c(0, 0) = "1";
  d(0, 0) = "0";
 }
 else
 {
  tmpsignificand = "000000000000000000000000";
  tmpsignificand(MAXFRBITS32 - 1,
                         MAXFRBITS32 - 1) = "1";
  tmpsignificand(MAXFRBITS32 - 2, 1) =
                         b(MAXFRBITS32 - 2, 1);
  shiftlen -= 1;
  tmpsignificand = tmpsignificand >> shiftlen;
  a(MAXFRBITS32 - 1, 0) =
    tmpsignificand(MAXFRBITS32 - 1, 0);
  ee(EXPSIZE32 - 1, 0) = be(EXPSIZE32 - 1, 0);
  c(0, 0) = "0";
  d(0, 0) = "1";
 }
}
else
{
 if(pwr1 < pwr0)
 {
   tmpsignificand = "000000000000000000000000";
   tmpsignificand(MAXFRBITS32 - 1,
                         MAXFRBITS32 - 1) = "1";
```

**Fig. 6.30** (continued)

```
      tmpsignificand(MAXFRBITS32 - 2, 1) =
                              b(MAXFRBITS32 - 2, 1);
      shiftlen -= 1;

      tmpsignificand = tmpsignificand >> shiftlen;
      b(MAXFRBITS32 - 1, 0) =
         tmpsignificand(MAXFRBITS32 - 1, 0);
      ee(EXPSIZE32 - 1, 0) = ae(EXPSIZE32 - 1, 0);
      c(0, 0) = "1";
      d(0, 0) = "0";
    }
    else
    {
      tmpsignificand = "00000000000000000000000";
      tmpsignificand(MAXFRBITS32 - 1,
                              MAXFRBITS32 - 1) = "1";
      tmpsignificand(MAXFRBITS32 - 2, 1) =
                              b(MAXFRBITS32 - 2, 1);
      shiftlen -= 1;
      tmpsignificand = tmpsignificand >> shiftlen;
      a(MAXFRBITS32 - 1, 0) =
         tmpsignificand(MAXFRBITS32 - 1, 0);
      ee(EXPSIZE32 - 1, 0) = be(EXPSIZE32 - 1, 0);
      c(0, 0) = "0";
      d(0, 0) = "1";
    }
    if(sign0(0, 0) == "1" && sign1(0, 0) == "0")
      { twocomplement(a);}
    else if(sign0(0, 0) == "0" && sign1(0, 0) == "1")
      { twocomplement(b);}
    else if(sign0(0, 0) == "1" && sign1(0, 0) == "1")
    {
      twocomplement(a);
      twocomplement(b);
    }
  }
  exponentcheck = true;
}
```

**Fig. 6.30** (continued)

```
}

/* Method for addition of two IEEE-754-2008 32
 bit floating point numbers */
void add(sc_dt::sc_bv<MAXFRBITS32> &a,
        sc_dt::sc_bv<MAXFRBITS32> &b,
        sc_dt::sc_bv<MAXFRBITS32> &r,
        sc_dt::sc_bv<EXPSIZE32> &ee,
        sc_dt::sc_bv<1> &c,
        sc_dt::sc_bv<1> &cb,
        sc_dt::sc_bv<1> &d,
        sc_dt::sc_bv<2> &e)
{
  sc_dt::sc_bv<1> tmp0;
  sc_dt::sc_bv<1> tmp1;
  unsigned int i = 0;
  r = "00000000000000000000000";
  cb = "0";

  if(exponentcheck == true)
  {
    for(i = 0; i < MAXFRBITS32; i++)
    {
      tmp0 = a(i,i);
      tmp1 = b(i,i);
      if(i == 0)
      {
        if(tmp0 == "0" && tmp1 == "1")
        { cb = 0; r(i,i) = "1"; }
        else if(tmp0 == "1" && tmp1 == "0")
        { cb = 0; r(i,i) = "1"; }
        else if(tmp0 == "1" && tmp1 == "1")
        { cb = 1; r(i,i) = "0"; }
      }
      else
      {
        if(tmp0 == "0" && tmp1 == "0" && carrybit == "0")
        { cb = 0; r(i,i) = "0"; }
        else if(tmp0 == "0" && tmp1 == "0" && carrybit == "1")
```

**Fig. 6.30** (continued)

```
        { cb = 0; r(i,i) = "1"; }
        else if(tmp0 == "0" && tmp1 == "1" && carrybit == "0")
        { cb = 0; r(i,i) = "1"; }
        else if(tmp0 == "1" && tmp1 == "0" && carrybit == "0")
        { cb = 0; r(i,i) = "1"; }
        else if(tmp0 == "0" && tmp1 == "1" && carrybit == "1")
        { cb = 1; r(i,i) = "0"; }
        else if(tmp0 == "1" && tmp1 == "0" && carrybit == "1")
        { cb = 1; r(i,i) = "0"; }
        else if(tmp0 == "1" && tmp1 == "1" && carrybit == "0")
        { cb = 1; r(i,i) = "0"; }
        else if(tmp0 == "1" && tmp1 == "1" && carrybit == "1")
        { cb = 1; r(i,i) = "1"; }
      }
    }
    if(cb(0, 0) == "1" &&
        c(0, 0) == "1" &&
        d == "1") e = "11";
    else if(cb(0, 0) == "1" &&
            c(0, 0) == "1" &&
            d == "0") e = "10";
    else if(cb(0, 0) == "1" &&
            c(0, 0) == "0" &&
            d == "1") e = "10";
    else if(cb(0, 0) == "0" &&
            c(0, 0) == "1" &&
            d == "1") e = "10";
    else if(cb(0, 0) == "0" &&
            c(0, 0) == "0" &&
            d == "1") e = "01";
    else if(cb(0, 0) == "0" &&
            c(0, 0) == "1" &&
            d == "0") e = "01";
    else if(cb(0, 0) == "1" &&
            c(0, 0) == "0" &&
            d == "0") e = "01";
    else if(cb(0, 0) == "0" &&
        c(0, 0) == "0" &&
        d == "0") e = "00";
```

**Fig. 6.30** (continued)

```
      normalize(r, ee, e);
      std::cout<<name()<<" "<<cb<<" "<<c<<" "<<d<<" "<<e<<" "<<ee<<"
"<<r<<std::endl;
    }
  }

public:
  /* Declare/define input/output ports */
  sc_core::sc_in< sc_dt::sc_bv<MAXFRBITS32> > significandin0;
  sc_core::sc_in< sc_dt::sc_bv<MAXFRBITS32> > significandin1;
  sc_core::sc_in< sc_dt::sc_bv<EXPSIZE32> > exponentin0;
  sc_core::sc_in< sc_dt::sc_bv<EXPSIZE32> > exponentin1;
  sc_core::sc_in< sc_dt::sc_bv<1> > signin0;
  sc_core::sc_in< sc_dt::sc_bv<1> > signin1;

  /* Declare/define data members */
  sc_dt::sc_bv<MAXFRBITS32 + EXPSIZE32> num0;
  sc_dt::sc_bv<MAXFRBITS32 + EXPSIZE32> num1;
  sc_dt::sc_bv<MAXFRBITS32M> multresf;
  sc_dt::sc_bv<MAXFRBITS32> significand0;
  sc_dt::sc_bv<MAXFRBITS32> significand1;
  sc_dt::sc_bv<MAXFRBITS32> significandresadd;
  sc_dt::sc_bv<MAXFRBITS32> significandresmult;
  sc_dt::sc_bv<MAXFRBITS32> significandressub;
  sc_dt::sc_bv<EXPSIZE32> exponent0;
  sc_dt::sc_bv<EXPSIZE32> exponent1;
  sc_dt::sc_bv<EXPSIZE32> nexponent;
  sc_dt::sc_bv<EXPSIZE32> mexponent;
  sc_dt::sc_bv<EXPSIZE32> sexponent;
  sc_dt::sc_bv<1> sign0;
  sc_dt::sc_bv<1> sign1;
  sc_dt::sc_bv<1> signadd;
  sc_dt::sc_bv<1> carrybit;
  sc_dt::sc_bv<1> hiddenbit0;
  sc_dt::sc_bv<1> hiddenbit1;
  sc_dt::sc_bv<2> hiddenbits;
  bool exponentcheck;
  unsigned int oper;
```

**Fig. 6.30** (continued)

```
void ieee754_32_arithmetic_all_bin_proc0()
{
  while(1)
  {
    wait();
    readinput();
      checkexponentandshift(significand0,
                         significand1,
                         exponent0,
                         exponent1,
                         nexponent,
                         hiddenbit0,
                         hiddenbit1, oper);
      if(oper == 1)
      {              add(significand0,
                     significand1,
                     significandresadd,
                     nexponent,
                     hiddenbit0,
                     carrybit,
                     hiddenbit1,
                     hiddenbits);
      }
    }
  }

  /* Constructor - initialize variables */
  SC_CTOR(ieee754_32_arithmetic_all_bin):
                num0("00000000000000000000000000000000"),
                num1("00000000000000000000000000000000"),
                significandresadd("00000000000000000000000"),
                significand0("00000000000000000000000"),
                significand1("00000000000000000000000"),
multresf("0000000000000000000000000000000000000000000000000"),
                         exponent0("00000000"),
                         exponent1("00000000"),
                         nexponent("00000000"),
                         carrybit("0"),
```

**Fig. 6.30** (continued)

```
                          sign0("0"),
                          sign1("0"),
                          signadd("0"),
                          signmult("0"),
                          signsub("0"),
                          hiddenbit0("0"),
                          hiddenbit1("0"),
                          exponentcheck(false), oper(0)
    {
      /* Declare/assign thread and sensitivity list */
      SC_THREAD(ieee754_32_arithmetic_all_bin_proc0);
      sensitive << exponentin0 << exponentin1 << significandin0 <<
  significandin1 << signin0 << signin1;
    }

    ~ieee754_32_arithmetic_all_bin(){ }
  };
```

**Fig. 6.30** (continued)

```
#include "dec2ieee754.h"

const unsigned int SIZE = 10;

int sc_main(int argc, char **argv)
{
  if(argc < 3)
  {
    std::cout<<"Insufficient parameters ... "<<std::endl;
    std::cout<<"usage ./sim <floating point number> <floating point number>
"<<std::endl;
    exit(0);
  }

  /* Declare/define channesl/signals */
  sc_core::sc_signal<float> numsig0;
  sc_core::sc_signal<float> numsig1;
  sc_core::sc_signal< sc_dt::sc_bv<EXPO> > opersig;
```

**Fig. 6.31** Test harness for IEEE 754-2008 format 32-bit floating-point number adder

```
sc_core::sc_signal< sc_dt::sc_bv<1> > sigf;
sc_core::sc_signal< sc_dt::sc_bv<1> > signsig0;
sc_core::sc_signal< sc_dt::sc_bv<1> > signsig1;
sc_core::sc_signal< sc_dt::sc_bv<MAXFRBITS32> > significandsig0;
sc_core::sc_signal< sc_dt::sc_bv<EXPSIZE32> > exponentsig0;
sc_core::sc_signal< sc_dt::sc_bv<MAXFRBITS32> > significandsig1;
sc_core::sc_signal< sc_dt::sc_bv<EXPSIZE32> > exponentsig1;
sc_core::sc_signal< sc_dt::sc_bv<MAXFRBITS32> > significandsigf;
sc_core::sc_signal< sc_dt::sc_bv<EXPSIZE32> > exponentsigf;

/* Declare/efine modules */
dec2ieee754_32_conv ieee_convt("ieee_convt");
ieee754_32_arithmetic_all_bin arithallbin("arith_all_bin");
/* Declare/define variables */
char number0[SIZE];
char number1[SIZE];
float num0;
float num1;
unsigned int i;
unsigned int j;
unsigned int len0;
unsigned int len1;

/* Connect module ports and channels */
ieee_convt.floatin0(numsig0);
ieee_convt.floatin1(numsig1);
ieee_convt.exponentout0(exponentsig0);
ieee_convt.significandout0(significandsig0);
ieee_convt.signout0(signsig0);
ieee_convt.exponentout1(exponentsig1);
ieee_convt.significandout1(significandsig1);
ieee_convt.signout1(signsig1);

arithallbin.significandin0(significandsig0);
arithallbin.significandin1(significandsig1);
arithallbin.exponentin0(exponentsig0);
arithallbin.exponentin1(exponentsig1);
arithallbin.signin0(signsig0);
arithallbin.signin1(signsig1);
```

**Fig. 6.31** (continued)

```
arithallbin.oper = 1;

len0 = strlen(argv[1]);
len1 = strlen(argv[2]);

if(len0 > 0 && len1 > 0)
{
  strcpy(number0, argv[1]);
  strcpy(number1, argv[2]);
}
else
{
  std::cout<<"invalid input ... exiting ..."<<std::endl;
    exit(0);
  }

  /* Generate new data values  */
  num0 = strtof(number0, NULL);
  num1 = strtof(number1, NULL);
  numsig0.write(0.0);
  numsig1.write(0.0);
 /* Run simulator for pre-defind time period */
  sc_core::sc_start(10.0, sc_core::SC_NS);

 /* Generate new data values  */
  numsig0.write(num0);
  numsig1.write(num1);
  /* Run simulator for pre-defind time period */
  sc_core::sc_start(10.0, sc_core::SC_NS);

  numsig0.write(num0);
  numsig1.write(num1);
  /* Run simulator for pre-defind time period */
  sc_core::sc_start(10.0, sc_core::SC_NS);

  /* Stop simulator */
  sc_core::sc_stop();
  return 0;
}
```

**Fig. 6.31** (continued)

test harness, respectively. As in the previous example, we use only SystemC's bit-vector manipulation for the addition module.

The output at the console for these two inputs is as follows:

./sim 100.0 0.25
SystemC 2.2.0—August 30, 2011, 19:39:07
Copyright (c) 1996–2006 by all contributors,
ALL RIGHTS RESERVED

100 6 1.5625
0.25 −2 1
arith_all_bin 0 10000101 10010001000000000000000
SystemC: simulation stopped by user.

SystemC 2.3.0-ASI—July 19, 2012, 18:53:11
Copyright (c) 1996–2012 by all contributors,
ALL RIGHTS RESERVED

100 6 1.5625
0.25 −2 1
arith_all_bin 0 10000101 10010001000000000000000

The first two lines of output display the two input values expressed in decimal IEEE 754-2008 format before conversion to binary. The third line shows the addition result in IEEE 754-2008 32-bit binary format. For the first two lines of output, the second digit represents the exponent of the largest power of 2 that is less than or equal to the input number, and the floating-point number represents the decimal representation of the input number in the IEEE 754-2008 format, *before* conversion to binary. For the result, the true exponent is obtained by first converting the 8-bit value to an unsigned number and subtracting 127 from it.

## 6.10 IEEE 754-2008 Format 32-Bit Floating-Point Number Multiplication

IEEE 754-2008 [4, 7] 32-bit multiplication is a simpler process than 32-bit addition, because for multiplication, it does not require any radix alignment. It involves the 'hidden' bits of both the multiplier and multiplicand and consists of a series of shift and addition operations. A simple example in Fig. 6.32 illustrates this.

Figure 6.33 contains the multiplier code and test harness. It must be noted that the multiplication code is a part of the same SystemC source file as the addition code, and so only the method is listed here.

The console output is as

./sim 100.0 0.25
SystemC 2.2.0—August 30, 2011, 19:39:07
Copyright (c) 1996–2006 by all contributors,
ALL RIGHTS RESERVED

```
Multiply  0 10000100 01000000000000000000000 by
1 00111100 11000000000000000000000
where the left most bit for each number is the sign bit followed by 8 exponent
bits, and finally 23 mantissa/significand bits.

Mantissa multiplication taking into account the hidden bit and ignoring the
lower order 0s in each mantissa:       1.0100
                                     x 1.1100
                                        00000
                                       00000
                                     10100
                                    10100
                                  10100
                                  ---------
                                 1000110000
becomes   10.00110000

add true exponents (otherwise the bias gets added in twice)
biased addtion case:
    10000100
  + 00111100
  ----------


 10000100      01111111  (switch the order of the subtraction,
 to get a negative value) - 01111111    - 00111100

 00000101  - true exponent is 5
 01000011 = true exponent is -67

add true exponents       5 + (-67) is -62.

re-bias exponent:     -62 + 127 is 65.
 unsigned representation for 65 is  01000001.

put the result back together (and add sign bit).
 1 01000001  10.00110000
 normalize the result:
 (moving the radix point one place to the left increases the exponent by 1.)
1 01000001  10.00110000000000000000000
```

**Fig. 6.32** IEEE 754-2008 32-bit floating-point number multiplication example

```
becomes
1 01000010  1.00011000000000000000000

value stored (no hidden bit):
1 01000010  00011000000000000000000
```

**Fig. 6.32** (continued)

```
/* Method for bit-by-bit multiplication of multiplicand by multiplier */
void multadd(sc_dt::sc_bv<MAXFRBITS32M> &a,
             sc_dt::sc_bv<MAXFRBITS32M> &b,
             sc_dt::sc_bv<1> &c)
  {
    sc_dt::sc_bv<1> tmp0 = "0";
    sc_dt::sc_bv<1> tmp1 = "0";
    sc_dt::sc_bv<1> tmp2 = "0";

    tmp2(0, 0) = c(0, 0);
    unsigned int i = 0;
    for(i = 0; i < MAXFRBITS32M; i++)
    {
      tmp0 = a(i, i);
      tmp1 = b(i, i);
      if(tmp0 == "1" && tmp1 == "1" && tmp2 == "1")
      { a(i, i) = "1"; tmp2(0, 0) = "1"; }
      else if(tmp0 == "1" && tmp1 == "1" && tmp2 == "0")
      { a(i, i) = "0"; tmp2(0, 0) = "1"; }
      else if(tmp0 == "0" && tmp1 == "1" && tmp2 == "1")
      { a(i, i) = "0"; tmp2(0, 0) = "1"; }
      else if(tmp0 == "0" && tmp1 == "1" && tmp2 == "0")
      { a(i, i) = "1"; tmp2(0, 0) = "0"; }
      else if(tmp0 == "1" && tmp1 == "0" && tmp2 == "1")
      { a(i, i) = "0"; tmp2(0, 0) = "1"; }
      else if(tmp0 == "1" && tmp1 == "0" && tmp2 == "0")
      { a(i, i) = "1"; tmp2(0, 0) = "0"; }
      else if(tmp0 == "0" && tmp1 == "0" && tmp2 == "1")
      { a(i, i) = "1"; tmp2(0, 0) = "0"; }
      else if(tmp0 == "0" && tmp1 == "0" && tmp2 == "0")
      { a(i, i) = "0"; tmp2(0, 0) = "0"; }
    }
```

**Fig. 6.33** IEEE 754-2008 32-bit floating-point number multiplier source code

```
   c(0, 0) = tmp2(0, 0);
  }

/* Main multiplication method - includes shifting and addition operations as
described in example */
void mult(sc_dt::sc_bv<MAXFRBITS32> &a,
        sc_dt::sc_bv<MAXFRBITS32> &b,
        sc_dt::sc_bv<EXPSIZE32> &c,
        sc_dt::sc_bv<EXPSIZE32> &d,
        sc_dt::sc_bv<1> &e,
        sc_dt::sc_bv<1> &f,
        sc_dt::sc_bv<MAXFRBITS32M> &r,
        sc_dt::sc_bv<EXPSIZE32> &ef,
        sc_dt::sc_bv<MAXFRBITS32> &sf,
        sc_dt::sc_bv<1> &sif)
  {
   sc_dt::sc_bv<MAXFRBITS32>  tmp0;
   sc_dt::sc_bv<MAXFRBITS32>  tmp1;
   sc_dt::sc_bv<MAXFRBITS32P> tmp2;
   sc_dt::sc_bv<MAXFRBITS32P> tmp3;
   sc_dt::sc_bv<MAXFRBITS32M> tmp4;
   sc_dt::sc_bv<MAXFRBITS32M> tmp5;
   sc_dt::sc_bv<EXPSIZE32>    tmp6;
   sc_dt::sc_bv<EXPSIZE32>    tmp7;
   sc_dt::sc_bv<1> tmp;
   sc_dt::sc_bv<1> mcarry;
   int expf;
   unsigned int i;

   tmp0 = "00000000000000000000000";
   tmp1 = "00000000000000000000000";
   tmp2 = "000000000000000000000000";
   tmp3 = "000000000000000000000000";
   tmp4 = "000000000000000000000000000000000000000000000000";
   tmp5 = "000000000000000000000000000000000000000000000000";
   mcarry = "0";

   tmp0(MAXFRBITS32 - 1, 0) = a(MAXFRBITS32 - 1, 0);
   tmp1(MAXFRBITS32 - 1, 0) = b(MAXFRBITS32 - 1, 0);
```

**Fig. 6.33** (continued)

```
    tmp2(MAXFRBITS32P - 1, MAXFRBITS32P - 1) = "1";
    tmp3(MAXFRBITS32P - 1, MAXFRBITS32P - 1) = "1";
    tmp2(MAXFRBITS32P - 2, 0) = tmp0(MAXFRBITS32 - 1, 0);
    tmp3(MAXFRBITS32P - 2, 0) = tmp1(MAXFRBITS32 - 1, 0);
    tmp4(MAXFRBITS32P - 1, 0) = tmp2(MAXFRBITS32P - 1, 0);

    expf = (bv8_2int(c) - 127) + (bv8_2int(d) - 127);
    for(i = 0; i < MAXFRBITS32P; i++)
    {
      tmp(0, 0) = tmp3(i, i);
      if(tmp(0, 0) == "0") { if(i > 0) tmp4 = tmp4 << 1; }
      else if(tmp(0, 0) == "1")
    {
      tmp4 = tmp4 << 1;
      multadd(tmp5, tmp4, mcarry);
    }
  }

  if(tmp5(MAXFRBITS32M - 1,
            MAXFRBITS32M - 1) == "0") tmp5 = tmp5 << 1;
  expf += 127;
  decexp2bin((unsigned int)expf, ef);
  if(e(0, 0) == "0" && f(0, 0) == "1") sif(0, 0) = "1";
  else if(e(0, 0) == "1" && f(0, 0) == "0") sif(0, 0) = "1";
  else if(e(0, 0) == "1" && f(0, 0) == "1") sif(0, 0) = "0";
  else if(e(0, 0) == "0" && f(0, 0) == "0") sif(0, 0) = "0";
  r(MAXFRBITS32M - 1, 0) = tmp5(MAXFRBITS32M - 1, 0);
  sf(MAXFRBITS32 - 1, 0) =
    tmp5(MAXFRBITS32M - 2, MAXFRBITS32M - 24);
}
```

**Fig. 6.33** (continued)

```
100 6 1.5625
0.25 −2 1
0 00000101 10010000000000000000000
ieee_dec_conv : 24.3886

    SystemC 2.3.0-ASI—July 19, 2012, 18:53:11
    Copyright (c) 1996–2012 by all contributors,
    ALL RIGHTS RESERVED
```

```
100 6 1.5625
0.25 −2 1
100 6 1.5625
0.25 −2 1
0 00000101 10010000000000000000000

   ./sim 100.0 −0.25
   SystemC 2.2.0—August 30, 2011, 19:39:07
   Copyright (c) 1996–2006 by all contributors,
   ALL RIGHTS RESERVED

100 6 1.5625
−0.25 −2 −1
1 00000101 10010000000000000000000
ieee_dec_conv : 24.3886

   SystemC 2.3.0-ASI—July 19, 2012, 18:53:11
   Copyright (c) 1996–2012 by all contributors,
   ALL RIGHTS RESERVED

100 6 1.5625
0.25 −2 1
1 00000101 10010000000000000000000
```

The number in the last line shows the binary multiplication result reconverted to decimal.

## 6.11 Simple 4-Bit Dual-Purpose Addition/Subtraction Module

While dedicated addition/subtraction modules are readily available, a dual-purpose module may be designed using properties of only the XOR gate and 2-bit adders. For illustrative purposes only, a simple 4-bit version of such a module is shown in Fig. 6.34, while the source code, test harness, and input/output traces are shown in

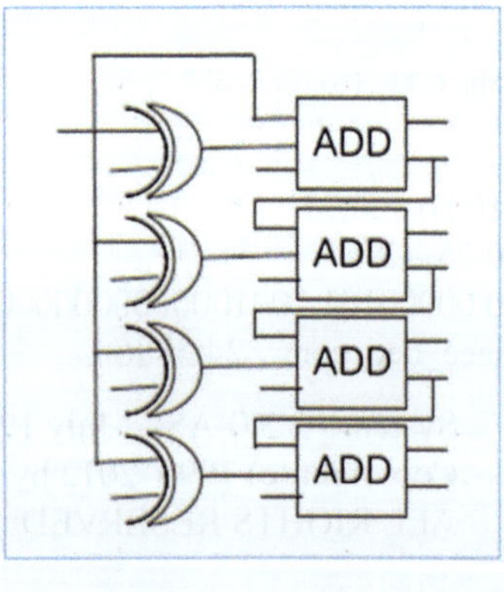


**Fig. 6.34** 4-bit dual-purpose addition/subtraction module 4-bit dual-purpose addition/subtraction module

```
#ifndef ADDSUB_H
#define ADDSUB_H
#include <systemc>

/* Two bit adder */
SC_MODULE(adder2bit)
{
  /* Declare/define input/output ports */
  sc_core::sc_in<bool> in0;
  sc_core::sc_in<bool> in1;
  sc_core::sc_in<bool> in2;
  sc_core::sc_out<bool> out0;
  sc_core::sc_out<bool> out1;

  /* Declare/define members */
  bool b0;
  bool b1;
  bool b2;
  bool b3;
  bool b4;

  /* Main adder thread */
  void adder2bit_proc0()
  {
    while(1)
    {
  wait();
  b0 = in0.read();
  b1 = in1.read();
  b2 = in2.read();

  if(b0 == false && b1 == false && b2 == false)
  {
    b3 = false;
    b4 = false;
  }
  else if(b0 == true && b1 == false && b2 == false)
  {
    b3 = true;
    b4 = false;
  }
```

**Fig. 6.35** Source code for 4-bit dual-purpose addition/subtraction module 4-bit dual-purpose addition/subtraction module

```
else if(b0 == false && b1 == true && b2 == false)
{
  b3 = true;
  b4 = false;
}
else if(b0 == false && b1 == false && b2 == true)
{
  b3 = true;
  b4 = false;
}
else if(b0 == true && b1 == true && b2 == false)
{
  b3 = false;
  b4 = true;
}
else if(b0 == true && b1 == false && b2 == true)
{
  b3 = false;
  b4 = true;
}
else if(b0 == false && b1 == true && b2 == true)
{
  b3 = false;
  b4 = true;
}
else if(b0 == true && b1 == true && b2 == true)

    {
      b3 = true;
      b4 = true;
    }
    out0.write(b3);
    out1.write(b4);
  }
}

/* Constructor */
SC_CTOR(adder2bit)
{
```

**Fig. 6.35** (continued)

```
    /* Declare/assign adder thread and activate sensitivity list */
    SC_THREAD(adder2bit_proc0);
    sensitive << in0 << in1 << in2;
  }

  ~adder2bit(){ }
 /* Destructor */
};

/* Simple XOR gate */
SC_MODULE(xorgate)
{
  /* Declare/define input/output ports */
  sc_core::sc_in<bool> in0;
  sc_core::sc_in<bool> in1;
  sc_core::sc_out<bool> out0;
  /* Declare/define members */
  bool b0;
  bool b1;
  bool b2;

  /* Main XOR gate thread */
  void xorgate_proc0()
  {
    while(1)
    {
      wait();
      b0 = in0.read();
      b1 = in1.read();
      b2 = ((b0 & !b1) | (!b0 & b1));
      out0.write(b2);
    }
  }

  /* Constructor */
  SC_CTOR(xorgate)
  {
    /* Declare/assign thread and activate sensitivity list */
    SC_THREAD(xorgate_proc0);
```

**Fig. 6.35** (continued)

```
    sensitive << in0 << in1;
  }

  ~xorgate(){ }
  /*Destructor */
};

/* 4 bit dual purpose adder/subtractor */
SC_MODULE(addsub4bit)
{
  /* Declare/define input/output ports */
  sc_core::sc_in<bool> in0;
  sc_core::sc_in<bool> in1;
  sc_core::sc_in<bool> in2;
  sc_core::sc_in<bool> in3;
  sc_core::sc_in<bool> in4;
  sc_core::sc_in<bool> in5;
  sc_core::sc_in<bool> in6;
  sc_core::sc_in<bool> in7;
  sc_core::sc_in<bool> in8;
  sc_core::sc_out<bool> out0;
  sc_core::sc_out<bool> out1;
  sc_core::sc_out<bool> out2;
  sc_core::sc_out<bool> out3;
  sc_core::sc_out<bool> out4;

/* Declare/define internal signals */
  sc_core::sc_signal<bool> sig0;
  sc_core::sc_signal<bool> sig1;
  sc_core::sc_signal<bool> sig2;
  sc_core::sc_signal<bool> sig3;

  sc_core::sc_signal<bool> sig4;
  sc_core::sc_signal<bool> sig5;
  sc_core::sc_signal<bool> sig6;
  sc_core::sc_signal<bool> sig7;

  /* Declare members */
  adder2bit ad0;
  adder2bit ad1;
```

**Fig. 6.35** (continued)

```
adder2bit ad2;
adder2bit ad3;

xorgate xog0;
xorgate xog1;
xorgate xog2;
xorgate xog3;

/* Constructor - initialize members and unternal channels */
SC_CTOR(addsub4bit):ad0("ad0"),
                    ad1("ad1"),
                    ad2("ad2"),
                    ad3("ad3"),
                    xog0("xog0"),
                    xog1("xog1"),
                    xog2("xog2"),
                    xog3("xog3"),
          sig0("sig0"), sig1("sig1"), sig2("sig2"), sig3("sig3"
          sig4("sig4"), sig5("sig5"), sig6("sig6"), sig7("sig7"
{
  /* Connect member module ports,
     signal channels and esternal ports */
  ad0.in0(in8);
  ad0.in1(in0);
  ad0.in2(sig0);
  ad0.out0(sig4);
  ad0.out1(out0);

  ad1.in0(in2);
  ad1.in1(sig1);
  ad1.in2(sig4);
  ad1.out0(sig5);
  ad1.out1(out1);
  ad2.in0(in4);
  ad2.in1(sig2);
  ad2.in2(sig5);
  ad2.out0(sig6);
  ad2.out1(out2);

  ad3.in0(in6);
```

**Fig. 6.35** (continued)

```
        ad3.in1(sig3);
        ad3.in2(sig6);
        ad3.out0(out3);
        ad3.out1(out4);

        xog0.in0(in8);
        xog0.in1(in1);
        xog0.out0(sig0);

        xog1.in0(in8);
        xog1.in1(in3);
        xog1.out0(sig1);

        xog2.in0(in8);
        xog2.in1(in5);
        xog2.out0(sig2);

        xog3.in0(in8);
        xog3.in1(in7);
        xog3.out0(sig3);

      }

      ~addsub4bit(){ }
     /* Destructor */
    };

    #endif
```

Fig. 6.35 (continued)

```
#include "addsub.h"
#include "systemc.h"
int sc_main(int argc, char **argv)
{
  /* Declare/define signal channels */
  sc_core::sc_signal<bool> insig0;
  sc_core::sc_signal<bool> insig1;
  sc_core::sc_signal<bool> insig2;
  sc_core::sc_signal<bool> insig3;
  sc_core::sc_signal<bool> insig4;
```

**Fig. 6.36** Test harness for 4-bit dual-purpose addition/subtraction module 4-bit dual-purpose addition/subtraction module

```
sc_core::sc_signal<bool> insig5;
sc_core::sc_signal<bool> insig6;
sc_core::sc_signal<bool> insig7;
sc_core::sc_signal<bool> modesig;
sc_core::sc_signal<bool> outsig0;
sc_core::sc_signal<bool> outsig1;
sc_core::sc_signal<bool> outsig2;
sc_core::sc_signal<bool> outsig3;
sc_core::sc_signal<bool> carryout;

/* Declare/define internal signals */
bool b0;
bool b1;
bool b2;
bool b3;
bool b4;
bool b5;
bool b6;
bool b7;
bool modebit;

/* Declare/define 4 bit dual-purpose adder/subtractor and trace file */
addsub4bit add_sub_4_bit("add_sub_4_bit");
sc_core::sc_trace_file *fp;
fp = sc_create_vcd_trace_file("tr_addsub4bit");
fp->set_time_unit(1.0, sc_core::SC_NS);

/* Connect module ports and siganal channels */
add_sub_4_bit.in0(insig0);
add_sub_4_bit.in1(insig1);
add_sub_4_bit.in2(insig2);
add_sub_4_bit.in3(insig3);
add_sub_4_bit.in4(insig4);
add_sub_4_bit.in5(insig5);
add_sub_4_bit.in6(insig6);
add_sub_4_bit.in7(insig7);
add_sub_4_bit.in8(modesig);
add_sub_4_bit.out0(outsig0);
add_sub_4_bit.out1(outsig1);
add_sub_4_bit.out2(outsig2);
```

**Fig. 6.36** (continued)

```
add_sub_4_bit.out3(outsig3);
add_sub_4_bit.out4(carryout);

/* Connect signal channels to trace file */
sc_core::sc_trace(fp, insig0, "in0");
sc_core::sc_trace(fp, insig1, "in1");
sc_core::sc_trace(fp, insig2, "in2");
sc_core::sc_trace(fp, insig3, "in3");
sc_core::sc_trace(fp, insig4, "in4");
sc_core::sc_trace(fp, insig5, "in5");
sc_core::sc_trace(fp, insig6, "in6");
sc_core::sc_trace(fp, insig7, "in7");
sc_core::sc_trace(fp, modesig, "mode");
sc_core::sc_trace(fp, outsig0, "out0");
sc_core::sc_trace(fp, outsig1, "out1");
sc_core::sc_trace(fp, outsig2, "out2");
sc_core::sc_trace(fp, outsig3, "out3");
sc_core::sc_trace(fp, carryout, "carryout");

/* Assign values to local variables, and write to input signal channels */
b0 = false;
b1 = false;
b2 = false;
b3 = false;
b4 = false;
b5 = false;
b6 = false;
b7 = false;
modebit = false;

insig0.write(b0);
insig1.write(b1);
insig2.write(b2);
insig3.write(b3);
insig4.write(b4);
insig5.write(b5);
insig6.write(b6);
insig7.write(b7);
modesig.write(modebit);
```

**Fig. 6.36** (continued)

```
/* Run simulation for pre-defined time interval */
sc_core::sc_start(4.0, sc_core::SC_NS);

/* Re-assign values to local variables, and write to input signal channels */
b0 = true;
b1 = false;
b2 = true;
b3 = false;
b4 = true;
b5 = false;
b6 = true;
b7 = false;
modebit = false;

insig0.write(b0);
insig1.write(b1);
insig2.write(b2);
insig3.write(b3);
insig4.write(b4);
insig5.write(b5);
insig6.write(b6);
insig7.write(b7);
modesig.write(modebit);
/* Run simulation for pre-defined time interval */
sc_core::sc_start(4.0, sc_core::SC_NS);

/* Re-assign values to local variables, and write to input signal channels */
b0 = true;
b1 = true;
b2 = true;
b3 = false;
b4 = false;
b5 = false;
b6 = true;

b7 = true;
modebit = true;

insig0.write(b0);
insig1.write(b1);
```

**Fig. 6.36** (continued)

```
insig2.write(b2);
insig3.write(b3);
insig4.write(b4);
insig5.write(b5);
insig6.write(b6);
insig7.write(b7);
modesig.write(modebit);

/* Run simulation for pre-defined time interval */
sc_core::sc_start(4.0, sc_core::SC_NS);

/* Re-assign values to local variables, and write to input signal channels */
b0 = false;
b1 = true;
b2 = false;
b3 = true;
b4 = false;
b5 = true;
b6 = true;
b7 = true;
modebit = true;

insig0.write(b0);
insig1.write(b1);
insig2.write(b2);
insig3.write(b3);
insig4.write(b4);
insig5.write(b5);
insig6.write(b6);
insig7.write(b7);
modesig.write(modebit);

/* Run simulation for pre-defined time interval */
sc_core::sc_start(4.0, sc_core::SC_NS);

/* Re-assign values to local variables, and write to input signal channels */
b0 = true;
b1 = true;
b2 = true;
b3 = true;
```

**Fig. 6.36** (continued)

```
b4 = false;
b5 = false;
b6 = true;
b7 = true;
modebit = true;

insig0.write(b0);
insig1.write(b1);
insig2.write(b2);
insig3.write(b3);
insig4.write(b4);
insig5.write(b5);
insig6.write(b6);
insig7.write(b7);
modesig.write(modebit);

/* Run simulation for pre-defined time interval */
sc_core::sc_start(4.0, sc_core::SC_NS);

/* Re-assign values to local variables, and write to input signal channels */
b0 = true;
b1 = true;
b2 = true;
b3 = false;
b4 = false;
b5 = false;
b6 = true;
b7 = true;
modebit = false;

insig0.write(b0);
insig1.write(b1);
insig2.write(b2);
insig3.write(b3);
insig4.write(b4);
insig5.write(b5);
insig6.write(b6);
insig7.write(b7);
modesig.write(modebit);
```

**Fig. 6.36** (continued)

```
/* Run simulation for pre-defined time interval */
sc_core::sc_start(4.0, sc_core::SC_NS);

/* Re-assign values to local variables, and write to input signal channels */
b0 = true;
b1 = true;
b2 = false;
b3 = false;
b4 = true;
b5 = false;
b6 = false;
b7 = true;
modebit = true;

insig0.write(b0);
insig1.write(b1);
insig2.write(b2);
insig3.write(b3);
insig4.write(b4);
insig5.write(b5);
insig6.write(b6);
insig7.write(b7);
modesig.write(modebit);

/* Run simulation for pre-defined time interval */
sc_core::sc_start(4.0, sc_core::SC_NS);

/* Re-assign values to local variables, and write to input signal channels */
b0 = true;
b1 = true;
b2 = true;
b3 = false;
b4 = false;
b5 = false;
b6 = true;
b7 = true;

modebit = true;
insig0.write(b0);
insig1.write(b1);
```

**Fig. 6.36** (continued)

```
insig2.write(b2);
insig3.write(b3);
insig4.write(b4);
insig5.write(b5);
insig6.write(b6);
insig7.write(b7);
modesig.write(modebit);

/* Run simulation for pre-defined time interval */
sc_core::sc_start(4.0, sc_core::SC_NS);

/* Re-assign values to local variables, and write to input signal channels */
b0 = false;
b1 = true;
b2 = false;
b3 = false;
b4 = true;
b5 = false;
b6 = false;
b7 = true;
modebit = false;

insig0.write(b0);
insig1.write(b1);
insig2.write(b2);
insig3.write(b3);
insig4.write(b4);

insig5.write(b5);
insig6.write(b6);
insig7.write(b7);
modesig.write(modebit);

/* Run simulation for pre-defined time interval */
sc_core::sc_start(4.0, sc_core::SC_NS);


/* Re-assign values to local variables, and write to input signal channels */
b0 = true;
```

**Fig. 6.36** (continued)

```
b1 = true;
b2 = false;
b3 = false;
b4 = true;
b5 = false;
b6 = true;
b7 = true;
modebit = false;

insig0.write(b0);
insig1.write(b1);
insig2.write(b2);
insig3.write(b3);
insig4.write(b4);
insig5.write(b5);
insig6.write(b6);
insig7.write(b7);
modesig.write(modebit);

/* Run simulation for pre-defined time interval */
sc_core::sc_start(4.0, sc_core::SC_NS);

/* Re-assign values to local variables, and write to input signal channels */
b0 = false;
b1 = false;
b2 = false;
b3 = false;
b4 = false;
b5 = false;
b6 = false;
b7 = true;
modebit = true;

insig0.write(b0);
insig1.write(b1);
insig2.write(b2);
insig3.write(b3);
insig4.write(b4);
insig5.write(b5);
insig6.write(b6);
insig7.write(b7);
```

**Fig. 6.36** (continued)

```
    modesig.write(modebit);

    /* Run simulation for pre-defined time interval */
    sc_core::sc_start(4.0, sc_core::SC_NS);

    /* Re-assign values to local variables, and write to input signal channels */
    b0 = true;
    b1 = true;
    b2 = false;
    b3 = false;
    b4 = true;
    b5 = true;
    b6 = false;
    b7 = true;
    modebit = true;
    insig0.write(b0);
    insig1.write(b1);
    insig2.write(b2);
    insig3.write(b3);
    insig4.write(b4);
    insig5.write(b5);
    insig6.write(b6);
    insig7.write(b7);
    modesig.write(modebit);

    /* Run simulation for pre-defined time interval */
    sc_core::sc_start(4.0, sc_core::SC_NS);

    /* Stop simulation and close trace file */
    sc_core::sc_stop();
    sc_core::sc_close_vcd_trace_file(fp);
    return 0;
}
```

**Fig. 6.36** (continued)

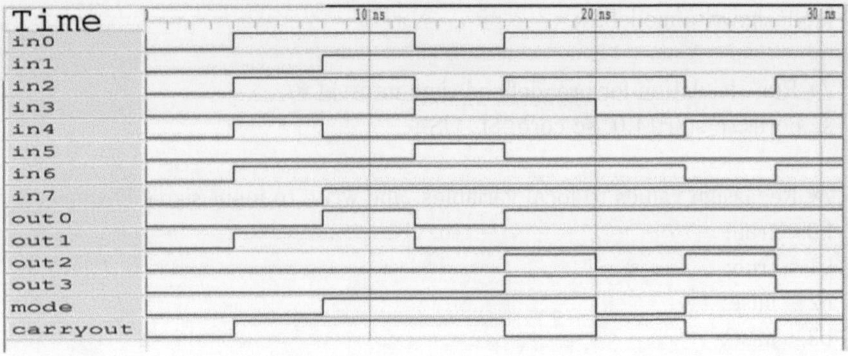


**Fig. 6.37** Input/output traces from 4-bit dual-purpose addition/subtraction module SystemC 2.2.0

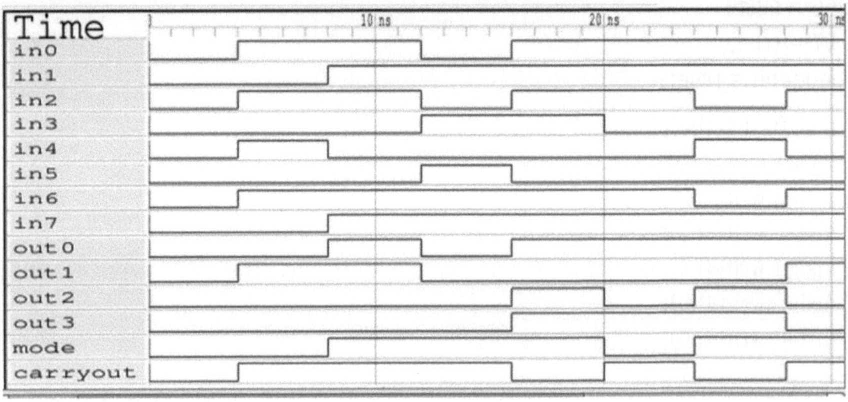


**Fig. 6.38** Input/output traces from 4-bit dual-purpose addition/subtraction module SystemC 2.3.0

Figs. 6.35, 6.36, 6.37 (SystemC 2.2.0), and 6.38 (SystemC 2.3.0). The key input signal is 'MODE,' which if set to logic '1' makes the module an adder and when set to logic '0' makes the module a subtractor.

## References

1. Sun Microsystems's open-source multi-core and multi-threading CPU design source code and related design documentation may be obtained from http://www.opensparc.net.
2. Charles, Roth H., and Kinney I. Larry, *Fundamentals of Logic Design*. Sixth ed. Cengage Learning, 2010. Print.
3. Kenneth, Martin. *Digital Integrated Circuit Design*. Oxford University Press, 2000. Print.
4. IEEE Computer Society. IEEE Xplore Digital Library 2008 *754-2008 IEEE Standard for Floating Point Arithmetic* http://ieeexplore.ieee.org/xpl/mostRecentIssue.jsp?punumber=4610933.
5. Ryan, William E., and Shu I. Lin. *Channel Codes: Classical and Modern*. Cambridge University Press, 2009. Print.

6. IEEE Ethernet Working Group. IEEE Standard Association - IEEE Get Program. IEEE, 2010. *IEEE Standard for Information Technology-Telecommunications and Information Exchange between Systems - Local and Metropolitan area Networks-* http://standards.ieee.org/getieee802/download/802.3ba-2010.pdf.
7. Hollasch, Steven. IEEE Standard 765 Floating Point. Ed. Steven Hollasch. 24 Feb. 2005 http://steve.hollasch.net/cgindex/coding/ieeefloat.html.
8. IEEE Standards Board, IEEE Standards Association Standards Board IEEE-SA - IEEE Get Program. IEEE, 2011 *IEEE Standard 1666 Open SystemC Language Reference Manual (LRM)* http://standards.ieee.org/getieee/1666/download/1666-2011.pdf.

# Chapter 7
# Explicit SystemC Events: Notify–Wait

**Abstract** For both combinational and sequential logic examples examined earlier, SystemC's [1] event-driven simulation model was used implicitly. In combinational logic examples, we used the *sensitivity list* concept, while for sequential logic we used both sensitivity list and clocked thread (*SC_CTHREAD*). There is another way to use SystemC's event-driven simulation model—by explicit use of SystemC events (*sc_core::sc_event*) with the library functions *notify()* and *wait()*. The wait() function may be used both implicitly or explicitly. Implicit SystemC events and their use are given in Table 7.1.

## 7.1 Explicit SystemC Events and wait()/notify()

Explicit SystemC [1] events are examined in this simple example. It consists of a stimulus object that triggers a master object and several slave objects, which in turn exchange data with the master. Once the triggering is complete, the stimulus object does nothing. Before the triggering event, the stimulus object can do other tasks not related to either master or the slaves. This example uses both implicit and explicit events (Table 7.1). Note that *wait()* may be used in only two ways from inside a SC_CTHREAD—wait() (wait for the next negative or positive clock edge) and wait(N) (N positive integer—wait for N positive or negative edges of clock). Waiting for events from inside clocked threads is deprecated with current versions of SystemC. The notification can be *immediate* (*event.notify()*), or *delta cycle* later (*event.notify(sc_core::SC_ZERO_TIME)*) or after a finite time interval (*event.notify(n, sc_core:: <SC_TIME_UNIT>)*—n positive number). Figures 7.1, 7.2, 7.3, and 7.4 contain components of our explicit event (notify()–wait()) example.

A. Banerjee and B. Sur, *SystemC and SystemC-AMS in Practice*,
DOI: 10.1007/978-3-319-01147-9_7,

**Table 7.1** Implicit SystemC events and their use

| | |
|---|---|
| SC_CTHREAD | Wait implicitly on positive or negative edges of clock (clock.pos()/neg()) |
| SC_THREAD | Wait implicitly on input port values changing (sensitivity list) |
| SC_METHOD | Wait implicitly on events specified in sensitivity list |

```
#ifndef EXPLICITEVENT_H
#define EXPLICITEVENT_H

#include <systemc>
sc_core::sc_event ev0;
SC_MODULE(stimulus)
{
 private:
 bool flag;
 /* stimulus class's private method that does not trigger master or slaves
*/
 void do_task()
  {
   unsigned int i;
   for(i = 0; i < 20; i++)
   {
     std::cout<<name()<<" do_task "<<std::endl;
   }
 }
 public:
 sc_core::sc_in<bool> clk;
/*Input port for external clock -- implicit event */
 void stimulus_proc0()
 {
   while(1)
   {
     wait();
      if(flag == false)
      {
        /* If flag is false, perform own task, notify master and slaves, set flag
to true*/
       do_task();
       ev0.notify(10.0, sc_core::SC_NS);
         /* Wait pre-defined time interval before notification */
       flag = true;
```

**Fig. 7.1** Explicit SystemC event model—stimulus class

```
        /*Set flag to true */
      std::cout<<"Notification sent ..."<<std::endl;
    }
  }
}

/* Constructor */
SC_CTOR(stimulus)
 {
  SC_CTHREAD(stimulus_proc0, clk.pos());
  /* Implicit event, sensitize process to clock positive edge */
}
/*Destructor */
~stimulus() { }
};
```

**Fig. 7.1** (continued)

```
SC_MODULE(master)
{
  private:
   bool flag;
  public:
  /* Declare/define input/outpit ports */
  sc_core::sc_in<bool> clk;
  sc_core::sc_out<double> dout;
  double dvalue;
  /* Master main operation  process*/
  void master_proc0()
  {
    while(1)
    {
      wait();
     /* Wait for implicit event - input clock posirive edge */
      if(flag == false)
      {
```

**Fig. 7.2** Explicit SystemC event model—master class

```
        wait(ev0);
        /* If flag is not set wait for explicit event - event notification from
stimulus */
        flag = true;
        /* Set flag */
        std::cout<<"master receives notification"<<std::endl;
      }
      else
      {
        dvalue = drand48();
        dout.write(dvalue);
       /*Send data to slaves */
      }
    }
  }
  /* Constructor */
  SC_CTOR(master):flag(false)
  {
    SC_THREAD(master_proc0);
    /*Declare/assign thread triggered by implicit event */
    sensitive << clk.pos();
    /*Sensitize thread to implicit event */
  }
  /*Destructor */
  ~master(){ }
};
```

**Fig. 7.2** (continued)

```
SC_MODULE(slave)
{
  private:
   bool flag;
  public:
  /* Declare/define input/outpit ports */
  sc_core::sc_in<bool> clk;
  sc_core::sc_in<double> din;
  double d;
  /* Slave main operation  process*/
  void slave_proc0()
```

**Fig. 7.3** Explicit SystemC event model—slave class

```
{
  while(1)
  {
   wait();
   /* Wait for implicit event - input clock posirive edge */
   if(flag == false)
   {
     wait(ev0);
     /* If flag is not set wait for explicit event -
          event notification from stimulus */
     flag = true;
     /* Set flag */
     std::cout<<name()<<" receives notification"<<std::endl;
   }
   else
   {
      d = din.read();
     /*Receive data from master  */
     std::cout<<name()<<" rec'd "<<d<<std::endl;
   }
  }
}
/*Constructor */
SC_CTOR(slave):flag(false)
{
  SC_THREAD(slave_proc0)
/*Declare/assign thread triggered by implicit event */
  sensitive << clk.pos();
/*Sensitize thread to implicit event */
}
/* Destructor */
~slave(){ }
};
#endif
```

**Fig. 7.3** (continued)

```
#include "explicitevent.h"
int sc_main(int argc, char **argv)
{
  /* Declare/define channel */
  sc_core::sc_signal<double> data;
  /* Declare/define clock for implicit event synchronization */
  sc_core::sc_clock clk("clock", 2.0, sc_core::SC_NS, 0.5);
  /* Declare/define components */
  stimulus stim("stim");
  master mas_ter("mas_ter");
  slave slave_0("slave_0");
  slave slave_1("slave_1");

  /* Connect modules and channels */
  stim.clk(clk);
  mas_ter.clk(clk);
  mas_ter.dout(data);
  slave_0.clk(clk);
  slave_0.din(data);
  slave_1.clk(clk);
  slave_1.din(data);
  /* Run simulation for pre-defined time interval and then stop */
  sc_core::sc_start(250.0, sc_core::SC_NS);
  sc_core::sc_stop();
  return 0;
}
```

**Fig. 7.4** Test harness for explicit SystemC event model—stimulus, master, and two slaves

The output at the console looks like:

```
$ ./sim
        SystemC 2.2.0 --- Aug 30 2011 19:39:07
     Copyright (c) 1996-2006 by all Contributors
              ALL RIGHTS RESERVED

stim do_task
stim do_task
....
....
stim do_task
stim do_task
stim do_task
stim do_task
stim do_task
stim do_task
stim do_task
Notification sent ...
slave_1 receives notification
slave_0 receives notification
master receives notification
slave_0 rec'd 0
slave_1 rec'd 0
slave_0 rec'd 3.90799e-14
slave_1 rec'd 3.90799e-14
slave_0 rec'd 0.000985395
slave_1 rec'd 0.000985395

slave_0 rec'd 0.931731
slave_1 rec'd 0.931731
slave_0 rec'd 0.56806
slave_1 rec'd 0.56806
slave_0 rec'd 0.556094
slave_1 rec'd 0.556094
slave_0 rec'd 0.0508319
slave_1 rec'd 0.0508319
slave_0 rec'd 0.767051
slave_1 rec'd 0.767051
```

```
.....
.....
.....
.....
slave_1 rec'd 0.466057
slave_0 rec'd 0.868966
slave_1 rec'd 0.868966
slave_0 rec'd 0.134039
SystemC: simulation stopped by user.

        SystemC 2.3.0-ASI --- Jul 19 2012 18:53:11
    Copyright (c) 1996-2012 by all Contributors,
    ALL RIGHTS RESERVED
stim do_task
stim do_task
stim do_task
stim do_task
.....
.....
stim do_task
stim do_task

Notification sent ...
master receives notification
slave_1 receives notification
slave_0 receives notification
slave_0 rec'd 0
slave_1 rec'd 0
slave_0 rec'd 3.90799e-14
slave_1 rec'd 3.90799e-14
slave_0 rec'd 0.000985395
slave_1 rec'd 0.000985395
.....
.....
slave_0 rec'd 0.984891
slave_1 rec'd 0.984891
.....
.....
```

```
slave_1 rec'd 0.701703
slave_0 rec'd 0.367247
slave_1 rec'd 0.367247
slave_0 rec'd 0.876312
.....
.....
slave_1 rec'd 0.577967
slave_0 rec'd 0.964704
slave_1 rec'd 0.964704
.....
.....
slave_1 rec'd 0.670082
slave_0 rec'd 0.799972
slave_1 rec'd 0.799972
Info: /OSCI/SystemC: Simulation stopped by user.
```

## Reference

1. IEEE Standards Board, IEEE Standards Association Standards Board IEEE-SA—IEEE Get Program. IEEE, 2011 *IEEE Standard 1666 Open SystemC Language Reference Manual (LRM)* http://standards.ieee.org/getieee/1666/download/1666-2011.pdf

# Chapter 8
# Hierarchical Combinational–Sequential System Design

**Abstract** Real-world digital circuits consist of combinations of combinational and sequential sub-circuits, connected and configured in a hierarchy. Having examined the behavior of some of these sub-circuits (both combinational and sequential) in detail earlier, we now examine their hierarchical combinations, as in real-world systems. All sequential circuits are clocked, while no combinational circuit is clocked. *The integration of these two seemingly contradictory design philosophies is possible* via *sensitivity lists—i.e., a combinational circuit block is sensitized to each of its input ports, while the sequential circuit block responds to its clock.* As before, these design examples are graded with increasing levels of complexity, ending with some widely used digital sub-circuits for integrated circuit testing.

## 8.1 Pseudo-Random Number Generator

A simple hierarchical combinational–sequential logic system is a pseudo-random number generator [1]. It consists of a register whose cells are typically D flip-flops, triggered on a clock. Taps at various cells of the shift register are fed into XOR gates, whose output is fed back into the shift register—Fig. 8.1. So, while clock triggered flip-flops are purely sequential, the XOR gate is a purely combinational circuit. The output is 'pseudo-random,' because a real random number would have an infinite number of bits in it, while all computer memory can only store a finite number of bits. The source code for the XOR gate, D flip-flop, and their combination (pseudo-random number generator is in Figs. 8.2, 8.3, and 8.4. The test harness and output trace are in Figs. 8.5, 8.6 (SystemC-2.2.0) and 8.7 (SystemC-2.3.0), respectively.

A. Banerjee and B. Sur, *SystemC and SystemC-AMS in Practice* ,
DOI: 10.1007/978-3-319-01147-9_8,

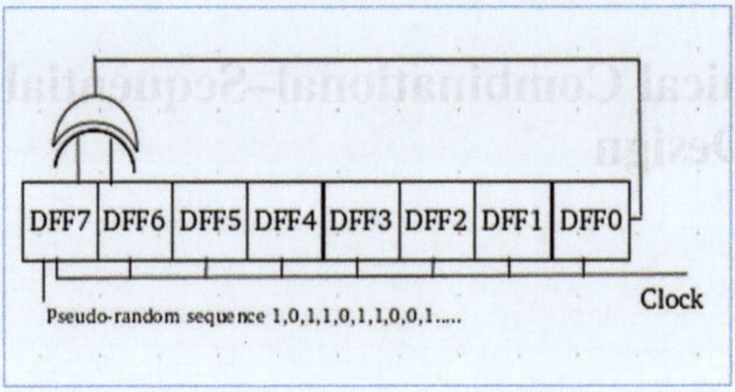


**Fig. 8.1** Simple eight-cell shift register. Each cell is a D flip-flop, and the output of the XOR gate is fed back into the shift register. The output of the last register cell is the pseudo-random bit stream. The register cells need to be preset to some initial value (random number seed)

```
#ifndef PRN_H
#define PRN_H

#include <systemc>

SC_MODULE(x_or)
{
 /* Declare/define input output ports */
 sc_core::sc_in<bool> xorin0;
 sc_core::sc_in<bool> xorin1;
 sc_core::sc_out<bool> xorout;

 /* Declare/define internal variables */
 bool b0;
 bool b1;
 bool b2;

/* XOR processing */
 void x_or_proc0()
 {
   while(1)
   {
     wait();
     b0 = xorin0.read();
     b1 = xorin1.read();
     b2 = (((!b0) & b1) | (b0 & (!b1)));
     xorout.write(b2);
```

**Fig. 8.2** Pseudo-random number generator component XOR gate—pure combinational logic

```
    }
  }

  /* Constructor */
  SC_CTOR(x_or):b0(false), b1(false), b2(false)
  {
    /* Declare/assign thread */
    SC_THREAD(x_or_proc0);
    sensitive << xorin0 << xorin1;
 /*Declare/define thread sensitivity list */
  }

/* Destructor */
  ~x_or(){}
};
```

**Fig. 8.2** (continued)

```
SC_MODULE(dff)
{
  /* Decllare/define input/output ports */
  sc_core::sc_in<bool> clk;
  sc_core::sc_in<bool> din;
  sc_core::sc_out<bool> dout;

  /* declare/define internal variables */
  bool b0;
  bool b1;
  /* Memory element */

  /* D flip-flop operation thread */
  void dff_proc0()
  {
    while(1)
    {
      wait();
      b0 = din.read();
      dout.write(b1);
      if(b0 != b1) b1 = b0;
    }
  }
```

**Fig. 8.3** D flip-flop shift register cell for pseudo-random number generator—pure sequential logic

```
  /* Constructor */
  SC_CTOR(dff)
  {
     /*Declare/assign clocked thread */
    SC_CTHREAD(dff_proc0, clk.pos());
  }
  ~dff(){ }
 /* Destructor */
};
```

**Fig. 8.3** (continued)

```
SC_MODULE(prn)
{
  /* Declare/define input/output ports */
  sc_core::sc_in<bool> clk;
  sc_core::sc_out<bool> prnout;

  /* Declare/define internal channels */
  sc_core::sc_signal<bool> sig0;
  sc_core::sc_signal<bool> sig1;
  sc_core::sc_signal<bool> sig2;
  sc_core::sc_signal<bool> sig3;
  sc_core::sc_signal<bool> sig4;
  sc_core::sc_signal<bool> sig5;
  sc_core::sc_signal<bool> sig6;
  sc_core::sc_signal<bool> sig7;
  sc_core::sc_signal<bool> sig8;

  /* Declare/define D flip-flops and XOR gate */
  dff dff_0;
  dff dff_1;
  dff dff_2;
  dff dff_3;
  dff dff_4;
  dff dff_5;
  dff dff_6;
  dffdualout dff_7;
  x_or x_o_r;
```

**Fig. 8.4** Pseudo-random number generator

```
/* Constructor - initialize D flip-flops and XOR gate */
SC_CTOR(prn):dff_0("dff_0"), dff_1("dff_1"), dff_2("dff_2"),
          dff_3("dff_3"), dff_4("dff_4"), dff_5("dff_5"),
          dff_6("dff_6"), dff_7("dff_7"), x_o_r("x_o_r")
{
  /* Connect D flip-flops and XOR gate ports with internal
     and external channels */
  dff_0.clk(clk);
  dff_0.din(sig8);
  dff_0.dout(sig0);

  dff_1.clk(clk);
  dff_1.din(sig0);
  dff_1.dout(sig1);

  dff_2.clk(clk);
  dff_2.din(sig1);
  dff_2.dout(sig2);

  dff_3.clk(clk);
  dff_3.din(sig2);
  dff_3.dout(sig3);

  dff_4.clk(clk);
  dff_4.din(sig3);
  dff_4.dout(sig4);

  dff_5.clk(clk);
  dff_5.din(sig4);
  dff_5.dout(sig5);

  dff_6.clk(clk);
  dff_6.din(sig5);
  dff_6.dout(sig6);

  dff_7.clk(clk);
  dff_7.din(sig6);
  dff_7.dout(sig7);
  dff_7.dualout(prnout);
```

**Fig. 8.4** (continued)

```
    x_o_r.xorin0(sig6);
    x_o_r.xorin1(sig7);
    x_o_r.xorout(sig8);

    /* Initialize memory bits in pseudo random number generator D flip-flops */
    dff_0.b1 = true;
    dff_1.b1 = false;
    dff_2.b1 = true;
    dff_3.b1 = false;
    dff_4.b1 = true;
    dff_5.b1 = false;
    dff_6.b1 = true;
    dff_7.b1 = false;
  }

  /* Destructor */
  ~prn(){ }
};

#endif
```

**Fig. 8.4** (continued)

```
#include "prn.h"

int sc_main(int argc, char **argv)
{
  /* Declare/define channels */
  sc_core::sc_signal<bool> prnout;
 /* Declare/define clock, pseudo-random number
    generator and trace file */
  sc_core::sc_clock prn_clk("prn_clk", 5.0, sc_core::SC_NS, 0.5);
  prn p_r_n("p_r_n");
  sc_core::sc_trace_file *fp =
                    sc_core::sc_create_vcd_trace_file("tr_prn");
  fp->set_time_unit(1.0, sc_core::SC_NS);
  /* Connect all module ports and channels */
  p_r_n.clk(prn_clk);
  p_r_n.prnout(prnout);
```

**Fig. 8.5** Test harness for pseudo-random number generator

```
  /* Connect trace file with clock output and data channels */
  sc_core::sc_trace(fp, prn_clk, "prn_clk");
  sc_core::sc_trace(fp, prnout, "prn_out");

/* Run simulation for pre-defind time inetrval, then stop and close trace file */
  sc_core::sc_start(5000.0, sc_core::SC_NS);
  sc_core::sc_stop();
  sc_core::sc_close_vcd_trace_file(fp);
  return 0;
}
```

**Fig. 8.5** (continued)

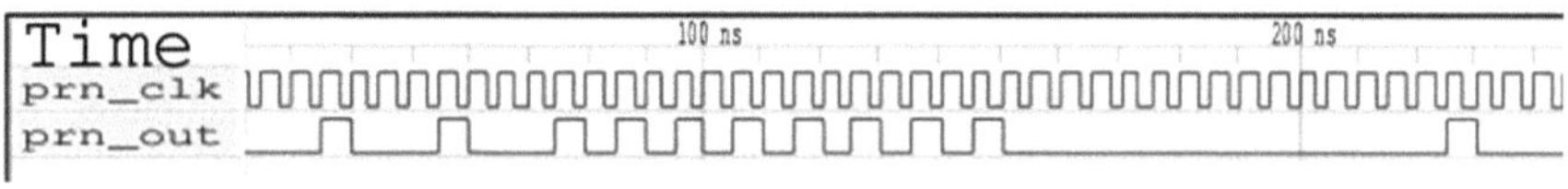


**Fig. 8.6** Output trace from pseudo-random number generator SystemC-2.2.0

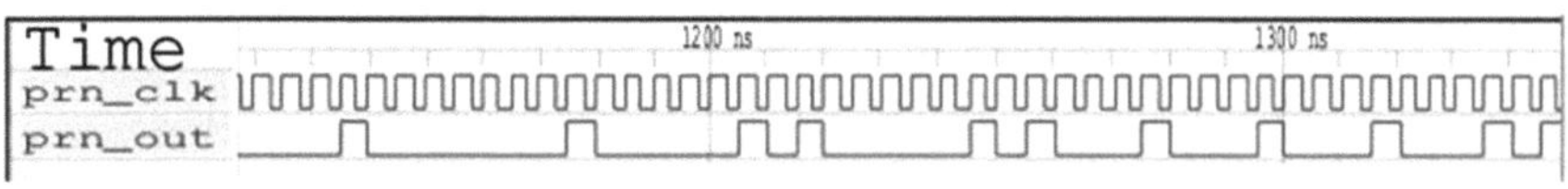


**Fig. 8.7** Output trace from pseudo-random number generator SystemC-2.3.0

## 8.2 Instruction Register-Level Scoreboard

Hierarchical designs/systems are very common in real world. A shift register has cells, and each cell is a flip-flop (D or JK). Here, we examine a more complicated design—a *scoreboard* [2]. The scoreboard consists of thirty-two 64-cell registers, and each cell consists of a *wakeup multiplexer*, Fig. 8.8. We focus on the register. Each multiplexer takes 5-bit input and 5-bit select input. Of the 5-bit input, 4 bits are supplied externally, while the 5th bit is a feedback bit from the output latch (latch B). The latches are triggered on two phases of the same clock. The wakeup multiplexer is a mixed combinational, sequential logic design as the core multiplexer is combinational, the latches sequential. Each member of the 64 cells of the register shares the common clock and the common select inputs, but the external data input is different. The output of each register cell (wakeup multiplexer) is fed into a 64-bit NOR gate, whose output is AND'ed with a bit that indicates whether that register is available or not. In addition, each of the 4 input bits to a register cell is the combination of bits from other registers in series. For example, cell '0' of the 2nd register takes as its input the 0th bits of the contents of registers 3, 4, 5, and 6. It is worth noting that

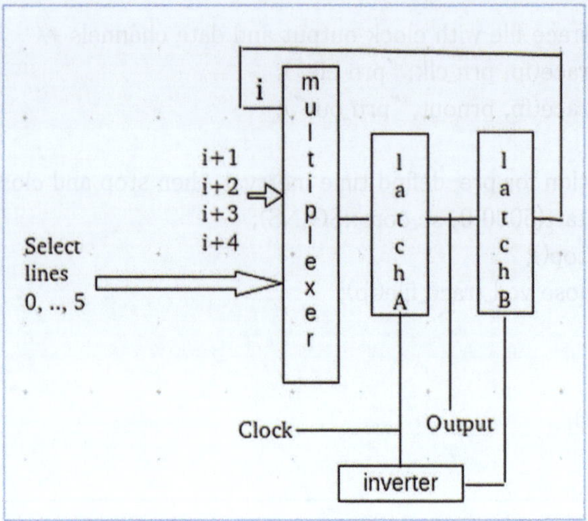


**Fig. 8.8** Wakeup multiplexer. Each cell in the 64-cell register is a wakeup multiplexer. The output of latch A is the first input to the multiplexer. The other inputs are the corresponding ith input of the succeeding or preceding registers, depending on whether a 'top-down' or 'bottom-up' approach is followed

the scoreboard 'register' in this design is not the same as a register in the traditional sense. Each scoreboard register datum is obtained from four preceding or succeeding traditional registers (each cell of a traditional register contains one bit of data).

The scoreboard register (Figs. 8.9, 8.10) is unlike any standard register, as each of its 64 cells is a wakeup multiplexer, and each multiplexer uses the following input—4-bit bit-vector data, a 5-bit bit-vector select, a clock for the two latches and a bit indication if the corresponding register (not a scoreboard register) is available for use or not. The 5-bit select input can have at most one 1 in it and the default case is when each bit is zero. The 4-bit data input is unique, as each of the 4 bits come from a different non-scoreboard register. Consider cell 0 of scoreboard register 2. Assuming top-down ordering, the four-bit data input to cell 0 of scoreboard register 2 will be as follows:

- Bit 0 of non-scoreboard register 3
- Bit 0 of non-scoreboard register 4
- Bit 0 of non-scoreboard register 5
- Bit 0 of non-scoreboard register 6.

Thus, each scoreboard register uses bits from four 64-bit non-scoreboard registers following it (top-down ordering) or preceding it (bottom-up ordering). To extract the bits from non-scoreboard registers and format them accurately to input to each cell of the scoreboard register, a special module is needed—Fig. 8.11. This module accepts the contents of four 64-bit non-scoreboard registers and generates sixty-four 4-bit bit-vectors that are input to the 64 cells of a scoreboard register.

Corresponding to each register there is a splitter, and to accurately direct the correct input to each cell of the scoreboard register, a wrapper module is

```
SC_MODULE(wakeup_mux)
{
 /*Declare/define input/output ports */
 sc_core::sc_in<bool> clk;
 sc_core::sc_in< sc_dt::sc_bv<4> > din;
/*Data input port */
 sc_core::sc_in< sc_dt::sc_bv<5> > sel;
/* Select input port */
 sc_core::sc_out< sc_dt::sc_bv<1> > new_reg;
/* Output port */
 /*Declare/define internal bit vectors */
 sc_dt::sc_bv<5> tmp0;
 sc_dt::sc_bv<5> tmp1;
 sc_dt::sc_bv<5> sel_in;
 sc_dt::sc_bv<4> din_in;
 sc_dt::sc_bv<1> tmp2;
 sc_dt::sc_bv<1> tmp3;

 /* Wakeup multiplexer - main multiplexing operation */
 void wakeup_mux_proc0()
 {
  while(1)
  {
    wait();
 /* Wait for new input data or new select bits */
    din_in = din.read();
    sel_in = sel.read();
    tmp0.range(4,4) = tmp2.range(0,0);
    tmp0.range(3,0) = din_in.range(3,0);
    tmp1 = tmp0 & sel_in;
    tmp3 = tmp1.or_reduce();
  }
 }

 /* Wakeup multiplexer latch A operation */
 void wakeup_mux_proc1()
 {
  while(1)
```

**Fig. 8.9** Wakeup multiplexer—each cell in the scoreboard register is a wakeup multiplexer

```
      {
       wait();
      /* Wait for positive clock edge */
       new_reg.write(tmp3);
      }
     }

 /* Wakeup register latch B operation */
 void wakeup_mux_proc2()
 {
  while(1)
  {
   wait();
  /*Wait for negative clock edge and feedback latch output */
   tmp2 = tmp3;
  /* to first input of multiplexer */
  }
 }

 /* Constructor  -- Initialize internal variables first */
 SC_CTOR(wakeup_mux):tmp0("00000"),
                     tmp1("00000"),
                     tmp2("0"),
                     tmp3("0")
 {
  /* Declare/define threads and sensitize them */
  SC_THREAD(wakeup_mux_proc0);
  sensitive << din << sel;
  SC_CTHREAD(wakeup_mux_proc1, clk.pos());
 /* Clocked thread */
  SC_CTHREAD(wakeup_mux_proc2, clk.neg());
 /* Clocked thread */
 }

 /* Destructor */
 ~wakeup_mux(){ }
}
```

**Fig. 8.9** (continued)

```
SC_MODULE(register64cell)
{
  /* Declare/define common clock port and data and select ports for each
register cell */
  /* Contents of each select input for each cell of the register is the same */
  sc_core::sc_in<bool> clk;
  sc_core::sc_in< sc_dt::sc_bv<4> > din_63;
  sc_core::sc_in< sc_dt::sc_bv<5> > sel_63;
  sc_core::sc_out< sc_dt::sc_bv<1> > new_reg_63;

  sc_core::sc_in< sc_dt::sc_bv<4> > din_62;
  sc_core::sc_in< sc_dt::sc_bv<5> > sel_62;
  sc_core::sc_out< sc_dt::sc_bv<1> > new_reg_62;
  sc_core::sc_in< sc_dt::sc_bv<4> > din_61;
  sc_core::sc_in< sc_dt::sc_bv<5> > sel_61;
  sc_core::sc_out< sc_dt::sc_bv<1> > new_reg_61;
  sc_core::sc_in< sc_dt::sc_bv<4> > din_60;
  sc_core::sc_in< sc_dt::sc_bv<5> > sel_60;
  sc_core::sc_out< sc_dt::sc_bv<1> > new_reg_60;
  sc_core::sc_in< sc_dt::sc_bv<4> > din_59;
  sc_core::sc_in< sc_dt::sc_bv<5> > sel_59;
  sc_core::sc_out< sc_dt::sc_bv<1> > new_reg_59;
  sc_core::sc_in< sc_dt::sc_bv<4> > din_58;
  sc_core::sc_in< sc_dt::sc_bv<5> > sel_58;
  sc_core::sc_out< sc_dt::sc_bv<1> > new_reg_58;
  sc_core::sc_in< sc_dt::sc_bv<4> > din_57;
  sc_core::sc_in< sc_dt::sc_bv<5> > sel_57;
  sc_core::sc_out< sc_dt::sc_bv<1> > new_reg_57;
  sc_core::sc_in< sc_dt::sc_bv<4> > din_56;
  sc_core::sc_in< sc_dt::sc_bv<5> > sel_56;
  sc_core::sc_out< sc_dt::sc_bv<1> > new_reg_56;
  sc_core::sc_in< sc_dt::sc_bv<4> > din_55;
  sc_core::sc_in< sc_dt::sc_bv<5> > sel_55;
  sc_core::sc_out< sc_dt::sc_bv<1> > new_reg_55;
  sc_core::sc_in< sc_dt::sc_bv<4> > din_54;
  sc_core::sc_in< sc_dt::sc_bv<5> > sel_54;
  sc_core::sc_out< sc_dt::sc_bv<1> > new_reg_54;
  sc_core::sc_in< sc_dt::sc_bv<4> > din_53;
  sc_core::sc_in< sc_dt::sc_bv<5> > sel_53;
```

**Fig. 8.10** 64-cell scoreboard register—each cell is a wakeup multiplexer

```
sc_core::sc_out< sc_dt::sc_bv<1> > new_reg_53;
sc_core::sc_in< sc_dt::sc_bv<4> > din_52;
sc_core::sc_in< sc_dt::sc_bv<5> > sel_52;
sc_core::sc_out< sc_dt::sc_bv<1> > new_reg_52;
sc_core::sc_in< sc_dt::sc_bv<4> > din_51;
sc_core::sc_in< sc_dt::sc_bv<5> > sel_51;
sc_core::sc_out< sc_dt::sc_bv<1> > new_reg_51;
sc_core::sc_in< sc_dt::sc_bv<4> > din_50;
sc_core::sc_in< sc_dt::sc_bv<5> > sel_50;
sc_core::sc_out< sc_dt::sc_bv<1> > new_reg_50;
sc_core::sc_in< sc_dt::sc_bv<4> > din_49;
sc_core::sc_in< sc_dt::sc_bv<5> > sel_49;
sc_core::sc_out< sc_dt::sc_bv<1> > new_reg_49;
sc_core::sc_in< sc_dt::sc_bv<4> > din_48;
sc_core::sc_in< sc_dt::sc_bv<5> > sel_48;
sc_core::sc_out< sc_dt::sc_bv<1> > new_reg_48;
sc_core::sc_in< sc_dt::sc_bv<4> > din_47;
sc_core::sc_in< sc_dt::sc_bv<5> > sel_47;
sc_core::sc_out< sc_dt::sc_bv<1> > new_reg_47;
sc_core::sc_in< sc_dt::sc_bv<4> > din_46;
sc_core::sc_in< sc_dt::sc_bv<5> > sel_46;
sc_core::sc_out< sc_dt::sc_bv<1> > new_reg_46;
sc_core::sc_in< sc_dt::sc_bv<4> > din_45;
sc_core::sc_in< sc_dt::sc_bv<5> > sel_45;
sc_core::sc_out< sc_dt::sc_bv<1> > new_reg_45;
sc_core::sc_in< sc_dt::sc_bv<4> > din_44;
sc_core::sc_in< sc_dt::sc_bv<5> > sel_44;
sc_core::sc_out< sc_dt::sc_bv<1> > new_reg_44;

sc_core::sc_in< sc_dt::sc_bv<4> > din_43;
sc_core::sc_in< sc_dt::sc_bv<5> > sel_43;
sc_core::sc_out< sc_dt::sc_bv<1> > new_reg_43;
sc_core::sc_in< sc_dt::sc_bv<4> > din_42;
sc_core::sc_in< sc_dt::sc_bv<5> > sel_42;
sc_core::sc_out< sc_dt::sc_bv<1> > new_reg_42;
sc_core::sc_in< sc_dt::sc_bv<4> > din_41;
sc_core::sc_in< sc_dt::sc_bv<5> > sel_41;
sc_core::sc_out< sc_dt::sc_bv<1> > new_reg_41;
```

**Fig. 8.10** (continued)

```
sc_core::sc_in< sc_dt::sc_bv<4> > din_40;
sc_core::sc_in< sc_dt::sc_bv<5> > sel_40;
sc_core::sc_out< sc_dt::sc_bv<1> > new_reg_40;
sc_core::sc_in< sc_dt::sc_bv<4> > din_39;
sc_core::sc_in< sc_dt::sc_bv<5> > sel_39;
sc_core::sc_out< sc_dt::sc_bv<1> > new_reg_39;
sc_core::sc_in< sc_dt::sc_bv<4> > din_38;
sc_core::sc_in< sc_dt::sc_bv<5> > sel_38;
sc_core::sc_out< sc_dt::sc_bv<1> > new_reg_38;
sc_core::sc_in< sc_dt::sc_bv<4> > din_37;
sc_core::sc_in< sc_dt::sc_bv<5> > sel_37;
sc_core::sc_out< sc_dt::sc_bv<1> > new_reg_37;
sc_core::sc_in< sc_dt::sc_bv<4> > din_36;
sc_core::sc_in< sc_dt::sc_bv<5> > sel_36;
sc_core::sc_out< sc_dt::sc_bv<1> > new_reg_36;
sc_core::sc_in< sc_dt::sc_bv<4> > din_35;
sc_core::sc_in< sc_dt::sc_bv<5> > sel_35;
sc_core::sc_out< sc_dt::sc_bv<1> > new_reg_35;
sc_core::sc_in< sc_dt::sc_bv<4> > din_34;
sc_core::sc_in< sc_dt::sc_bv<5> > sel_34;
sc_core::sc_out< sc_dt::sc_bv<1> > new_reg_34;
sc_core::sc_in< sc_dt::sc_bv<4> > din_33;
sc_core::sc_in< sc_dt::sc_bv<5> > sel_33;
sc_core::sc_out< sc_dt::sc_bv<1> > new_reg_33;
sc_core::sc_in< sc_dt::sc_bv<4> > din_32;
sc_core::sc_in< sc_dt::sc_bv<5> > sel_32;
sc_core::sc_out< sc_dt::sc_bv<1> > new_reg_32;
sc_core::sc_in< sc_dt::sc_bv<4> > din_31;
sc_core::sc_in< sc_dt::sc_bv<5> > sel_31;
sc_core::sc_out< sc_dt::sc_bv<1> > new_reg_31;
sc_core::sc_in< sc_dt::sc_bv<4> > din_30;
sc_core::sc_in< sc_dt::sc_bv<5> > sel_30;
sc_core::sc_out< sc_dt::sc_bv<1> > new_reg_30;
sc_core::sc_in< sc_dt::sc_bv<4> > din_29;
sc_core::sc_in< sc_dt::sc_bv<5> > sel_29;
sc_core::sc_out< sc_dt::sc_bv<1> > new_reg_29;
sc_core::sc_in< sc_dt::sc_bv<4> > din_28;
sc_core::sc_in< sc_dt::sc_bv<5> > sel_28;
sc_core::sc_out< sc_dt::sc_bv<1> > new_reg_28;
```

**Fig. 8.10** (continued)

```
sc_core::sc_in< sc_dt::sc_bv<4> > din_27;
sc_core::sc_in< sc_dt::sc_bv<5> > sel_27;
sc_core::sc_out< sc_dt::sc_bv<1> > new_reg_27;
sc_core::sc_in< sc_dt::sc_bv<4> > din_26;
sc_core::sc_in< sc_dt::sc_bv<5> > sel_26;
sc_core::sc_out< sc_dt::sc_bv<1> > new_reg_26;
sc_core::sc_in< sc_dt::sc_bv<4> > din_25;
sc_core::sc_in< sc_dt::sc_bv<5> > sel_25;
sc_core::sc_out< sc_dt::sc_bv<1> > new_reg_25;
sc_core::sc_in< sc_dt::sc_bv<4> > din_24;
sc_core::sc_in< sc_dt::sc_bv<5> > sel_24;
sc_core::sc_out< sc_dt::sc_bv<1> > new_reg_24;
sc_core::sc_in< sc_dt::sc_bv<4> > din_23;
sc_core::sc_in< sc_dt::sc_bv<5> > sel_23;
sc_core::sc_out< sc_dt::sc_bv<1> > new_reg_23;
sc_core::sc_in< sc_dt::sc_bv<4> > din_22;
sc_core::sc_in< sc_dt::sc_bv<5> > sel_22;
sc_core::sc_out< sc_dt::sc_bv<1> > new_reg_22;
sc_core::sc_in< sc_dt::sc_bv<4> > din_21;
sc_core::sc_in< sc_dt::sc_bv<5> > sel_21;
sc_core::sc_out< sc_dt::sc_bv<1> > new_reg_21;
sc_core::sc_in< sc_dt::sc_bv<4> > din_20;
sc_core::sc_in< sc_dt::sc_bv<5> > sel_20;
sc_core::sc_out< sc_dt::sc_bv<1> > new_reg_20;
sc_core::sc_in< sc_dt::sc_bv<4> > din_19;
sc_core::sc_in< sc_dt::sc_bv<5> > sel_19;
sc_core::sc_out< sc_dt::sc_bv<1> > new_reg_19;
sc_core::sc_in< sc_dt::sc_bv<4> > din_18;
sc_core::sc_in< sc_dt::sc_bv<5> > sel_18;
sc_core::sc_out< sc_dt::sc_bv<1> > new_reg_18;
sc_core::sc_in< sc_dt::sc_bv<4> > din_17;
sc_core::sc_in< sc_dt::sc_bv<5> > sel_17;
sc_core::sc_out< sc_dt::sc_bv<1> > new_reg_17;
sc_core::sc_in< sc_dt::sc_bv<4> > din_16;
sc_core::sc_in< sc_dt::sc_bv<5> > sel_16;
sc_core::sc_out< sc_dt::sc_bv<1> > new_reg_16;
sc_core::sc_in< sc_dt::sc_bv<4> > din_15;
sc_core::sc_in< sc_dt::sc_bv<5> > sel_15;
sc_core::sc_out< sc_dt::sc_bv<1> > new_reg_15;
```

**Fig. 8.10** (continued)

```
sc_core::sc_in< sc_dt::sc_bv<4> > din_14;
sc_core::sc_in< sc_dt::sc_bv<5> > sel_14;
sc_core::sc_out< sc_dt::sc_bv<1> > new_reg_14;
sc_core::sc_in< sc_dt::sc_bv<4> > din_13;
sc_core::sc_in< sc_dt::sc_bv<5> > sel_13;
sc_core::sc_out< sc_dt::sc_bv<1> > new_reg_13;
sc_core::sc_in< sc_dt::sc_bv<4> > din_12;
sc_core::sc_in< sc_dt::sc_bv<5> > sel_12;
sc_core::sc_out< sc_dt::sc_bv<1> > new_reg_12;
sc_core::sc_in< sc_dt::sc_bv<4> > din_11;
sc_core::sc_in< sc_dt::sc_bv<5> > sel_11;
sc_core::sc_out< sc_dt::sc_bv<1> > new_reg_11;
sc_core::sc_in< sc_dt::sc_bv<4> > din_10;
sc_core::sc_in< sc_dt::sc_bv<5> > sel_10;
sc_core::sc_out< sc_dt::sc_bv<1> > new_reg_10;
sc_core::sc_in< sc_dt::sc_bv<4> > din_9;
sc_core::sc_in< sc_dt::sc_bv<5> > sel_9;
sc_core::sc_out< sc_dt::sc_bv<1> > new_reg_9;
sc_core::sc_in< sc_dt::sc_bv<4> > din_8;
sc_core::sc_in< sc_dt::sc_bv<5> > sel_8;
sc_core::sc_out< sc_dt::sc_bv<1> > new_reg_8;
sc_core::sc_in< sc_dt::sc_bv<4> > din_7;
sc_core::sc_in< sc_dt::sc_bv<5> > sel_7;
sc_core::sc_out< sc_dt::sc_bv<1> > new_reg_7;
sc_core::sc_in< sc_dt::sc_bv<4> > din_6;
sc_core::sc_in< sc_dt::sc_bv<5> > sel_6;
sc_core::sc_out< sc_dt::sc_bv<1> > new_reg_6;
sc_core::sc_in< sc_dt::sc_bv<4> > din_5;
sc_core::sc_in< sc_dt::sc_bv<5> > sel_5;
sc_core::sc_out< sc_dt::sc_bv<1> > new_reg_5;
sc_core::sc_in< sc_dt::sc_bv<4> > din_4;
sc_core::sc_in< sc_dt::sc_bv<5> > sel_4;
sc_core::sc_out< sc_dt::sc_bv<1> > new_reg_4;
sc_core::sc_in< sc_dt::sc_bv<4> > din_3;
sc_core::sc_in< sc_dt::sc_bv<5> > sel_3;
sc_core::sc_out< sc_dt::sc_bv<1> > new_reg_3;
sc_core::sc_in< sc_dt::sc_bv<4> > din_2;
sc_core::sc_in< sc_dt::sc_bv<5> > sel_2;
sc_core::sc_out< sc_dt::sc_bv<1> > new_reg_2;
```

**Fig. 8.10** (continued)

```
sc_core::sc_in< sc_dt::sc_bv<4> > din_1;
sc_core::sc_in< sc_dt::sc_bv<5> > sel_1;
sc_core::sc_out< sc_dt::sc_bv<1> > new_reg_1;
sc_core::sc_in< sc_dt::sc_bv<4> > din_0;
sc_core::sc_in< sc_dt::sc_bv<5> > sel_0;
sc_core::sc_out< sc_dt::sc_bv<1> > new_reg_0;

/* Declare/define 64 wakeup multiplexers that form the register cells */
wakeup_mux w_m_63;
wakeup_mux w_m_62;
wakeup_mux w_m_61;
wakeup_mux w_m_60;
wakeup_mux w_m_59;
wakeup_mux w_m_58;
wakeup_mux w_m_57;
wakeup_mux w_m_56;
wakeup_mux w_m_55;
wakeup_mux w_m_54;
wakeup_mux w_m_53;
wakeup_mux w_m_52;
wakeup_mux w_m_51;
wakeup_mux w_m_50;
wakeup_mux w_m_49;
wakeup_mux w_m_48;
wakeup_mux w_m_47;
wakeup_mux w_m_46;
wakeup_mux w_m_45;
wakeup_mux w_m_44;
wakeup_mux w_m_43;
wakeup_mux w_m_42;
wakeup_mux w_m_41;
wakeup_mux w_m_40;
wakeup_mux w_m_39;
wakeup_mux w_m_38;
wakeup_mux w_m_37;
wakeup_mux w_m_36;
wakeup_mux w_m_35;
wakeup_mux w_m_34;
wakeup_mux w_m_33;
```

**Fig. 8.10** (continued)

```
wakeup_mux w_m_32;
wakeup_mux w_m_31;
wakeup_mux w_m_30;
wakeup_mux w_m_29;
wakeup_mux w_m_28;
wakeup_mux w_m_27;
wakeup_mux w_m_26;
wakeup_mux w_m_25;
wakeup_mux w_m_24;
wakeup_mux w_m_23;
wakeup_mux w_m_22;
wakeup_mux w_m_21;
wakeup_mux w_m_20;
wakeup_mux w_m_19;
wakeup_mux w_m_18;
wakeup_mux w_m_17;
wakeup_mux w_m_16;
wakeup_mux w_m_15;
wakeup_mux w_m_14;
wakeup_mux w_m_13;
wakeup_mux w_m_12;
wakeup_mux w_m_11;
wakeup_mux w_m_10;
wakeup_mux w_m_9;
wakeup_mux w_m_8;
wakeup_mux w_m_7;
wakeup_mux w_m_6;
wakeup_mux w_m_5;
wakeup_mux w_m_4;
wakeup_mux w_m_3;
wakeup_mux w_m_2;
wakeup_mux w_m_1;
wakeup_mux w_m_0;

/* Constructor - inotialize the wakeup multiplexer cella */
SC_CTOR(register64cell):w_m_63("w_m_63"),w_m_62("w_m_62"),
                        w_m_61("w_m_61"),w_m_60("w_m_60"),
                        w_m_59("w_m_59"),w_m_58("w_m_58"),
```

**Fig. 8.10** (continued)

```
            w_m_57("w_m_57"),w_m_56("w_m_56"),
            w_m_55("w_m_55"),w_m_54("w_m_54"),
            w_m_53("w_m_53"),w_m_52("w_m_52"),
            w_m_51("w_m_51"),w_m_50("w_m_50"),
            w_m_49("w_m_49"),w_m_48("w_m_48"),
            w_m_47("w_m_47"),w_m_46("w_m_46"),
            w_m_45("w_m_45"),w_m_44("w_m_44"),
            w_m_43("w_m_43"),w_m_42("w_m_42"),
            w_m_41("w_m_41"),w_m_40("w_m_40"),
            w_m_39("w_m_39"),w_m_38("w_m_38"),
            w_m_37("w_m_37"),w_m_36("w_m_36"),
            w_m_35("w_m_35"),w_m_34("w_m_34"),
            w_m_33("w_m_33"),w_m_32("w_m_32"),
            w_m_31("w_m_31"),w_m_30("w_m_30"),
            w_m_29("w_m_29"),w_m_28("w_m_28"),
            w_m_27("w_m_27"),w_m_26("w_m_26"),
            w_m_25("w_m_25"),w_m_24("w_m_24"),
            w_m_23("w_m_23"),w_m_22("w_m_22"),
            w_m_21("w_m_21"),w_m_20("w_m_20"),
            w_m_19("w_m_19"),w_m_18("w_m_18"),
            w_m_17("w_m_17"),w_m_16("w_m_16"),
            w_m_15("w_m_15"),w_m_14("w_m_14"),
            w_m_13("w_m_13"),w_m_12("w_m_12"),
            w_m_11("w_m_11"),w_m_10("w_m_10"),
            w_m_9("w_m_9"),w_m_8("w_m_8"),
            w_m_7("w_m_7"),w_m_6("w_m_6"),
            w_m_5("w_m_5"),w_m_4("w_m_4"),
            w_m_3("w_m_3"),w_m_2("w_m_2"),
            w_m_1("w_m_1"),w_m_0("w_m_0")
{
  /* Connect each wakeup multiplexer cell with its input/output ports */
  w_m_63.clk(clk);
  w_m_63.din(din_63);
  w_m_63.sel(sel_63);
  w_m_63.new_reg(new_reg_63);

  w_m_62.clk(clk);
  w_m_62.din(din_62);
  w_m_62.sel(sel_62);
```

**Fig. 8.10** (continued)

```
w_m_62.new_reg(new_reg_62);

w_m_61.clk(clk);
w_m_61.din(din_61);
w_m_61.sel(sel_61);
w_m_61.new_reg(new_reg_61);

w_m_60.clk(clk);
w_m_60.din(din_60);
w_m_60.sel(sel_60);
w_m_60.new_reg(new_reg_60);

w_m_59.clk(clk);
w_m_59.din(din_59);
w_m_59.sel(sel_59);
w_m_59.new_reg(new_reg_59);

w_m_58.clk(clk);
w_m_58.din(din_58);
w_m_58.sel(sel_58);
w_m_58.new_reg(new_reg_58);

w_m_57.clk(clk);
w_m_57.din(din_57);
w_m_57.sel(sel_57);
w_m_57.new_reg(new_reg_57);

w_m_56.clk(clk);
w_m_56.din(din_56);
w_m_56.sel(sel_56);
w_m_56.new_reg(new_reg_56);

w_m_55.clk(clk);
w_m_55.din(din_55);
w_m_55.sel(sel_55);
w_m_55.new_reg(new_reg_55);

w_m_54.clk(clk);
```

**Fig. 8.10** (continued)

```
w_m_54.din(din_54);
w_m_54.sel(sel_54);
w_m_54.new_reg(new_reg_54);

w_m_53.clk(clk);
w_m_53.din(din_53);
w_m_53.sel(sel_53);
w_m_53.new_reg(new_reg_53);

w_m_52.clk(clk);
w_m_52.din(din_52);
w_m_52.sel(sel_52);
w_m_52.new_reg(new_reg_52);

w_m_51.clk(clk);
w_m_51.din(din_51);
w_m_51.sel(sel_51);
w_m_51.new_reg(new_reg_51);

w_m_50.clk(clk);
w_m_50.din(din_50);
w_m_50.sel(sel_50);
w_m_50.new_reg(new_reg_50);

w_m_49.clk(clk);
w_m_49.din(din_49);
w_m_49.sel(sel_49);
w_m_49.new_reg(new_reg_49);

w_m_48.clk(clk);
w_m_48.din(din_48);
w_m_48.sel(sel_48);
w_m_48.new_reg(new_reg_48);

w_m_47.clk(clk);
w_m_47.din(din_47);
w_m_47.sel(sel_47);
w_m_47.new_reg(new_reg_47);
```

**Fig. 8.10** (continued)

```
w_m_46.clk(clk);
w_m_46.din(din_46);
w_m_46.sel(sel_46);
w_m_46.new_reg(new_reg_46);

w_m_45.clk(clk);
w_m_45.din(din_45);
w_m_45.sel(sel_45);
w_m_45.new_reg(new_reg_45);

w_m_44.clk(clk);
w_m_44.din(din_44);
w_m_44.sel(sel_44);
w_m_44.new_reg(new_reg_44);

w_m_43.clk(clk);
w_m_43.din(din_43);
w_m_43.sel(sel_43);
w_m_43.new_reg(new_reg_43);

w_m_42.clk(clk);
w_m_42.din(din_42);
w_m_42.sel(sel_42);
w_m_42.new_reg(new_reg_42);

w_m_41.clk(clk);
w_m_41.din(din_41);
w_m_41.sel(sel_41);
w_m_41.new_reg(new_reg_41);

w_m_40.clk(clk);
w_m_40.din(din_40);
w_m_40.sel(sel_40);
w_m_40.new_reg(new_reg_40);

w_m_39.clk(clk);
w_m_39.din(din_39);
w_m_39.sel(sel_39);
w_m_39.new_reg(new_reg_39);
```

**Fig. 8.10** (continued)

```
w_m_38.clk(clk);
w_m_38.din(din_38);
w_m_38.sel(sel_38);
w_m_38.new_reg(new_reg_38);

w_m_37.clk(clk);
w_m_37.din(din_37);
w_m_37.sel(sel_37);
w_m_37.new_reg(new_reg_37);
w_m_36.clk(clk);
w_m_36.din(din_36);
w_m_36.sel(sel_36);
w_m_36.new_reg(new_reg_36);

w_m_35.clk(clk);
w_m_35.din(din_35);
w_m_35.sel(sel_35);
w_m_35.new_reg(new_reg_35);

w_m_34.clk(clk);
w_m_34.din(din_34);
w_m_34.sel(sel_34);
w_m_34.new_reg(new_reg_34);

w_m_33.clk(clk);
w_m_33.din(din_33);
w_m_33.sel(sel_33);
w_m_33.new_reg(new_reg_33);

w_m_32.clk(clk);
w_m_32.din(din_32);
w_m_32.sel(sel_32);
w_m_32.new_reg(new_reg_32);

w_m_31.clk(clk);
w_m_31.din(din_31);
w_m_31.sel(sel_31);
w_m_31.new_reg(new_reg_31);
```

**Fig. 8.10** (continued)

```
w_m_30.clk(clk);
w_m_30.din(din_30);
w_m_30.sel(sel_30);
w_m_30.new_reg(new_reg_30);

w_m_29.clk(clk);
w_m_29.din(din_29);
w_m_29.sel(sel_29);
w_m_29.new_reg(new_reg_29);

w_m_28.clk(clk);
w_m_28.din(din_28);
w_m_28.sel(sel_28);
w_m_28.new_reg(new_reg_28);

w_m_27.clk(clk);
w_m_27.din(din_27);
w_m_27.sel(sel_27);
w_m_27.new_reg(new_reg_27);

w_m_26.clk(clk);
w_m_26.din(din_26);
w_m_26.sel(sel_26);
w_m_26.new_reg(new_reg_26);

w_m_25.clk(clk);
w_m_25.din(din_25);
w_m_25.sel(sel_25);
w_m_25.new_reg(new_reg_25);

w_m_24.clk(clk);
w_m_24.din(din_24);
w_m_24.sel(sel_24);
w_m_24.new_reg(new_reg_24);

w_m_23.clk(clk);
w_m_23.din(din_23);
w_m_23.sel(sel_23);
```

**Fig. 8.10** (continued)

```
w_m_23.new_reg(new_reg_23);

w_m_22.clk(clk);
w_m_22.din(din_22);
w_m_22.sel(sel_22);
w_m_22.new_reg(new_reg_22);

w_m_21.clk(clk);
w_m_21.din(din_21);
w_m_21.sel(sel_21);
w_m_21.new_reg(new_reg_21);

w_m_20.clk(clk);
w_m_20.din(din_20);
w_m_20.sel(sel_20);
w_m_20.new_reg(new_reg_20);

w_m_19.clk(clk);
w_m_19.din(din_19);
w_m_19.sel(sel_19);
w_m_19.new_reg(new_reg_19);

w_m_18.clk(clk);
w_m_18.din(din_18);
w_m_18.sel(sel_18);
w_m_18.new_reg(new_reg_18);

w_m_17.clk(clk);
w_m_17.din(din_17);
w_m_17.sel(sel_17);
w_m_17.new_reg(new_reg_17);

w_m_16.clk(clk);
w_m_16.din(din_16);
w_m_16.sel(sel_16);
w_m_16.new_reg(new_reg_16);

w_m_15.clk(clk);
w_m_15.din(din_15);
```

**Fig. 8.10** (continued)

```
w_m_15.sel(sel_15);
w_m_15.new_reg(new_reg_15);

w_m_14.clk(clk);
w_m_14.din(din_14);
w_m_14.sel(sel_14);
w_m_14.new_reg(new_reg_14);

w_m_13.clk(clk);
w_m_13.din(din_13);
w_m_13.sel(sel_13);
w_m_13.new_reg(new_reg_13);

w_m_12.clk(clk);
w_m_12.din(din_12);
w_m_12.sel(sel_12);
w_m_12.new_reg(new_reg_12);
w_m_11.clk(clk);
w_m_11.din(din_11);
w_m_11.sel(sel_11);
w_m_11.new_reg(new_reg_11);

w_m_10.clk(clk);
w_m_10.din(din_10);
w_m_10.sel(sel_10);
w_m_10.new_reg(new_reg_10);

w_m_9.clk(clk);
w_m_9.din(din_9);
w_m_9.sel(sel_9);
w_m_9.new_reg(new_reg_9);

w_m_8.clk(clk);
w_m_8.din(din_8);
w_m_8.sel(sel_8);
w_m_8.new_reg(new_reg_8);
```

**Fig. 8.10** (continued)

```
w_m_7.clk(clk);
w_m_7.din(din_7);
w_m_7.sel(sel_7);
w_m_7.new_reg(new_reg_7);

w_m_6.clk(clk);
w_m_6.din(din_6);
w_m_6.sel(sel_6);
w_m_6.new_reg(new_reg_6);

w_m_5.clk(clk);
w_m_5.din(din_5);
w_m_5.sel(sel_5);
w_m_5.new_reg(new_reg_5);

w_m_4.clk(clk);
w_m_4.din(din_4);
w_m_4.sel(sel_4);
w_m_4.new_reg(new_reg_4);

w_m_3.clk(clk);
w_m_3.din(din_3);
w_m_3.sel(sel_3);
w_m_3.new_reg(new_reg_3);

w_m_2.clk(clk);
w_m_2.din(din_2);
w_m_2.sel(sel_2);
w_m_2.new_reg(new_reg_2);

w_m_1.clk(clk);
w_m_1.din(din_1);
w_m_1.sel(sel_1);
w_m_1.new_reg(new_reg_1);

w_m_0.clk(clk);
w_m_0.din(din_0);
```

**Fig. 8.10** (continued)

```
    w_m_0.sel(sel_0);
    w_m_0.new_reg(new_reg_0);
  }
  /* Destructor */
   ~register64cell(){ }
};
```

**Fig. 8.10** (continued)

```
SC_MODULE(splitter)
{
 /* Declare/define input ports for contents of 4 64-bit non-scoreboard
register */
  sc_core::sc_in< sc_dt::sc_bv<64> > in0;
  sc_core::sc_in< sc_dt::sc_bv<64> > in1;
  sc_core::sc_in< sc_dt::sc_bv<64> > in2;
  sc_core::sc_in< sc_dt::sc_bv<64> > in3;

 /* Declare/define output ports for each of 64 4 bit bit-vector outputs*/
 /* Each 4 bit bit-vector is the input for a wakeup multiplexer cell
    of the scoreboard register */
  sc_core::sc_out< sc_dt::sc_bv<4> > bit4out0;
  sc_core::sc_out< sc_dt::sc_bv<4> > bit4out1;
  sc_core::sc_out< sc_dt::sc_bv<4> > bit4out2;
  sc_core::sc_out< sc_dt::sc_bv<4> > bit4out3;
  sc_core::sc_out< sc_dt::sc_bv<4> > bit4out4;
  sc_core::sc_out< sc_dt::sc_bv<4> > bit4out5;
  sc_core::sc_out< sc_dt::sc_bv<4> > bit4out6;
  sc_core::sc_out< sc_dt::sc_bv<4> > bit4out7;
  sc_core::sc_out< sc_dt::sc_bv<4> > bit4out8;
  sc_core::sc_out< sc_dt::sc_bv<4> > bit4out9;
  sc_core::sc_out< sc_dt::sc_bv<4> > bit4out10;
  sc_core::sc_out< sc_dt::sc_bv<4> > bit4out11;
  sc_core::sc_out< sc_dt::sc_bv<4> > bit4out12;
  sc_core::sc_out< sc_dt::sc_bv<4> > bit4out13;
  sc_core::sc_out< sc_dt::sc_bv<4> > bit4out14;
  sc_core::sc_out< sc_dt::sc_bv<4> > bit4out15;
  sc_core::sc_out< sc_dt::sc_bv<4> > bit4out16;
  sc_core::sc_out< sc_dt::sc_bv<4> > bit4out17;
```

**Fig. 8.11** Module for extracting and formatting the bits of 4 non-scoreboard 64-bit registers and formatting the bits into 4-bit bit-vectors for each of the 64 cells of a scoreboard register

```
sc_core::sc_out< sc_dt::sc_bv<4> > bit4out18;
sc_core::sc_out< sc_dt::sc_bv<4> > bit4out19;
sc_core::sc_out< sc_dt::sc_bv<4> > bit4out20;
sc_core::sc_out< sc_dt::sc_bv<4> > bit4out21;
sc_core::sc_out< sc_dt::sc_bv<4> > bit4out22;
sc_core::sc_out< sc_dt::sc_bv<4> > bit4out23;
sc_core::sc_out< sc_dt::sc_bv<4> > bit4out24;
sc_core::sc_out< sc_dt::sc_bv<4> > bit4out25;
sc_core::sc_out< sc_dt::sc_bv<4> > bit4out26;
sc_core::sc_out< sc_dt::sc_bv<4> > bit4out27;
sc_core::sc_out< sc_dt::sc_bv<4> > bit4out28;
sc_core::sc_out< sc_dt::sc_bv<4> > bit4out29;
sc_core::sc_out< sc_dt::sc_bv<4> > bit4out30;
sc_core::sc_out< sc_dt::sc_bv<4> > bit4out31;
sc_core::sc_out< sc_dt::sc_bv<4> > bit4out32;
sc_core::sc_out< sc_dt::sc_bv<4> > bit4out33;
sc_core::sc_out< sc_dt::sc_bv<4> > bit4out34;
sc_core::sc_out< sc_dt::sc_bv<4> > bit4out35;
sc_core::sc_out< sc_dt::sc_bv<4> > bit4out36;
sc_core::sc_out< sc_dt::sc_bv<4> > bit4out37;
sc_core::sc_out< sc_dt::sc_bv<4> > bit4out38;
sc_core::sc_out< sc_dt::sc_bv<4> > bit4out39;
sc_core::sc_out< sc_dt::sc_bv<4> > bit4out40;
sc_core::sc_out< sc_dt::sc_bv<4> > bit4out41;
sc_core::sc_out< sc_dt::sc_bv<4> > bit4out42;
sc_core::sc_out< sc_dt::sc_bv<4> > bit4out43;
sc_core::sc_out< sc_dt::sc_bv<4> > bit4out44;
sc_core::sc_out< sc_dt::sc_bv<4> > bit4out45;
sc_core::sc_out< sc_dt::sc_bv<4> > bit4out46;
sc_core::sc_out< sc_dt::sc_bv<4> > bit4out47;
sc_core::sc_out< sc_dt::sc_bv<4> > bit4out48;
sc_core::sc_out< sc_dt::sc_bv<4> > bit4out49;
sc_core::sc_out< sc_dt::sc_bv<4> > bit4out50;
sc_core::sc_out< sc_dt::sc_bv<4> > bit4out51;
sc_core::sc_out< sc_dt::sc_bv<4> > bit4out52;
sc_core::sc_out< sc_dt::sc_bv<4> > bit4out53;
sc_core::sc_out< sc_dt::sc_bv<4> > bit4out54;
sc_core::sc_out< sc_dt::sc_bv<4> > bit4out55;
```

**Fig. 8.11** (continued)

```
sc_core::sc_out< sc_dt::sc_bv<4> > bit4out56;
sc_core::sc_out< sc_dt::sc_bv<4> > bit4out57;
sc_core::sc_out< sc_dt::sc_bv<4> > bit4out58;
sc_core::sc_out< sc_dt::sc_bv<4> > bit4out59;
sc_core::sc_out< sc_dt::sc_bv<4> > bit4out60;
sc_core::sc_out< sc_dt::sc_bv<4> > bit4out61;
sc_core::sc_out< sc_dt::sc_bv<4> > bit4out62;
sc_core::sc_out< sc_dt::sc_bv<4> > bit4out63;

/* Declare/define bit vectors for intermediate data storage */
sc_dt::sc_bv<64> in0_bv;
sc_dt::sc_bv<64> in1_bv;
sc_dt::sc_bv<64> in2_bv;
sc_dt::sc_bv<64> in3_bv;

sc_dt::sc_bv<4> bit4out0_bv;
sc_dt::sc_bv<4> bit4out1_bv;
sc_dt::sc_bv<4> bit4out2_bv;
sc_dt::sc_bv<4> bit4out3_bv;
sc_dt::sc_bv<4> bit4out4_bv;
sc_dt::sc_bv<4> bit4out5_bv;
sc_dt::sc_bv<4> bit4out6_bv;
sc_dt::sc_bv<4> bit4out7_bv;
sc_dt::sc_bv<4> bit4out8_bv;
sc_dt::sc_bv<4> bit4out9_bv;
sc_dt::sc_bv<4> bit4out10_bv;
sc_dt::sc_bv<4> bit4out11_bv;
sc_dt::sc_bv<4> bit4out12_bv;
sc_dt::sc_bv<4> bit4out13_bv;
sc_dt::sc_bv<4> bit4out14_bv;
sc_dt::sc_bv<4> bit4out15_bv;
sc_dt::sc_bv<4> bit4out16_bv;
sc_dt::sc_bv<4> bit4out17_bv;
sc_dt::sc_bv<4> bit4out18_bv;
sc_dt::sc_bv<4> bit4out19_bv;
sc_dt::sc_bv<4> bit4out20_bv;
sc_dt::sc_bv<4> bit4out21_bv;
sc_dt::sc_bv<4> bit4out22_bv;
sc_dt::sc_bv<4> bit4out23_bv;
```

**Fig. 8.11** (continued)

```
sc_dt::sc_bv<4> bit4out24_bv;
sc_dt::sc_bv<4> bit4out25_bv;
sc_dt::sc_bv<4> bit4out26_bv;
sc_dt::sc_bv<4> bit4out27_bv;
sc_dt::sc_bv<4> bit4out28_bv;
sc_dt::sc_bv<4> bit4out29_bv;
sc_dt::sc_bv<4> bit4out30_bv;
sc_dt::sc_bv<4> bit4out31_bv;
sc_dt::sc_bv<4> bit4out32_bv;
sc_dt::sc_bv<4> bit4out33_bv;
sc_dt::sc_bv<4> bit4out34_bv;
sc_dt::sc_bv<4> bit4out35_bv;
sc_dt::sc_bv<4> bit4out36_bv;
sc_dt::sc_bv<4> bit4out37_bv;
sc_dt::sc_bv<4> bit4out38_bv;
sc_dt::sc_bv<4> bit4out39_bv;
sc_dt::sc_bv<4> bit4out40_bv;
sc_dt::sc_bv<4> bit4out41_bv;
sc_dt::sc_bv<4> bit4out42_bv;
sc_dt::sc_bv<4> bit4out43_bv;
sc_dt::sc_bv<4> bit4out44_bv;
sc_dt::sc_bv<4> bit4out45_bv;
sc_dt::sc_bv<4> bit4out46_bv;
sc_dt::sc_bv<4> bit4out47_bv;
sc_dt::sc_bv<4> bit4out48_bv;
sc_dt::sc_bv<4> bit4out49_bv;
sc_dt::sc_bv<4> bit4out50_bv;
sc_dt::sc_bv<4> bit4out51_bv;
sc_dt::sc_bv<4> bit4out52_bv;
sc_dt::sc_bv<4> bit4out53_bv;
sc_dt::sc_bv<4> bit4out54_bv;
sc_dt::sc_bv<4> bit4out55_bv;
sc_dt::sc_bv<4> bit4out56_bv;
sc_dt::sc_bv<4> bit4out57_bv;
sc_dt::sc_bv<4> bit4out58_bv;
sc_dt::sc_bv<4> bit4out59_bv;
sc_dt::sc_bv<4> bit4out60_bv;
sc_dt::sc_bv<4> bit4out61_bv;
sc_dt::sc_bv<4> bit4out62_bv;
```

**Fig. 8.11** (continued)

```
sc_dt::sc_bv<4> bit4out63_bv;

void splitter_proc0()
/* Main splitter operation thread */
{
  while(1)
  {
    wait();
/* Wait for change in any input bit-vector contents */
    in0_bv = in0.read();
    in1_bv = in1.read();
    in2_bv = in2.read();
    in3_bv = in3.read();
    /* After extracting data from input data bit-vectors,
       format it for output bit-vectors */
    bit4out63_bv.range(3,3) = in3_bv.range(63,63);
    bit4out63_bv.range(2,2) = in2_bv.range(63,63);
    bit4out63_bv.range(1,1) = in1_bv.range(63,63);
    bit4out63_bv.range(0,0) = in0_bv.range(63,63);
    bit4out62_bv.range(3,3) = in3_bv.range(62,62);
    bit4out62_bv.range(2,2) = in2_bv.range(62,62);
    bit4out62_bv.range(1,1) = in1_bv.range(62,62);
    bit4out62_bv.range(0,0) = in0_bv.range(62,62);
    bit4out61_bv.range(3,3) = in3_bv.range(61,61);
    bit4out61_bv.range(2,2) = in2_bv.range(61,61);
    bit4out61_bv.range(1,1) = in1_bv.range(61,61);
    bit4out61_bv.range(0,0) = in0_bv.range(61,61);
    bit4out60_bv.range(3,3) = in3_bv.range(60,60);
    bit4out60_bv.range(2,2) = in2_bv.range(60,60);
    bit4out60_bv.range(1,1) = in1_bv.range(60,60);
    bit4out60_bv.range(0,0) = in0_bv.range(60,60);
    bit4out59_bv.range(3,3) = in3_bv.range(59,59);
    bit4out59_bv.range(2,2) = in2_bv.range(59,59);
    bit4out59_bv.range(1,1) = in1_bv.range(59,59);
    bit4out59_bv.range(0,0) = in0_bv.range(59,59);
    bit4out58_bv.range(3,3) = in3_bv.range(58,58);
    bit4out58_bv.range(2,2) = in2_bv.range(58,58);
    bit4out58_bv.range(1,1) = in1_bv.range(58,58);
    bit4out58_bv.range(0,0) = in0_bv.range(58,58);
```

**Fig. 8.11** (continued)

```
bit4out57_bv.range(3,3) = in3_bv.range(57,57);
bit4out57_bv.range(2,2) = in2_bv.range(57,57);
bit4out57_bv.range(1,1) = in1_bv.range(57,57);
bit4out57_bv.range(0,0) = in0_bv.range(57,57);
bit4out56_bv.range(3,3) = in3_bv.range(56,56);
bit4out56_bv.range(2,2) = in2_bv.range(56,56);
bit4out56_bv.range(1,1) = in1_bv.range(56,56);
bit4out56_bv.range(0,0) = in0_bv.range(56,56);
bit4out55_bv.range(3,3) = in3_bv.range(55,55);
bit4out55_bv.range(2,2) = in2_bv.range(55,55);
bit4out55_bv.range(1,1) = in1_bv.range(55,55);
bit4out55_bv.range(0,0) = in0_bv.range(55,55);
bit4out54_bv.range(3,3) = in3_bv.range(54,54);
bit4out54_bv.range(2,2) = in2_bv.range(54,54);
bit4out54_bv.range(1,1) = in1_bv.range(54,54);
bit4out54_bv.range(0,0) = in0_bv.range(54,54);
bit4out53_bv.range(3,3) = in3_bv.range(53,53);
bit4out53_bv.range(2,2) = in2_bv.range(53,53);
bit4out53_bv.range(1,1) = in1_bv.range(53,53);
bit4out53_bv.range(0,0) = in0_bv.range(53,53);

bit4out52_bv.range(3,3) = in3_bv.range(52,52);
bit4out52_bv.range(2,2) = in2_bv.range(52,52);
bit4out52_bv.range(1,1) = in1_bv.range(52,52);
bit4out52_bv.range(0,0) = in0_bv.range(52,52);
bit4out51_bv.range(3,3) = in3_bv.range(51,51);
bit4out51_bv.range(2,2) = in2_bv.range(51,51);
bit4out51_bv.range(1,1) = in1_bv.range(51,51);
bit4out51_bv.range(0,0) = in0_bv.range(51,51);
bit4out50_bv.range(3,3) = in3_bv.range(50,50);
bit4out50_bv.range(2,2) = in2_bv.range(50,50);
bit4out50_bv.range(1,1) = in1_bv.range(50,50);
bit4out50_bv.range(0,0) = in0_bv.range(50,50);
bit4out49_bv.range(3,3) = in3_bv.range(49,49);
bit4out49_bv.range(2,2) = in2_bv.range(49,49);
bit4out49_bv.range(1,1) = in1_bv.range(49,49);
bit4out49_bv.range(0,0) = in0_bv.range(49,49);
bit4out48_bv.range(3,3) = in3_bv.range(48,48);
bit4out48_bv.range(2,2) = in2_bv.range(48,48);
```

**Fig. 8.11** (continued)

```
bit4out48_bv.range(1,1) = in1_bv.range(48,48);
bit4out48_bv.range(0,0) = in0_bv.range(48,48);
bit4out47_bv.range(3,3) = in3_bv.range(47,47);
bit4out47_bv.range(2,2) = in2_bv.range(47,47);
bit4out47_bv.range(1,1) = in1_bv.range(47,47);
bit4out47_bv.range(0,0) = in0_bv.range(47,47);
bit4out46_bv.range(3,3) = in3_bv.range(46,46);
bit4out46_bv.range(2,2) = in2_bv.range(46,46);
bit4out46_bv.range(1,1) = in1_bv.range(46,46);
bit4out46_bv.range(0,0) = in0_bv.range(46,46);
bit4out45_bv.range(3,3) = in3_bv.range(45,45);
bit4out45_bv.range(2,2) = in2_bv.range(45,45);
bit4out45_bv.range(1,1) = in1_bv.range(45,45);
bit4out45_bv.range(0,0) = in0_bv.range(45,45);
bit4out44_bv.range(3,3) = in3_bv.range(44,44);
bit4out44_bv.range(2,2) = in2_bv.range(44,44);
bit4out44_bv.range(1,1) = in1_bv.range(44,44);
bit4out44_bv.range(0,0) = in0_bv.range(44,44);
bit4out43_bv.range(3,3) = in3_bv.range(43,43);
bit4out43_bv.range(2,2) = in2_bv.range(43,43);
bit4out43_bv.range(1,1) = in1_bv.range(43,43);
bit4out43_bv.range(0,0) = in0_bv.range(43,43);

bit4out42_bv.range(3,3) = in3_bv.range(42,42);
bit4out42_bv.range(2,2) = in2_bv.range(42,42);
bit4out42_bv.range(1,1) = in1_bv.range(42,42);
bit4out42_bv.range(0,0) = in0_bv.range(42,42);
bit4out41_bv.range(3,3) = in3_bv.range(41,41);
bit4out41_bv.range(2,2) = in2_bv.range(41,41);
bit4out41_bv.range(1,1) = in1_bv.range(41,41);
bit4out41_bv.range(0,0) = in0_bv.range(41,41);
bit4out40_bv.range(3,3) = in3_bv.range(40,40);
bit4out40_bv.range(2,2) = in2_bv.range(40,40);
bit4out40_bv.range(1,1) = in1_bv.range(40,40);
bit4out40_bv.range(0,0) = in0_bv.range(40,40);
bit4out39_bv.range(3,3) = in3_bv.range(39,39);
bit4out39_bv.range(2,2) = in2_bv.range(39,39);
bit4out39_bv.range(1,1) = in1_bv.range(39,39);
bit4out39_bv.range(0,0) = in0_bv.range(39,39);
```

**Fig. 8.11** (continued)

```
bit4out38_bv.range(3,3) = in3_bv.range(38,38);
bit4out38_bv.range(2,2) = in2_bv.range(38,38);
bit4out38_bv.range(1,1) = in1_bv.range(38,38);
bit4out38_bv.range(0,0) = in0_bv.range(38,38);
bit4out37_bv.range(3,3) = in3_bv.range(37,37);
bit4out37_bv.range(2,2) = in2_bv.range(37,37);
bit4out37_bv.range(1,1) = in1_bv.range(37,37);
bit4out37_bv.range(0,0) = in0_bv.range(37,37);
bit4out36_bv.range(3,3) = in3_bv.range(36,36);
bit4out36_bv.range(2,2) = in2_bv.range(36,36);
bit4out36_bv.range(1,1) = in1_bv.range(36,36);
bit4out36_bv.range(0,0) = in0_bv.range(36,36);
bit4out35_bv.range(3,3) = in3_bv.range(35,35);
bit4out35_bv.range(2,2) = in2_bv.range(35,35);
bit4out35_bv.range(1,1) = in1_bv.range(35,35);
bit4out35_bv.range(0,0) = in0_bv.range(35,35);
bit4out34_bv.range(3,3) = in3_bv.range(34,34);
bit4out34_bv.range(2,2) = in2_bv.range(34,34);
bit4out34_bv.range(1,1) = in1_bv.range(34,34);
bit4out34_bv.range(0,0) = in0_bv.range(34,34);
bit4out33_bv.range(3,3) = in3_bv.range(33,33);
bit4out33_bv.range(2,2) = in2_bv.range(33,33);
bit4out33_bv.range(1,1) = in1_bv.range(33,33);
bit4out33_bv.range(0,0) = in0_bv.range(33,33);

bit4out32_bv.range(3,3) = in3_bv.range(32,32);
bit4out32_bv.range(2,2) = in2_bv.range(32,32);
bit4out32_bv.range(1,1) = in1_bv.range(32,32);
bit4out32_bv.range(0,0) = in0_bv.range(32,32);
bit4out31_bv.range(3,3) = in3_bv.range(31,31);
bit4out31_bv.range(2,2) = in2_bv.range(31,31);
bit4out31_bv.range(1,1) = in1_bv.range(31,31);
bit4out31_bv.range(0,0) = in0_bv.range(31,31);
bit4out30_bv.range(3,3) = in3_bv.range(30,30);
bit4out30_bv.range(2,2) = in2_bv.range(30,30);
bit4out30_bv.range(1,1) = in1_bv.range(30,30);
bit4out30_bv.range(0,0) = in0_bv.range(30,30);
bit4out29_bv.range(3,3) = in3_bv.range(29,29);
bit4out29_bv.range(2,2) = in2_bv.range(29,29);
```

**Fig. 8.11** (continued)

```
bit4out29_bv.range(1,1) = in1_bv.range(29,29);
bit4out29_bv.range(0,0) = in0_bv.range(29,29);
bit4out28_bv.range(3,3) = in3_bv.range(28,28);
bit4out28_bv.range(2,2) = in2_bv.range(28,28);
bit4out28_bv.range(1,1) = in1_bv.range(28,28);
bit4out28_bv.range(0,0) = in0_bv.range(28,28);
bit4out27_bv.range(3,3) = in3_bv.range(27,27);
bit4out27_bv.range(2,2) = in2_bv.range(27,27);
bit4out27_bv.range(1,1) = in1_bv.range(27,27);
bit4out27_bv.range(0,0) = in0_bv.range(27,27);
bit4out26_bv.range(3,3) = in3_bv.range(26,26);
bit4out26_bv.range(2,2) = in2_bv.range(26,26);
bit4out26_bv.range(1,1) = in1_bv.range(26,26);
bit4out26_bv.range(0,0) = in0_bv.range(26,26);
bit4out25_bv.range(3,3) = in3_bv.range(25,25);
bit4out25_bv.range(2,2) = in2_bv.range(25,25);
bit4out25_bv.range(1,1) = in1_bv.range(25,25);
bit4out25_bv.range(0,0) = in0_bv.range(25,25);
bit4out24_bv.range(3,3) = in3_bv.range(24,24);
bit4out24_bv.range(2,2) = in2_bv.range(24,24);
bit4out24_bv.range(1,1) = in1_bv.range(24,24);
bit4out24_bv.range(0,0) = in0_bv.range(24,24);
bit4out23_bv.range(3,3) = in3_bv.range(23,23);
bit4out23_bv.range(2,2) = in2_bv.range(23,23);
bit4out23_bv.range(1,1) = in1_bv.range(23,23);
bit4out23_bv.range(0,0) = in0_bv.range(23,23);

bit4out22_bv.range(3,3) = in3_bv.range(22,22);
bit4out22_bv.range(2,2) = in2_bv.range(22,22);
bit4out22_bv.range(1,1) = in1_bv.range(22,22);
bit4out22_bv.range(0,0) = in0_bv.range(22,22);
bit4out21_bv.range(3,3) = in3_bv.range(21,21);
bit4out21_bv.range(2,2) = in2_bv.range(21,21);
bit4out21_bv.range(1,1) = in1_bv.range(21,21);
bit4out21_bv.range(0,0) = in0_bv.range(21,21);
bit4out20_bv.range(3,3) = in3_bv.range(20,20);
bit4out20_bv.range(2,2) = in2_bv.range(20,20);
bit4out20_bv.range(1,1) = in1_bv.range(20,20);
```

**Fig. 8.11** (continued)

```
bit4out20_bv.range(0,0) = in0_bv.range(20,20);
bit4out19_bv.range(3,3) = in3_bv.range(19,19);
bit4out19_bv.range(2,2) = in2_bv.range(19,19);
bit4out19_bv.range(1,1) = in1_bv.range(19,19);
bit4out19_bv.range(0,0) = in0_bv.range(19,19);
bit4out18_bv.range(3,3) = in3_bv.range(18,18);
bit4out18_bv.range(2,2) = in2_bv.range(18,18);
bit4out18_bv.range(1,1) = in1_bv.range(18,18);
bit4out18_bv.range(0,0) = in0_bv.range(18,18);
bit4out17_bv.range(3,3) = in3_bv.range(17,17);
bit4out17_bv.range(2,2) = in2_bv.range(17,17);
bit4out17_bv.range(1,1) = in1_bv.range(17,17);
bit4out17_bv.range(0,0) = in0_bv.range(17,17);
bit4out16_bv.range(3,3) = in3_bv.range(16,16);
bit4out16_bv.range(2,2) = in2_bv.range(16,16);
bit4out16_bv.range(1,1) = in1_bv.range(16,16);
bit4out16_bv.range(0,0) = in0_bv.range(16,16);
bit4out15_bv.range(3,3) = in3_bv.range(15,15);
bit4out15_bv.range(2,2) = in2_bv.range(15,15);
bit4out15_bv.range(1,1) = in1_bv.range(15,15);
bit4out15_bv.range(0,0) = in0_bv.range(15,15);
bit4out14_bv.range(3,3) = in3_bv.range(14,14);
bit4out14_bv.range(2,2) = in2_bv.range(14,14);
bit4out14_bv.range(1,1) = in1_bv.range(14,14);
bit4out14_bv.range(0,0) = in0_bv.range(14,14);
bit4out13_bv.range(3,3) = in3_bv.range(13,13);
bit4out13_bv.range(2,2) = in2_bv.range(13,13);
bit4out13_bv.range(1,1) = in1_bv.range(13,13);
bit4out13_bv.range(0,0) = in0_bv.range(13,13);

bit4out12_bv.range(3,3) = in3_bv.range(12,12);
bit4out12_bv.range(2,2) = in2_bv.range(12,12);
bit4out12_bv.range(1,1) = in1_bv.range(12,12);
bit4out12_bv.range(0,0) = in0_bv.range(12,12);
bit4out11_bv.range(3,3) = in3_bv.range(11,11);
bit4out11_bv.range(2,2) = in2_bv.range(11,11);
bit4out11_bv.range(1,1) = in1_bv.range(11,11);
bit4out11_bv.range(0,0) = in0_bv.range(11,11);
bit4out10_bv.range(3,3) = in3_bv.range(10,10);
bit4out10_bv.range(2,2) = in2_bv.range(10,10);
bit4out10_bv.range(1,1) = in1_bv.range(10,10);
```

**Fig. 8.11** (continued)

```
bit4out10_bv.range(0,0) = in0_bv.range(10,10);
bit4out9_bv.range(3,3) = in3_bv.range(9,9);
bit4out9_bv.range(2,2) = in2_bv.range(9,9);
bit4out9_bv.range(1,1) = in1_bv.range(9,9);
bit4out9_bv.range(0,0) = in0_bv.range(9,9);
bit4out8_bv.range(3,3) = in3_bv.range(8,8);
bit4out8_bv.range(2,2) = in2_bv.range(8,8);
bit4out8_bv.range(1,1) = in1_bv.range(8,8);
bit4out8_bv.range(0,0) = in0_bv.range(8,8);
bit4out7_bv.range(3,3) = in3_bv.range(7,7);
bit4out7_bv.range(2,2) = in2_bv.range(7,7);
bit4out7_bv.range(1,1) = in1_bv.range(7,7);
bit4out7_bv.range(0,0) = in0_bv.range(7,7);
bit4out6_bv.range(3,3) = in3_bv.range(6,6);
bit4out6_bv.range(2,2) = in2_bv.range(6,6);
bit4out6_bv.range(1,1) = in1_bv.range(6,6);
bit4out6_bv.range(0,0) = in0_bv.range(6,6);
bit4out5_bv.range(3,3) = in3_bv.range(5,5);
bit4out5_bv.range(2,2) = in2_bv.range(5,5);
bit4out5_bv.range(1,1) = in1_bv.range(5,5);
bit4out5_bv.range(0,0) = in0_bv.range(5,5);
bit4out4_bv.range(3,3) = in3_bv.range(4,4);
bit4out4_bv.range(2,2) = in2_bv.range(4,4);
bit4out4_bv.range(1,1) = in1_bv.range(4,4);
bit4out4_bv.range(0,0) = in0_bv.range(4,4);
bit4out3_bv.range(3,3) = in3_bv.range(3,3);
bit4out3_bv.range(2,2) = in2_bv.range(3,3);
bit4out3_bv.range(1,1) = in1_bv.range(3,3);
bit4out3_bv.range(0,0) = in0_bv.range(3,3);

bit4out2_bv.range(3,3) = in3_bv.range(2,2);
bit4out2_bv.range(2,2) = in2_bv.range(2,2);
bit4out2_bv.range(1,1) = in1_bv.range(2,2);
bit4out2_bv.range(0,0) = in0_bv.range(2,2);
bit4out1_bv.range(3,3) = in3_bv.range(1,1);
bit4out1_bv.range(2,2) = in2_bv.range(1,1);
bit4out1_bv.range(1,1) = in1_bv.range(1,1);
bit4out1_bv.range(0,0) = in0_bv.range(1,1);
bit4out0_bv.range(3,3) = in3_bv.range(0,0);
bit4out0_bv.range(2,2) = in2_bv.range(0,0);
bit4out0_bv.range(1,1) = in1_bv.range(0,0);
```

**Fig. 8.11** (continued)

```
    bit4out0_bv.range(0,0) = in0_bv.range(0,0);

/* Write 4 bit bit-vectors to output ports */
    bit4out0.write(bit4out0_bv);
    bit4out1.write(bit4out1_bv);
    bit4out2.write(bit4out2_bv);
    bit4out3.write(bit4out3_bv);
    bit4out4.write(bit4out4_bv);
    bit4out5.write(bit4out5_bv);
    bit4out6.write(bit4out6_bv);
    bit4out7.write(bit4out7_bv);
    bit4out8.write(bit4out8_bv);
    bit4out9.write(bit4out9_bv);
    bit4out10.write(bit4out10_bv);
    bit4out11.write(bit4out11_bv);
    bit4out12.write(bit4out12_bv);
    bit4out13.write(bit4out13_bv);
    bit4out14.write(bit4out14_bv);
    bit4out15.write(bit4out15_bv);
    bit4out16.write(bit4out16_bv);
    bit4out17.write(bit4out17_bv);
    bit4out18.write(bit4out18_bv);
    bit4out19.write(bit4out19_bv);
    bit4out20.write(bit4out20_bv);
    bit4out21.write(bit4out21_bv);
    bit4out22.write(bit4out22_bv);
    bit4out23.write(bit4out23_bv);
    bit4out24.write(bit4out24_bv);
    bit4out25.write(bit4out25_bv);
    bit4out26.write(bit4out26_bv);
    bit4out27.write(bit4out27_bv);
    bit4out28.write(bit4out28_bv);
    bit4out29.write(bit4out29_bv);
    bit4out30.write(bit4out30_bv);
    bit4out31.write(bit4out31_bv);
    bit4out32.write(bit4out32_bv);
    bit4out33.write(bit4out33_bv);
    bit4out34.write(bit4out34_bv);
    bit4out35.write(bit4out35_bv);
    bit4out36.write(bit4out36_bv);
    bit4out37.write(bit4out37_bv);
```

**Fig. 8.11** (continued)

```
        bit4out38.write(bit4out38_bv);
        bit4out39.write(bit4out39_bv);
        bit4out40.write(bit4out40_bv);
        bit4out41.write(bit4out41_bv);
        bit4out42.write(bit4out42_bv);
        bit4out43.write(bit4out43_bv);
        bit4out44.write(bit4out44_bv);
        bit4out45.write(bit4out45_bv);
        bit4out46.write(bit4out46_bv);
        bit4out47.write(bit4out47_bv);
        bit4out48.write(bit4out48_bv);
        bit4out49.write(bit4out49_bv);
        bit4out50.write(bit4out50_bv);
        bit4out51.write(bit4out51_bv);
        bit4out52.write(bit4out52_bv);
        bit4out53.write(bit4out53_bv);
        bit4out54.write(bit4out54_bv);
        bit4out55.write(bit4out55_bv);
        bit4out56.write(bit4out56_bv);
        bit4out57.write(bit4out57_bv);
        bit4out58.write(bit4out58_bv);
        bit4out59.write(bit4out59_bv);
        bit4out60.write(bit4out60_bv);
        bit4out61.write(bit4out61_bv);
        bit4out62.write(bit4out62_bv);
        bit4out63.write(bit4out63_bv);
      }
    }

    /* Constructor */
    SC_CTOR(splitter)
    {
     /* Declare/assign thread */
      SC_THREAD(splitter_proc0);
      sensitive << in0 << in1 << in2 << in3; /* Sensitivity list for thread */
      dont_initialize();
    }

   /* Destructor */
    ~splitter(){ }
};
```

**Fig. 8.11** (continued)

defined—Fig. 8.12. This means that for each scoreboard register, two helper modules (splitter and the wrapper) must be used for accurate register operation.

The test harness for the scoreboard register is in Fig. 8.13, and the input to and output from a single 64-cell scoreboard register is in Figs. 8.14 (SystemC-2.2.0) and 8.15 (SystemC-2.3.0).

## 8.3 T2 256 × 132 Asynchronous Memory Array

A memory block presents a curious issue to the designer. Two essential input signals—*read clock* and *write clock*—are sequential. However, the read/write enable signals are asynchronous, in that data is read in/out only when the processor requires it, and there could be conflicts—concurrent read/write requests to same memory location, which the memory controller must handle correctly to maintain data integrity. A 256 x 132 asynchronous memory block from the open-source T2 [3] microprocessor is examined here. As mentioned earlier, judicious use of sensitivity lists allows the designer to reconcile the conflicting demands of the combinational and sequential sub-circuits. Figures 8.16, 8.17 and 8.18 (SystemC-2.2.0) and 8.19 (SystemC-2.3.0) contain the source code, test harness, and partial memory timing diagram, respectively.

## 8.4 T2 64 × 45 Content Addressable Memory Array

A common memory component associated with all modern microprocessors is the content addressable memory (CAM)—also known as associative array/memory/storage. It is used for very high-speed searches. Unlike the RAM in which the user supplies a memory address and the RAM returns the data word stored at that address, for a CAM, the user supplies a data word and the CAM searches its entire memory to see if that data word is stored anywhere in it. If the data word is found, the CAM returns a list of one or more storage addresses where the word was found (and in some architectures, it also returns the data word, or other associated pieces of data). A CAM is the hardware embodiment of a software *hashtable*. The operation of a simple CAM block of the T2 [3] processor is analyzed here. This CAM block can contain 64 entries, each 45 bits wide. Figures 8.20, 8.21 and 8.22 (SystemC-2.2.0) and 8.23 (SystemC-2.3.0) contain the source code, the test harness, and sample input/output traces for this CAM.

## 8.5 Triangle Wave Carrier, DC Modulator Pulse Width Modulation

Pulse width modulation (PWM) is a very efficient and popular control scheme used in a wide variety of electronic equipment as switched mode power supplies (SMPS), DC–AC inverters, industrial electric motor control, monolithic buck/

```
SC_MODULE(register64celldriver)
{
  /* Declare/define all input/output ports  for 64-cell scoreboard register */
  sc_core::sc_in<bool> clk;
  /* Clock input for latches and common select input for each regiister cell */
  sc_core::sc_in< sc_dt::sc_bv<5> > sel;
  sc_core::sc_in< sc_dt::sc_bv<1> > reg_avail;
  sc_core::sc_out< sc_dt::sc_bv<1> > rdy;
  /* Scoreboard register output */
  /* Input port for 4 bit data input for each cell, and intermediate data
channels */
  sc_core::sc_in< sc_dt::sc_bv<4> > din_63;
  sc_core::sc_signal< sc_dt::sc_bv<1> > new_reg_63;

  sc_core::sc_in< sc_dt::sc_bv<4> > din_62;
  sc_core::sc_signal< sc_dt::sc_bv<1> > new_reg_62;

  sc_core::sc_in< sc_dt::sc_bv<4> > din_61;
  sc_core::sc_signal< sc_dt::sc_bv<1> > new_reg_61;

  sc_core::sc_in< sc_dt::sc_bv<4> > din_60;
  sc_core::sc_signal< sc_dt::sc_bv<1> > new_reg_60;

  sc_core::sc_in< sc_dt::sc_bv<4> > din_59;
  sc_core::sc_signal< sc_dt::sc_bv<1> > new_reg_59;

  sc_core::sc_in< sc_dt::sc_bv<4> > din_58;
  sc_core::sc_signal< sc_dt::sc_bv<1> > new_reg_58;

  sc_core::sc_in< sc_dt::sc_bv<4> > din_57;
  sc_core::sc_signal< sc_dt::sc_bv<1> > new_reg_57;

  sc_core::sc_in< sc_dt::sc_bv<4> > din_56;
  sc_core::sc_signal< sc_dt::sc_bv<1> > new_reg_56;

  sc_core::sc_in< sc_dt::sc_bv<4> > din_55;
  sc_core::sc_signal< sc_dt::sc_bv<1> > new_reg_55;
```

**Fig. 8.12** Wrapper module for each 64-cell scoreboard register. Each scoreboard register has an associated splitter and wrapper module

```
sc_core::sc_in< sc_dt::sc_bv<4> > din_54;
sc_core::sc_signal< sc_dt::sc_bv<1> > new_reg_54;

sc_core::sc_in< sc_dt::sc_bv<4> > din_53;
sc_core::sc_signal< sc_dt::sc_bv<1> > new_reg_53;

sc_core::sc_in< sc_dt::sc_bv<4> > din_52;
sc_core::sc_signal< sc_dt::sc_bv<1> > new_reg_52;

sc_core::sc_in< sc_dt::sc_bv<4> > din_51;
sc_core::sc_signal< sc_dt::sc_bv<1> > new_reg_51;

sc_core::sc_in< sc_dt::sc_bv<4> > din_50;
sc_core::sc_signal< sc_dt::sc_bv<1> > new_reg_50;

sc_core::sc_in< sc_dt::sc_bv<4> > din_49;
sc_core::sc_signal< sc_dt::sc_bv<1> > new_reg_49;

sc_core::sc_in< sc_dt::sc_bv<4> > din_48;
sc_core::sc_signal< sc_dt::sc_bv<1> > new_reg_48;

sc_core::sc_in< sc_dt::sc_bv<4> > din_47;
sc_core::sc_signal< sc_dt::sc_bv<1> > new_reg_47;

sc_core::sc_in< sc_dt::sc_bv<4> > din_46;
sc_core::sc_signal< sc_dt::sc_bv<1> > new_reg_46;

sc_core::sc_in< sc_dt::sc_bv<4> > din_45;
sc_core::sc_signal< sc_dt::sc_bv<1> > new_reg_45;

sc_core::sc_in< sc_dt::sc_bv<4> > din_44;
sc_core::sc_signal< sc_dt::sc_bv<1> > new_reg_44;

sc_core::sc_in< sc_dt::sc_bv<4> > din_43;
sc_core::sc_signal< sc_dt::sc_bv<1> > new_reg_43;

sc_core::sc_in< sc_dt::sc_bv<4> > din_42;
sc_core::sc_signal< sc_dt::sc_bv<1> > new_reg_42;

sc_core::sc_in< sc_dt::sc_bv<4> > din_41;
```

**Fig. 8.12** (continued)

```
sc_core::sc_signal< sc_dt::sc_bv<1> > new_reg_41;

sc_core::sc_in< sc_dt::sc_bv<4> > din_40;
sc_core::sc_signal< sc_dt::sc_bv<1> > new_reg_40;

sc_core::sc_in< sc_dt::sc_bv<4> > din_39;
sc_core::sc_signal< sc_dt::sc_bv<1> > new_reg_39;

sc_core::sc_in< sc_dt::sc_bv<4> > din_38;
sc_core::sc_signal< sc_dt::sc_bv<1> > new_reg_38;

sc_core::sc_in< sc_dt::sc_bv<4> > din_37;
sc_core::sc_signal< sc_dt::sc_bv<1> > new_reg_37;

sc_core::sc_in< sc_dt::sc_bv<4> > din_36;
sc_core::sc_signal< sc_dt::sc_bv<1> > new_reg_36;

sc_core::sc_in< sc_dt::sc_bv<4> > din_35;
sc_core::sc_signal< sc_dt::sc_bv<1> > new_reg_35;

sc_core::sc_in< sc_dt::sc_bv<4> > din_34;
sc_core::sc_signal< sc_dt::sc_bv<1> > new_reg_34;

sc_core::sc_in< sc_dt::sc_bv<4> > din_33;
sc_core::sc_signal< sc_dt::sc_bv<1> > new_reg_33;

sc_core::sc_in< sc_dt::sc_bv<4> > din_32;
sc_core::sc_signal< sc_dt::sc_bv<1> > new_reg_32;

sc_core::sc_in< sc_dt::sc_bv<4> > din_31;
sc_core::sc_signal< sc_dt::sc_bv<1> > new_reg_31;

sc_core::sc_in< sc_dt::sc_bv<4> > din_30;
sc_core::sc_signal< sc_dt::sc_bv<1> > new_reg_30;

sc_core::sc_in< sc_dt::sc_bv<4> > din_29;
sc_core::sc_signal< sc_dt::sc_bv<1> > new_reg_29;

sc_core::sc_in< sc_dt::sc_bv<4> > din_28;
sc_core::sc_signal< sc_dt::sc_bv<1> > new_reg_28;
```

**Fig. 8.12** (continued)

```
sc_core::sc_in< sc_dt::sc_bv<4> > din_27;
sc_core::sc_signal< sc_dt::sc_bv<1> > new_reg_27;

sc_core::sc_in< sc_dt::sc_bv<4> > din_26;
sc_core::sc_signal< sc_dt::sc_bv<1> > new_reg_26;

sc_core::sc_in< sc_dt::sc_bv<4> > din_25;
sc_core::sc_signal< sc_dt::sc_bv<1> > new_reg_25;

sc_core::sc_in< sc_dt::sc_bv<4> > din_24;
sc_core::sc_signal< sc_dt::sc_bv<1> > new_reg_24;

sc_core::sc_in< sc_dt::sc_bv<4> > din_23;
sc_core::sc_signal< sc_dt::sc_bv<1> > new_reg_23;

sc_core::sc_in< sc_dt::sc_bv<4> > din_22;
sc_core::sc_signal< sc_dt::sc_bv<1> > new_reg_22;

sc_core::sc_in< sc_dt::sc_bv<4> > din_21;
sc_core::sc_signal< sc_dt::sc_bv<1> > new_reg_21;

sc_core::sc_in< sc_dt::sc_bv<4> > din_20;
sc_core::sc_signal< sc_dt::sc_bv<1> > new_reg_20;

sc_core::sc_in< sc_dt::sc_bv<4> > din_19;
sc_core::sc_signal< sc_dt::sc_bv<1> > new_reg_19;

sc_core::sc_in< sc_dt::sc_bv<4> > din_18;
sc_core::sc_signal< sc_dt::sc_bv<1> > new_reg_18;

sc_core::sc_in< sc_dt::sc_bv<4> > din_17;
sc_core::sc_signal< sc_dt::sc_bv<1> > new_reg_17;

sc_core::sc_in< sc_dt::sc_bv<4> > din_16;
sc_core::sc_signal< sc_dt::sc_bv<1> > new_reg_16;

sc_core::sc_in< sc_dt::sc_bv<4> > din_15;
sc_core::sc_signal< sc_dt::sc_bv<1> > new_reg_15;
```

**Fig. 8.12** (continued)

```
sc_core::sc_in< sc_dt::sc_bv<4> > din_14;
sc_core::sc_signal< sc_dt::sc_bv<1> > new_reg_14;

sc_core::sc_in< sc_dt::sc_bv<4> > din_13;
sc_core::sc_signal< sc_dt::sc_bv<1> > new_reg_13;

sc_core::sc_in< sc_dt::sc_bv<4> > din_12;
sc_core::sc_signal< sc_dt::sc_bv<1> > new_reg_12;

sc_core::sc_in< sc_dt::sc_bv<4> > din_11;
sc_core::sc_signal< sc_dt::sc_bv<1> > new_reg_11;

sc_core::sc_in< sc_dt::sc_bv<4> > din_10;
sc_core::sc_signal< sc_dt::sc_bv<1> > new_reg_10;

sc_core::sc_in< sc_dt::sc_bv<4> > din_9;
sc_core::sc_signal< sc_dt::sc_bv<1> > new_reg_9;

sc_core::sc_in< sc_dt::sc_bv<4> > din_8;
sc_core::sc_signal< sc_dt::sc_bv<1> > new_reg_8;

sc_core::sc_in< sc_dt::sc_bv<4> > din_7;
sc_core::sc_signal< sc_dt::sc_bv<1> > new_reg_7;

sc_core::sc_in< sc_dt::sc_bv<4> > din_6;
sc_core::sc_signal< sc_dt::sc_bv<1> > new_reg_6;

sc_core::sc_in< sc_dt::sc_bv<4> > din_5;
sc_core::sc_signal< sc_dt::sc_bv<1> > new_reg_5;

sc_core::sc_in< sc_dt::sc_bv<4> > din_4;
sc_core::sc_signal< sc_dt::sc_bv<1> > new_reg_4;

sc_core::sc_in< sc_dt::sc_bv<4> > din_3;
sc_core::sc_signal< sc_dt::sc_bv<1> > new_reg_3;

sc_core::sc_in< sc_dt::sc_bv<4> > din_2;
sc_core::sc_signal< sc_dt::sc_bv<1> > new_reg_2;
```

**Fig. 8.12** (continued)

```
  sc_core::sc_in< sc_dt::sc_bv<4> > din_1;
  sc_core::sc_signal< sc_dt::sc_bv<1> > new_reg_1;

  sc_core::sc_in< sc_dt::sc_bv<4> > din_0;
  sc_core::sc_signal< sc_dt::sc_bv<1> > new_reg_0;

/* Bit vectors for intermediate data storage */
  sc_dt::sc_bv<64> rdy_bv;
  sc_dt::sc_bv<1> rdy_1_bv;
  sc_dt::sc_bv<1> rdy_2_bv;
  sc_dt::sc_bv<1> reg_avail_bv;

  register64cell reg_64_cell;  /* Wrapped 64 cell scoreboard register */

  void register64celldriver_proc0()  /* Main operation thread */
  {
    while(1)
    {
     wait();
     reg_avail_bv = reg_avail.read();
    /* Concatenate outputs from each register cell and perform final NOR and
AND operations */
     rdy_bv = (new_reg_63.read(), new_reg_62.read(), new_reg_61.read(),
new_reg_60.read(), new_reg_59.read(), new_reg_58.read(), new_reg_57.read(),
new_reg_56.read(), new_reg_55.read(), new_reg_54.read(), new_reg_53.read(),
new_reg_52.read(), new_reg_51.read(), new_reg_50.read(), new_reg_49.read(),
new_reg_48.read(), new_reg_47.read(), new_reg_46.read(), new_reg_45.read(),
new_reg_44.read(), new_reg_43.read(), new_reg_42.read(), new_reg_41.read(),
new_reg_40.read(), new_reg_39.read(), new_reg_38.read(), new_reg_37.read(),
new_reg_36.read(), new_reg_36.read(), new_reg_35.read(), new_reg_34.read(),
new_reg_33.read(), new_reg_32.read(), new_reg_31.read(), new_reg_30.read(),
new_reg_29.read(), new_reg_28.read(), new_reg_27.read(), new_reg_26.read(),
new_reg_25.read(), new_reg_24.read(), new_reg_23.read(), new_reg_22.read(),
new_reg_21.read(), new_reg_20.read(), new_reg_19.read(), new_reg_18.read(),
new_reg_17.read(), new_reg_16.read(), new_reg_15.read(), new_reg_14.read(),
new_reg_13.read(), new_reg_12.read(), new_reg_11.read(), new_reg_10.read(),
new_reg_9.read(), new_reg_9.read(), new_reg_7.read(), new_reg_6.read(),
new_reg_5.read(), new_reg_4.read(), new_reg_3.read(), new_reg_2.read(),
new_reg_1.read(), new_reg_0.read());
```

**Fig. 8.12** (continued)

```
      rdy_bv &= statusreg;
      rdy_1_bv = rdy_bv.or_reduce();
      rdy_2_bv = rdy_1_bv[0] == "1" ? "1" : "0";
      rdy_2_bv &= reg_avail_bv;
      rdy.write(rdy_2_bv);
    }
  }

  /* Constructor */
  SC_CTOR(register64celldriver):reg_64_cell("reg_64_cell")
  {
    /* Connect input/output ports to/from wrapped register and internal
channels */
    reg_64_cell.clk(clk);
    reg_64_cell.din_63(din_63);
    reg_64_cell.din_62(din_62);
    reg_64_cell.din_61(din_61);
    reg_64_cell.din_60(din_60);
    reg_64_cell.din_59(din_59);
    reg_64_cell.din_58(din_58);
    reg_64_cell.din_57(din_57);
    reg_64_cell.din_56(din_56);
    reg_64_cell.din_55(din_55);
    reg_64_cell.din_54(din_54);
    reg_64_cell.din_53(din_53);
    reg_64_cell.din_52(din_52);
    reg_64_cell.din_51(din_51);
    reg_64_cell.din_50(din_50);
    reg_64_cell.din_49(din_49);
    reg_64_cell.din_48(din_48);
    reg_64_cell.din_47(din_47);
    reg_64_cell.din_46(din_46);
    reg_64_cell.din_45(din_45);
    reg_64_cell.din_44(din_44);
    reg_64_cell.din_43(din_43);
    reg_64_cell.din_42(din_42);
    reg_64_cell.din_41(din_41);
    reg_64_cell.din_40(din_40);
    reg_64_cell.din_39(din_39);
    reg_64_cell.din_38(din_38);
    reg_64_cell.din_37(din_37);
```

**Fig. 8.12** (continued)

```
    reg_64_cell.din_36(din_36);
    reg_64_cell.din_35(din_35);
    reg_64_cell.din_34(din_34);
    reg_64_cell.din_33(din_33);
    reg_64_cell.din_32(din_32);
    reg_64_cell.din_31(din_31);
    reg_64_cell.din_30(din_30);
    reg_64_cell.din_29(din_29);
    reg_64_cell.din_28(din_28);
    reg_64_cell.din_27(din_27);
    reg_64_cell.din_26(din_26);
    reg_64_cell.din_25(din_25);
    reg_64_cell.din_24(din_24);
    reg_64_cell.din_23(din_23);
    reg_64_cell.din_22(din_22);
    reg_64_cell.din_21(din_21);
    reg_64_cell.din_20(din_20);
    reg_64_cell.din_19(din_19);
    reg_64_cell.din_18(din_18);
    reg_64_cell.din_17(din_17);
    reg_64_cell.din_16(din_16);
    reg_64_cell.din_15(din_15);
    reg_64_cell.din_14(din_14);
    reg_64_cell.din_13(din_13);
    reg_64_cell.din_12(din_12);
    reg_64_cell.din_11(din_11);
    reg_64_cell.din_10(din_10);
    reg_64_cell.din_9(din_9);
    reg_64_cell.din_8(din_8);
    reg_64_cell.din_7(din_7);
    reg_64_cell.din_6(din_6);
    reg_64_cell.din_5(din_5);
    reg_64_cell.din_4(din_4);
    reg_64_cell.din_3(din_3);
    reg_64_cell.din_2(din_2);
    reg_64_cell.din_1(din_1);
    reg_64_cell.din_0(din_0);
    /* Connect select input port of 64 cell register to warpper module select
input port */
    reg_64_cell.sel(sel);
```

**Fig. 8.12** (continued)

```
/* Connect output of each cell to intermediate channel */
    reg_64_cell.new_reg_63(new_reg_63);
    reg_64_cell.new_reg_62(new_reg_62);
    reg_64_cell.new_reg_61(new_reg_61);
    reg_64_cell.new_reg_60(new_reg_60);
    reg_64_cell.new_reg_59(new_reg_59);
    reg_64_cell.new_reg_58(new_reg_58);
    reg_64_cell.new_reg_57(new_reg_57);
    reg_64_cell.new_reg_56(new_reg_56);
    reg_64_cell.new_reg_55(new_reg_55);
    reg_64_cell.new_reg_54(new_reg_54);
    reg_64_cell.new_reg_53(new_reg_53);
    reg_64_cell.new_reg_52(new_reg_52);
    reg_64_cell.new_reg_51(new_reg_51);
    reg_64_cell.new_reg_50(new_reg_50);
    reg_64_cell.new_reg_49(new_reg_49);
    reg_64_cell.new_reg_48(new_reg_48);
    reg_64_cell.new_reg_47(new_reg_47);
    reg_64_cell.new_reg_46(new_reg_46);
    reg_64_cell.new_reg_45(new_reg_45);
    reg_64_cell.new_reg_44(new_reg_44);
    reg_64_cell.new_reg_43(new_reg_43);
    reg_64_cell.new_reg_42(new_reg_42);
    reg_64_cell.new_reg_41(new_reg_41);
    reg_64_cell.new_reg_40(new_reg_40);
    reg_64_cell.new_reg_39(new_reg_39);
    reg_64_cell.new_reg_38(new_reg_38);
    reg_64_cell.new_reg_37(new_reg_37);
    reg_64_cell.new_reg_36(new_reg_36);
    reg_64_cell.new_reg_35(new_reg_35);
    reg_64_cell.new_reg_34(new_reg_34);
    reg_64_cell.new_reg_33(new_reg_33);
    reg_64_cell.new_reg_32(new_reg_32);
    reg_64_cell.new_reg_31(new_reg_31);
    reg_64_cell.new_reg_30(new_reg_30);
    reg_64_cell.new_reg_29(new_reg_29);
    reg_64_cell.new_reg_28(new_reg_28);
    reg_64_cell.new_reg_27(new_reg_27);
    reg_64_cell.new_reg_26(new_reg_26);
    reg_64_cell.new_reg_25(new_reg_25);
    reg_64_cell.new_reg_24(new_reg_24);
```

**Fig. 8.12** (continued)

```
    reg_64_cell.new_reg_23(new_reg_23);
    reg_64_cell.new_reg_22(new_reg_22);
    reg_64_cell.new_reg_21(new_reg_21);
    reg_64_cell.new_reg_20(new_reg_20);
    reg_64_cell.new_reg_19(new_reg_19);
    reg_64_cell.new_reg_18(new_reg_18);
    reg_64_cell.new_reg_17(new_reg_17);
    reg_64_cell.new_reg_16(new_reg_16);
    reg_64_cell.new_reg_15(new_reg_15);
    reg_64_cell.new_reg_14(new_reg_14);
    reg_64_cell.new_reg_13(new_reg_13);
    reg_64_cell.new_reg_12(new_reg_12);
    reg_64_cell.new_reg_11(new_reg_11);
    reg_64_cell.new_reg_10(new_reg_10);
    reg_64_cell.new_reg_9(new_reg_9);
    reg_64_cell.new_reg_8(new_reg_8);
    reg_64_cell.new_reg_7(new_reg_7);
    reg_64_cell.new_reg_6(new_reg_6);
    reg_64_cell.new_reg_5(new_reg_5);
    reg_64_cell.new_reg_4(new_reg_4);
    reg_64_cell.new_reg_3(new_reg_3);
    reg_64_cell.new_reg_2(new_reg_2);
    reg_64_cell.new_reg_1(new_reg_1);
    reg_64_cell.new_reg_0(new_reg_0);

    SC_THREAD(register64celldriver_proc0);
  /*Declare/assign thread and sensitivity list */
    sensitive << sel << din_0 << din_1 << din_2 << din_3 << din_4 << din_5 <<
  din_6 << din_7 << din_8 << din_9 << din_10 << din_11 << din_12 << din_13 <<
  din_14 << din_15 << din_16 << din_17 << din_18 << din_19 << din_20 << din_21
  << din_22 << din_23 << din_24 << din_25 << din_26 << din_27 << din_28 <<
  din_29 << din_30 << din_31 << din_32 << din_33 << din_34 << din_35 << din_36
  << din_37 << din_38 << din_39 << din_40 << din_41 << din_42 << din_43 <<
  din_44 << din_45 << din_46 << din_47 << din_48 << din_49 << din_50 << din_51
  << din_52 << din_53 << din_54 << din_55 << din_56 << din_57 << din_58 <<
  din_59 << din_60 << din_61 << din_62 << din_63 << clk.pos() << clk.neg();

  }
  /*Destructor */
  ~register64celldriver() { }
};
```

**Fig. 8.12** (continued)

```
#include "scoreboard.h"
#include <cstdlib>
#include <cstring>

int sc_main(int argc, char **argv)
{
  /* Declare/define internal channels */
  sc_core::sc_signal< sc_dt::sc_bv<64> > in0_63;
  sc_core::sc_signal< sc_dt::sc_bv<64> > in1_63;
  sc_core::sc_signal< sc_dt::sc_bv<64> > in2_63;
  sc_core::sc_signal< sc_dt::sc_bv<64> > in3_63;

  /*Declare/define channels to transfer 4 bit
    bit-vectors from splitter to scoreboard register wrapper */
  sc_core::sc_signal< sc_dt::sc_bv<4> > bit4out0_63;
  sc_core::sc_signal< sc_dt::sc_bv<4> > bit4out1_63;
  sc_core::sc_signal< sc_dt::sc_bv<4> > bit4out2_63;
  sc_core::sc_signal< sc_dt::sc_bv<4> > bit4out3_63;
  sc_core::sc_signal< sc_dt::sc_bv<4> > bit4out4_63;
  sc_core::sc_signal< sc_dt::sc_bv<4> > bit4out5_63;
  sc_core::sc_signal< sc_dt::sc_bv<4> > bit4out6_63;
  sc_core::sc_signal< sc_dt::sc_bv<4> > bit4out7_63;
  sc_core::sc_signal< sc_dt::sc_bv<4> > bit4out8_63;
  sc_core::sc_signal< sc_dt::sc_bv<4> > bit4out9_63;
  sc_core::sc_signal< sc_dt::sc_bv<4> > bit4out10_63;
  sc_core::sc_signal< sc_dt::sc_bv<4> > bit4out11_63;
  sc_core::sc_signal< sc_dt::sc_bv<4> > bit4out12_63;
  sc_core::sc_signal< sc_dt::sc_bv<4> > bit4out13_63;
  sc_core::sc_signal< sc_dt::sc_bv<4> > bit4out14_63;
  sc_core::sc_signal< sc_dt::sc_bv<4> > bit4out15_63;
  sc_core::sc_signal< sc_dt::sc_bv<4> > bit4out16_63;
  sc_core::sc_signal< sc_dt::sc_bv<4> > bit4out17_63;
  sc_core::sc_signal< sc_dt::sc_bv<4> > bit4out18_63;
  sc_core::sc_signal< sc_dt::sc_bv<4> > bit4out19_63;
  sc_core::sc_signal< sc_dt::sc_bv<4> > bit4out20_63;
  sc_core::sc_signal< sc_dt::sc_bv<4> > bit4out21_63;
  sc_core::sc_signal< sc_dt::sc_bv<4> > bit4out22_63;
  sc_core::sc_signal< sc_dt::sc_bv<4> > bit4out23_63;
  sc_core::sc_signal< sc_dt::sc_bv<4> > bit4out24_63;
  sc_core::sc_signal< sc_dt::sc_bv<4> > bit4out25_63;
```

**Fig. 8.13** Test harness for splitter, 64-cell register, 64-cell register wrapper

```
sc_core::sc_signal< sc_dt::sc_bv<4> > bit4out26_63;
sc_core::sc_signal< sc_dt::sc_bv<4> > bit4out27_63;
sc_core::sc_signal< sc_dt::sc_bv<4> > bit4out28_63;
sc_core::sc_signal< sc_dt::sc_bv<4> > bit4out29_63;
sc_core::sc_signal< sc_dt::sc_bv<4> > bit4out30_63;
sc_core::sc_signal< sc_dt::sc_bv<4> > bit4out31_63;
sc_core::sc_signal< sc_dt::sc_bv<4> > bit4out32_63;
sc_core::sc_signal< sc_dt::sc_bv<4> > bit4out33_63;
sc_core::sc_signal< sc_dt::sc_bv<4> > bit4out34_63;
sc_core::sc_signal< sc_dt::sc_bv<4> > bit4out35_63;
sc_core::sc_signal< sc_dt::sc_bv<4> > bit4out36_63;
sc_core::sc_signal< sc_dt::sc_bv<4> > bit4out37_63;
sc_core::sc_signal< sc_dt::sc_bv<4> > bit4out38_63;
sc_core::sc_signal< sc_dt::sc_bv<4> > bit4out39_63;
sc_core::sc_signal< sc_dt::sc_bv<4> > bit4out40_63;
sc_core::sc_signal< sc_dt::sc_bv<4> > bit4out41_63;
sc_core::sc_signal< sc_dt::sc_bv<4> > bit4out42_63;
sc_core::sc_signal< sc_dt::sc_bv<4> > bit4out43_63;
sc_core::sc_signal< sc_dt::sc_bv<4> > bit4out44_63;
sc_core::sc_signal< sc_dt::sc_bv<4> > bit4out45_63;
sc_core::sc_signal< sc_dt::sc_bv<4> > bit4out46_63;
sc_core::sc_signal< sc_dt::sc_bv<4> > bit4out47_63;
sc_core::sc_signal< sc_dt::sc_bv<4> > bit4out48_63;
sc_core::sc_signal< sc_dt::sc_bv<4> > bit4out49_63;
sc_core::sc_signal< sc_dt::sc_bv<4> > bit4out50_63;
sc_core::sc_signal< sc_dt::sc_bv<4> > bit4out51_63;
sc_core::sc_signal< sc_dt::sc_bv<4> > bit4out52_63;
sc_core::sc_signal< sc_dt::sc_bv<4> > bit4out53_63;
sc_core::sc_signal< sc_dt::sc_bv<4> > bit4out54_63;
sc_core::sc_signal< sc_dt::sc_bv<4> > bit4out55_63;
sc_core::sc_signal< sc_dt::sc_bv<4> > bit4out56_63;
sc_core::sc_signal< sc_dt::sc_bv<4> > bit4out57_63;
sc_core::sc_signal< sc_dt::sc_bv<4> > bit4out58_63;
sc_core::sc_signal< sc_dt::sc_bv<4> > bit4out59_63;
sc_core::sc_signal< sc_dt::sc_bv<4> > bit4out60_63;
sc_core::sc_signal< sc_dt::sc_bv<4> > bit4out61_63;
sc_core::sc_signal< sc_dt::sc_bv<4> > bit4out62_63;
sc_core::sc_signal< sc_dt::sc_bv<4> > bit4out63_63;
```

**Fig. 8.13** (continued)

```
    sc_core::sc_signal< sc_dt::sc_bv<5> > sel_63;
    sc_core::sc_signal< sc_dt::sc_bv<1> > rdy_out_63;
    sc_core::sc_signal< sc_dt::sc_bv<1> > reg_avail_63;
    sc_dt::sc_bv<5> sel_bv_63;
    sc_dt::sc_bv<1> reg_avail_bv_63;

  /* Declare/define internal bit-vectors */
    sc_dt::sc_bv<64> in0_bv;
    sc_dt::sc_bv<64> in1_bv;
    sc_dt::sc_bv<64> in2_bv;
    sc_dt::sc_bv<64> in3_bv;

    /* Declare/define clock, splitter and scoreboard register wrapper */
    sc_core::sc_clock clk("clk", 4.0, sc_core::SC_NS, 0.5);
    splitter splitter_63("splitter_63");
    register64celldriver reg_driver_63("reg_64_cell_driver_63");

    /* Declare/define trace file generator */
    sc_core::sc_trace_file *fp = sc_core::sc_create_vcd_trace_file("tr_splitter");
    fp->set_time_unit(1.0, sc_core::SC_NS);

    /* Connect splitter input ports and input channels */
    splitter_63.in0(in0_63);
    splitter_63.in1(in1_63);
    splitter_63.in2(in2_63);
    splitter_63.in3(in3_63);

   /* Connect splitter output ports and channels */
    splitter_63.bit4out0(bit4out0_63);
    splitter_63.bit4out1(bit4out1_63);
    splitter_63.bit4out2(bit4out2_63);
    splitter_63.bit4out3(bit4out3_63);
    splitter_63.bit4out4(bit4out4_63);
    splitter_63.bit4out5(bit4out5_63);
    splitter_63.bit4out6(bit4out6_63);
    splitter_63.bit4out7(bit4out7_63);
```

**Fig. 8.13** (continued)

```
splitter_63.bit4out8(bit4out8_63);
splitter_63.bit4out9(bit4out9_63);
splitter_63.bit4out10(bit4out10_63);
splitter_63.bit4out11(bit4out11_63);
splitter_63.bit4out12(bit4out12_63);
splitter_63.bit4out13(bit4out13_63);
splitter_63.bit4out14(bit4out14_63);
splitter_63.bit4out15(bit4out15_63);
splitter_63.bit4out16(bit4out16_63);
splitter_63.bit4out17(bit4out17_63);
splitter_63.bit4out18(bit4out18_63);
splitter_63.bit4out19(bit4out19_63);
splitter_63.bit4out20(bit4out20_63);
splitter_63.bit4out21(bit4out21_63);
splitter_63.bit4out22(bit4out22_63);
splitter_63.bit4out23(bit4out23_63);
splitter_63.bit4out24(bit4out24_63);
splitter_63.bit4out25(bit4out25_63);
splitter_63.bit4out26(bit4out26_63);
splitter_63.bit4out27(bit4out27_63);
splitter_63.bit4out28(bit4out28_63);
splitter_63.bit4out29(bit4out29_63);
splitter_63.bit4out30(bit4out30_63);
splitter_63.bit4out31(bit4out31_63);
splitter_63.bit4out32(bit4out32_63);
splitter_63.bit4out33(bit4out33_63);
splitter_63.bit4out34(bit4out34_63);
splitter_63.bit4out35(bit4out35_63);
splitter_63.bit4out36(bit4out36_63);
splitter_63.bit4out37(bit4out37_63);
splitter_63.bit4out38(bit4out38_63);
splitter_63.bit4out39(bit4out39_63);
splitter_63.bit4out40(bit4out40_63);
splitter_63.bit4out41(bit4out41_63);
splitter_63.bit4out42(bit4out42_63);
splitter_63.bit4out43(bit4out43_63);
splitter_63.bit4out44(bit4out44_63);
splitter_63.bit4out45(bit4out45_63);
splitter_63.bit4out46(bit4out46_63);
splitter_63.bit4out47(bit4out47_63);
splitter_63.bit4out48(bit4out48_63);
```

**Fig. 8.13** (continued)

```
  splitter_63.bit4out49(bit4out49_63);
  splitter_63.bit4out50(bit4out50_63);
  splitter_63.bit4out51(bit4out51_63);
  splitter_63.bit4out52(bit4out52_63);
  splitter_63.bit4out53(bit4out53_63);
  splitter_63.bit4out54(bit4out54_63);
  splitter_63.bit4out55(bit4out55_63);
  splitter_63.bit4out56(bit4out56_63);
  splitter_63.bit4out57(bit4out57_63);
  splitter_63.bit4out58(bit4out58_63);
  splitter_63.bit4out59(bit4out59_63);
  splitter_63.bit4out60(bit4out60_63);
  splitter_63.bit4out61(bit4out61_63);
  splitter_63.bit4out62(bit4out62_63);
  splitter_63.bit4out63(bit4out63_63);

  /* Connect scoreboard wrapper to input clock and data input channels from
splitter */
  reg_driver_63.clk(clk);
  reg_driver_63.din_63(bit4out63_63);
  reg_driver_63.din_62(bit4out62_63);
  reg_driver_63.din_61(bit4out61_63);
  reg_driver_63.din_60(bit4out60_63);
  reg_driver_63.din_59(bit4out59_63);
  reg_driver_63.din_58(bit4out58_63);
  reg_driver_63.din_57(bit4out57_63);
  reg_driver_63.din_56(bit4out56_63);
  reg_driver_63.din_55(bit4out55_63);
  reg_driver_63.din_54(bit4out54_63);
  reg_driver_63.din_53(bit4out53_63);
  reg_driver_63.din_52(bit4out52_63);
  reg_driver_63.din_51(bit4out51_63);
  reg_driver_63.din_50(bit4out50_63);
  reg_driver_63.din_49(bit4out49_63);
  reg_driver_63.din_48(bit4out48_63);
  reg_driver_63.din_47(bit4out47_63);
  reg_driver_63.din_46(bit4out46_63);
  reg_driver_63.din_45(bit4out45_63);
  reg_driver_63.din_44(bit4out44_63);
  reg_driver_63.din_43(bit4out43_63);
  reg_driver_63.din_42(bit4out42_63);
```

**Fig. 8.13** (continued)

```
reg_driver_63.din_41(bit4out41_63);
reg_driver_63.din_40(bit4out40_63);
reg_driver_63.din_39(bit4out39_63);
reg_driver_63.din_38(bit4out38_63);
reg_driver_63.din_37(bit4out37_63);
reg_driver_63.din_36(bit4out36_63);
reg_driver_63.din_35(bit4out35_63);
reg_driver_63.din_34(bit4out34_63);
reg_driver_63.din_33(bit4out33_63);
reg_driver_63.din_32(bit4out32_63);
reg_driver_63.din_31(bit4out31_63);
reg_driver_63.din_30(bit4out30_63);
reg_driver_63.din_29(bit4out29_63);
reg_driver_63.din_28(bit4out28_63);
reg_driver_63.din_27(bit4out27_63);
reg_driver_63.din_26(bit4out26_63);
reg_driver_63.din_25(bit4out25_63);
reg_driver_63.din_24(bit4out24_63);
reg_driver_63.din_23(bit4out23_63);
reg_driver_63.din_22(bit4out22_63);
reg_driver_63.din_21(bit4out21_63);
reg_driver_63.din_20(bit4out20_63);
reg_driver_63.din_19(bit4out19_63);
reg_driver_63.din_18(bit4out18_63);
reg_driver_63.din_17(bit4out17_63);
reg_driver_63.din_16(bit4out16_63);
reg_driver_63.din_15(bit4out15_63);
reg_driver_63.din_14(bit4out14_63);
reg_driver_63.din_13(bit4out13_63);
reg_driver_63.din_12(bit4out12_63);
reg_driver_63.din_11(bit4out11_63);
reg_driver_63.din_10(bit4out10_63);
reg_driver_63.din_9(bit4out9_63);
reg_driver_63.din_8(bit4out8_63);
reg_driver_63.din_7(bit4out7_63);
reg_driver_63.din_6(bit4out6_63);
reg_driver_63.din_5(bit4out5_63);
reg_driver_63.din_4(bit4out4_63);
reg_driver_63.din_3(bit4out3_63);
reg_driver_63.din_2(bit4out2_63);
reg_driver_63.din_1(bit4out1_63);
```

**Fig. 8.13** (continued)

```
  reg_driver_63.din_0(bit4out0_63);
  /* Connect other scoreboard regiater wrapper and other input/output
channels */
  reg_driver_63.sel(sel_63);
  reg_driver_63.reg_avail(reg_avail_63);
  reg_driver_63.rdy(rdy_out_63);

  /* Connect trace file to data channels for plotting */
  sc_core::sc_trace(fp, in0_63, "in0");
  sc_core::sc_trace(fp, in1_63, "in1");
  sc_core::sc_trace(fp, in2_63, "in2");
  sc_core::sc_trace(fp, in3_63, "in3");
  sc_core::sc_trace(fp, rdy_out_63, "rdy_63");
  sc_core::sc_trace(fp, bit4out62_63, "bit4out62_63");
  sc_core::sc_trace(fp, bit4out62_63, "bit4out62_63");

/* Generate 64 bit input bit vectors for splitter etc., */

in0_bv="000000000000000000000000000000000000000000000000000000000
0000010";

in1_bv="000000000000000000000000000000000000000000000000000000000
0000010";

in2_bv="000000000000000000000000000000000000000000000000000000000
0000010";

in3_bv="000000000000000000000000000000000000000000000000000000000
0000010";

  reg_avail_bv_63 = "1";
  sel_bv_63 = "00010";

  /* Write input data to correct channels */
  in0_63.write(in0_bv);
  in1_63.write(in1_bv);
  in2_63.write(in2_bv);
  in3_63.write(in3_bv);
  sel_63.write(sel_bv_63);
  reg_avail_63.write(reg_avail_bv_63);
```

**Fig. 8.13** (continued)

```
  /* Run simulation fro pre-defind time period */
  sc_core::sc_start(2.0, sc_core::SC_NS);

  /* Change 64 bit input bit vectors for splitter etc., */

in0_bv="1011001101110001101001010011100100110011001001100011110010100010";

in1_bv="1001011111010101001101011111101101110001101011101011110110101110";

in2_bv="1010001001111001001011010110010100100111001101101011010110110010";

in3_bv="1011011101110101101011000111100100110111011001101011100011100111";
  reg_avail_bv_63 = "1";
  sel_bv_63 = "00010";

  /* Write input data to correct channels */
  in0_63.write(in0_bv);
  in1_63.write(in1_bv);
  in2_63.write(in2_bv);
  in3_63.write(in3_bv);
  sel_63.write(sel_bv_63);
  reg_avail_63.write(reg_avail_bv_63);

  /* Run simulation fro pre-defind time period */
  sc_core::sc_start(2.0, sc_core::SC_NS);

  /* Change 64 bit input bit vectors for splitter etc., */

in0_bv="0011011101010011101011010011110101111011001001101011100110101010";

in1_bv="1001011001010101101101011001101101010101101010101010110010100110";

in2_bv="0000101001001001001011010110010100100101001101101011110110110010";
```

**Fig. 8.13** (continued)

```
in3_bv="1010011101010101101011010110100100110111011001101011100011100111";
  reg_avail_bv_63 = "1";
  sel_bv_63 = "01000";

  /* Write input data to correct channels */
  in0_63.write(in0_bv);
  in1_63.write(in1_bv);
  in2_63.write(in2_bv);
  in3_63.write(in3_bv);
  sel_63.write(sel_bv_63);
  reg_avail_63.write(reg_avail_bv_63);

 /* Run simulation fro pre-defind time period */
  sc_core::sc_start(2.0, sc_core::SC_NS);

   /* Change 64 bit input bit vectors for splitter etc., */
  in0_bv="0011011101010011101011010011110101

in0_bv="11110111010100111010110100111101011110110010011010111001101010100";

in1_bv="10010110110101011011010110011011010101011010101010101100101001100";

in2_bv="00001010010010010010110101100101001001010011011010111101101100100";

in3_bv="10100111010101011010110101101111001101110110011010111011111000111";
  reg_avail_bv_63 = "1";
  sel_bv_63 = "00010";

   /* Write input data to correct channels */
  in0_63.write(in0_bv);
  in1_63.write(in1_bv);
  in2_63.write(in2_bv);
  in3_63.write(in3_bv);
  sel_63.write(sel_bv_63);
```

**Fig. 8.13** (continued)

```
  reg_avail_63.write(reg_avail_bv_63);

  /* Run simulation fro pre-defind time period *
  sc_core::sc_start(2.0, sc_core::SC_NS);

  /* Stop simulation and close trace file */
  sc_core::sc_stop();
  sc_core::sc_close_vcd_trace_file(fp);
   return 0;
}
```

**Fig. 8.13** (Continued)

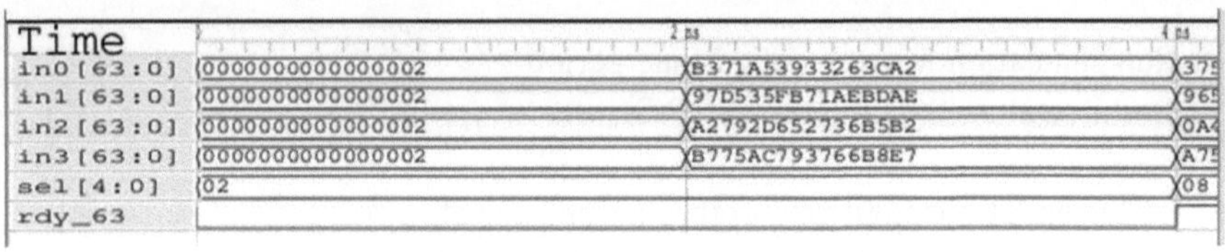


**Fig. 8.14** Trace output from scoreboard register test harness SystemC-2.2.0

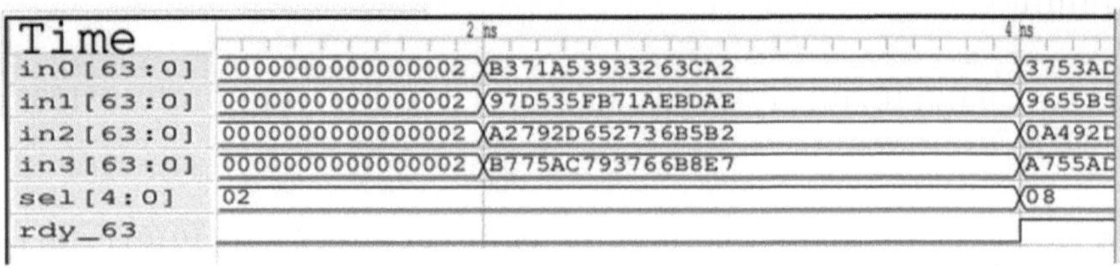


**Fig. 8.15** Trace output from scoreboard register test harness SystemC-2.3.0

```
#ifndef N2_COM_256X132ASYNC_DP_CUST_ARRAY_H
#define N2_COM_256X132ASYNC_DP_CUST_ARRAY_H

#include <systemc>

SC_MODULE(n2_com_256x132async_dp_cust_array)
{
  sc_core::sc_in<bool> wr_clk;
  /* write clk */
  sc_core::sc_in< sc_dt::sc_bv<8> > wr_addr_array;
```

**Fig. 8.16** T2 asynchronous 256 × 132 memory array

```
        /* write address port  */
        sc_core::sc_in<bool> wr_en_array;
        /* write port enable */
        sc_core::sc_in< sc_dt::sc_bv<132> > din_array;
        /* data in */
        sc_core::sc_in<bool> rd_clk;
        /* read clk */
        sc_core::sc_in< sc_dt::sc_bv<8> > rd_addr_array;
        /* read port address  */
        sc_core::sc_in<bool> rd_en_array;
        /* read port enable */
        sc_core::sc_out< sc_dt::sc_bv<132> > dout_array;
        /* data out */

        /* Declare/define memory array and data structures */
        sc_dt::sc_bv<132> array_ram[256];
        sc_dt::sc_bv<132> tmp_dout;
        sc_dt::sc_bv<132> tmp;
        bool notinitialized;
        unsigned int i;

       /* Initialize memory array and contents */
        void initialize_array()
  {
    unsigned int i;
    if(notinitialized == false)
    {

tmp="000000000000000000000000000000000000000000000000000000000000000000
000000000000000000000000000000000000000000000000000000000000000000";

tmp_dout="000000000000000000000000000000000000000000000000000000000000
00000000000000000000000000000000000000000000000000000000000000000000000
0";
      for(i = 0 ; i < 256; i++) { array_ram[i] = tmp; }
      notinitialized = true;
    }
  }
```

**Fig. 8.16** (continued)

```
/* Thread for reading data from a specified memory address */
  void n2_com_256x132async_dp_cust_array_proc0()
  {
    while(1)
    {
      wait();
      if(rd_clk.read() == true)
      {
        if(rd_en_array.read() == true)
        {
          tmp_dout = array_ram[rd_addr_array.read().to_int()];
        }
      }
    }
  }

  /* Thread to control concurrent read/write requests from same memory
address */
  void n2_com_256x132async_dp_cust_array_proc1()
  {
    while(1)
    {
      wait();
      if(rd_clk.read() == true)
      {
      if(rd_en_array.read() == true)
      {
        if((wr_en_array.read() == true) &&
        rd_addr_array.read() == wr_addr_array.read())
        {
          std::cout<<"Simultaneous read/write disallowed ... "<<std::endl;
        }
        else
        {
          tmp_dout = array_ram[rd_addr_array.read().to_int()];
        }
      }
    }
```

**Fig. 8.16** (continued)

```
    }
  }

  /* Thread to write data to specified memory address */
  void n2_com_256x132async_dp_cust_array_proc2()
  {
    while(1)
    {
      wait();
       if(wr_clk.read() == false)
       {
         if(wr_en_array.read() == true)
         {
           tmp = din_array.read();
           array_ram[wr_addr_array.read().to_int()] = tmp;
         }
       }
    }
  }

  /* Transfer output data from temporary store to output port */
  void n2_com_256x132async_dp_cust_array_proc3()
  {
    dout_array.write(tmp_dout);
  }

  /* Constructor - initialize members */
  SC_CTOR(n2_com_256x132async_dp_cust_array):i(0),

          notinitialized(false)
   {
     /* Declare/assign methods and threads and their sensitivity lists */
      initialize_array();
      SC_METHOD(n2_com_256x132async_dp_cust_array_proc3);
      SC_THREAD(n2_com_256x132async_dp_cust_array_proc0);
      sensitive << rd_clk << rd_en_array << rd_addr_array;
      SC_THREAD(n2_com_256x132async_dp_cust_array_proc1);
      sensitive << rd_clk << rd_en_array <<  rd_addr_array << wr_en_array <<
  wr_addr_array;
```

**Fig. 8.16** (continued)

```
    SC_THREAD(n2_com_256x132async_dp_cust_array_proc2);
    sensitive << wr_en_array << wr_addr_array << wr_clk << din_array;
  }

  ~n2_com_256x132async_dp_cust_array(){ } /*Destructor */
};

#endif
```

**Fig. 8.16** (continued)

```
#include "n2_com_256x132async_dp_cust_array.h"
#include "memmisc.h"
#include <cstdlib>
#include <cstring>

int sc_main(int argc, char **argv)
{
  /* Declare/define channels/signals */
  sc_core::sc_signal< sc_dt::sc_bv<8> > rd_addr_sig;
  sc_core::sc_signal< sc_dt::sc_bv<8> > wr_addr_sig;
  sc_core::sc_signal<bool> rd_en_array_sig;
  sc_core::sc_signal<bool> wr_en_array_sig;
  sc_core::sc_signal<bool> wclk;
  sc_core::sc_signal< sc_dt::sc_bv<132> > din_array_sig;
  sc_core::sc_signal< sc_dt::sc_bv<132> > dout_array_sig;
  /* Declare/define members */
  bool rd_en_array_value;
  bool wr_en_array_value;
  sc_dt::sc_bv<8> rd_array_addr;
  sc_dt::sc_bv<8> wr_array_addr;
  sc_dt::sc_bv<132> din_array_value;
  sc_dt::sc_bv<132> dout_array_value;

 /* Declare define master clock, memory array and trace file */
  sc_core::sc_clock rclk("read_clk", 2.0, sc_core::SC_NS, 0.5);
  inv_clk in_v_clk("inv_clk");
  n2_com_256x132async_dp_cust_array mem_array("mem_array");
  sc_core::sc_trace_file *fp =
```

**Fig. 8.17** Test harness for T2 asynchronous 256 × 132 memory array

```
sc_core::sc_create_vcd_trace_file("tr_mem_array");
 fp->set_time_unit(1.0, sc_core::SC_NS);
 in_v_clk.clk(rclk);
 in_v_clk.inv_clk_out(wclk);

 /* Connect module ports and channels */
 mem_array.rd_clk(rclk);
 mem_array.wr_clk(wclk);
 mem_array.rd_en_array(rd_en_array_sig);
 mem_array.wr_en_array(wr_en_array_sig);
 mem_array.rd_addr_array(rd_addr_sig);
 mem_array.wr_addr_array(wr_addr_sig);
 mem_array.din_array(din_array_sig);
 mem_array.dout_array(dout_array_sig);
/* Connect signals to trace file */
sc_core::sc_trace(fp, rclk, "rd_clk");
sc_core::sc_trace(fp, wclk, "wr_clk");
sc_core::sc_trace(fp, rd_en_array_sig, "rd_en");
sc_core::sc_trace(fp, wr_en_array_sig, "wr_en");
sc_core::sc_trace(fp, rd_addr_sig, "rd_addr");
sc_core::sc_trace(fp, wr_addr_sig, "wr_addr");
sc_core::sc_trace(fp, din_array_sig, "din");
sc_core::sc_trace(fp, dout_array_sig, "dout");

/* Assign values to members, and write to channels */
rd_en_array_value = false;
wr_en_array_value = false;
rd_array_addr="00000000";
wr_array_addr="00000000";

din_array_value="0000000000000000000000000000000000000000000000000
000000000000000000000000000000000000000000000000000000000000000
0000000000000000000";
dout_array_value="000000000000000000000000000000000000000000000000
000000000000000000000000000000000000000000000000000000000000000
00000000000000000000";
 rd_en_array_sig.write(rd_en_array_value);
 wr_en_array_sig.write(wr_en_array_value);
 rd_addr_sig.write(rd_array_addr);
```

**Fig. 8.17** (continued)

```
  wr_addr_sig.write(wr_array_addr);
  din_array_sig.write(din_array_value);
  dout_array_sig.write(dout_array_value);

  /* Run simulation for pre-defind time-interval */
  sc_core::sc_start(5.0, sc_core::SC_NS);

  /* Change member values and write to channels */
  rd_en_array_value = false;
  wr_en_array_value = true;
  rd_array_addr="00000000";
  wr_array_addr="00000000";
din_array_value="0100100100110101001101010010110101100111001010101010
1010111001010101011011100101010101010010111001110010011110010010
1101110011000101 0110";
  rd_en_array_sig.write(rd_en_array_value);
  wr_en_array_sig.write(wr_en_array_value);
  rd_addr_sig.write(rd_array_addr);
  wr_addr_sig.write(wr_array_addr);
  din_array_sig.write(din_array_value);

 /* Run simulation for pre-defind time-interval */
  sc_core::sc_start(5.0, sc_core::SC_NS);
  /* Change member values and write to channels */
  rd_en_array_value = false;
  wr_en_array_value = true;
  rd_array_addr="00000000";
  wr_array_addr="00000001";
din_array_value="0100100111110101001101011010110101100111001010101010
1010111001010101011011110101010101010010111101110010111110010010
11011100110001010110";
  rd_en_array_sig.write(rd_en_array_value);
  wr_en_array_sig.write(wr_en_array_value);
  rd_addr_sig.write(rd_array_addr);
  wr_addr_sig.write(wr_array_addr);
  din_array_sig.write(din_array_value);

  /* Run simulation for pre-defind time-interval */
  sc_core::sc_start(5.0, sc_core::SC_NS);
```

**Fig. 8.17** (continued)

```
  /* Change member values and write to channels */
  rd_en_array_value = false;
  wr_en_array_value = true;
  rd_array_addr="00000000";
  wr_array_addr="00000010";
din_array_value="01001101001000010011010100101101010001110010101010
1010111001010101011011100101010101010010111001110010010010010010
11011100110001010110";
  rd_en_array_sig.write(rd_en_array_value);
  wr_en_array_sig.write(wr_en_array_value);
  rd_addr_sig.write(rd_array_addr);
  wr_addr_sig.write(wr_array_addr);
  din_array_sig.write(din_array_value);

  /* Run simulation for pre-defind time-interval */
  sc_core::sc_start(5.0, sc_core::SC_NS);

  /* Change member values and write to channels */
  rd_en_array_value = false;
  wr_en_array_value = true;
  rd_array_addr="00000000";
  wr_array_addr="00000011";
din_array_value="01001101011011010011010101101101010101110010101010
1010111101010101011011100101010101010010111001110010010010010010
11011100110001010110";
  rd_en_array_sig.write(rd_en_array_value);
  wr_en_array_sig.write(wr_en_array_value);
  rd_addr_sig.write(rd_array_addr);
  wr_addr_sig.write(wr_array_addr);
  din_array_sig.write(din_array_value);
  sc_core::sc_start(5.0, sc_core::SC_NS);

  /* Change member values and write to channels */
  rd_en_array_value = true;
  wr_en_array_value = false;
  rd_array_addr="00000010";
  wr_array_addr="00000000";
din_array_value="01001101001000010011010100101101010001110010101010
1010111001010101011011100101010101010010111001110010010010010010
```

Fig. 8.17 (continued)

```
1101110011000101011 0";
  rd_en_array_sig.write(rd_en_array_value);
  wr_en_array_sig.write(wr_en_array_value);
  rd_addr_sig.write(rd_array_addr);
  wr_addr_sig.write(wr_array_addr);
  din_array_sig.write(din_array_value);

  /* Run simulation for pre-defind time-interval */
  sc_core::sc_start(5.0, sc_core::SC_NS);

  /* Stop simulation and close trace file */
  sc_core::sc_stop();
  sc_core::sc_close_vcd_trace_file(fp);
  return 0;
}
```

**Fig. 8.17** (continued)

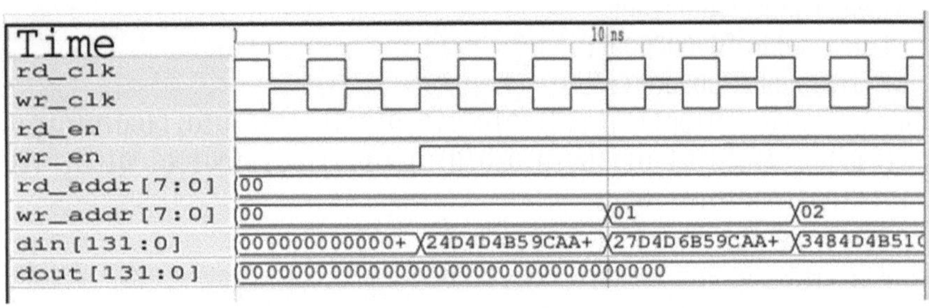


**Fig. 8.18** Partial signal trace for T2 asynchronous 256 × 132 memory array SystemC-2.2.0

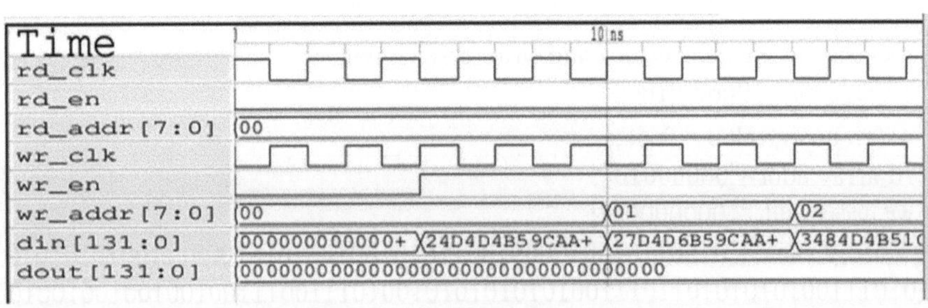


**Fig. 8.19** Partial signal trace for T2 asynchronous 256 × 132 memory array SystemC-2.3.0

```
#ifndef N2_STB_CM_64X45_ARRAY_H
#define N2_STB_CM_64X45_ARRAY_H

#include <systemc>

SC_MODULE(n2_stb_cm_64x45_array)
{
  /* Declare/define input/output ports */
   sc_core::sc_in<bool> clk;
  /*master clock */
   sc_core::sc_in<bool> cam_ldq ; /* quad-ld cam. */
   sc_core::sc_in<bool> tcu_array_wr_inhibit;
   sc_core::sc_in<bool> siclk;
  /* scan-in clock */
   sc_core::sc_in< sc_dt::sc_bv<3> > cam_rw_ptr ;
  /* wr pointer for single port. */
   sc_core::sc_in< sc_dt::sc_bv<3> > cam_rw_tid ;
  /* thread id for rw. */
   sc_core::sc_in<bool> wptr_vld ; /* write pointer vld */
   sc_core::sc_in<bool> rptr_vld ; /* read pointer vld */
   sc_core::sc_in< sc_dt::sc_bv<45> > camwr_data ;
   /* data for compare/write */
   sc_core::sc_in<bool> cam_vld ; /* cam is required. */
   sc_core::sc_in< sc_dt::sc_bv<3> > cam_cm_tid ;
   /* thread id for cam operation.*/
   sc_core::sc_in< sc_dt::sc_bv<8> > cam_line_en;
   /* mask for squashing cam */
   sc_core::sc_out< sc_dt::sc_bv<45> > stb_rdata;
   /* rd data from CAM. */
   sc_core::sc_out< bool> stb_ld_partial_raw ;
   /* ld with partial raw. */

   sc_core::sc_out< sc_dt::sc_bv<3> > stb_cam_hit_ptr;
   sc_core::sc_out< bool> stb_cam_hit ;  /* any hit in stb */
   sc_core::sc_out< bool> stb_cam_mhit ;  /* multiple hits in stb */
   /* Internal channels/signals */
   sc_core::sc_signal< sc_dt::sc_bv<6> > rw_addr;
   sc_core::sc_signal<bool> write_vld;
   sc_core::sc_signal<bool> read_vld;
```

**Fig. 8.20** T2 64 × 45 content addressable memory

```
sc_core::sc_signal<bool> read_vld;
sc_core::sc_signal< sc_dt::sc_bv<8> > byte_overlap_mx;
sc_core::sc_signal< sc_dt::sc_bv<8> > byte_match_mx;
sc_core::sc_signal< sc_dt::sc_bv<8> > ptag_hit_mx;
sc_core::sc_signal< sc_dt::sc_bv<8> > cam_hit;
unsigned int i;
unsigned int l;

/* Internal data structures and temporary members */
sc_dt::sc_bv<45> stb_ramc[64];
sc_dt::sc_bv<45> ramc_entry;
sc_dt::sc_bv<37> cam_tag;
sc_dt::sc_bv<8> cam_bmask;
sc_dt::sc_bv<64> ptag_hit;
sc_dt::sc_bv<64> byte_match;
sc_dt::sc_bv<64> byte_overlap;
sc_dt::sc_bv<1> tmp;
sc_dt::sc_bv<1> tmp1;
sc_dt::sc_bv<8> tmp2;
sc_dt::sc_bv<4> nibble0;
sc_dt::sc_bv<4> nibble1;

/* Initialize CAM contents */
void n2_stb_cm_64x45_array_init()
{
  unsigned int i;
  for(i = 0; i < 64; i++) stb_ramc[i]
="0000000000000000000000000000000000000000000000000";
}

/* Operations performed each time the CAM object is invoked */
void n2_stb_cm_64x45_array_proc0()
{
   bool tmp0;
   bool tmp1;
   sc_dt::sc_bv<6> tmp2;

   tmp0 = wptr_vld.read() & !(tcu_array_wr_inhibit.read());
   tmp1 = rptr_vld.read() & !(tcu_array_wr_inhibit.read());
```

**Fig. 8.20** (continued)

```
      tmp2 = (cam_rw_tid.read(), cam_rw_ptr.read());
      write_vld.write(tmp0);
      read_vld.write(tmp1);
      rw_addr.write(tmp2);
    }

    /* Thread to write new data to CAM or read data from it */
    void n2_stb_cm_64x45_array_proc1()
    {
      sc_dt::sc_bv<45> tmp =
   "000000000000000000000000000000000000000000000";
      while(1)
      {
     wait();
     if((clk.read() == true) & (write_vld.read() == true))
     {
       if(cam_vld.read() == true)
       {
        stb_ramc[rw_addr.read().to_int()] = tmp;
       }
       else
       {
         stb_ramc[rw_addr.read().to_int()] = camwr_data.read();
       }
     }
    }
  }

  /* Thread to check for/block concurrent CAM reads/writes *
  /* and read out CAM data otherwise */
  void n2_stb_cm_64x45_array_proc2()
  {
    while(1)
    {
      wait();
      if(clk.read() == true)
      {
        if((write_vld.read() | !(read_vld.read()) == true))
```

**Fig. 8.20** (continued)

```
      {
        std::cout<<"Concurrent CAM read/write disallowed ..."<<std::endl;
      }
      else
      {
        stb_rdata = stb_ramc[rw_addr.read().to_int()];
      }
    }
  }
}

/* Thread for maintaining CAM data integrity */
void n2_stb_cm_64x45_array_proc3()
{
  while(1)
  {
    wait();
    for(l = 0; l < 64; l++)
    {
      ramc_entry = stb_ramc[l];
      cam_tag = ramc_entry.range(44,8);
      cam_bmask = ramc_entry;
      if(cam_vld.read() == true) tmp[0] = "1";
      else tmp[0] = "0";
      ptag_hit[l] = (cam_tag.range(36,1) == camwr_data.read().range(44,9)) &
                (cam_tag[0] == camwr_data.read()[8]) &
                (!cam_ldq.read() | cam_ldq.read()) & cam_vld.read() & tmp[0];
      byte_match[l] = (cam_bmask.range(7,0) &
camwr_data.read().range(7,0)).or_reduce() & tmp[0];
      byte_overlap[l] = (~(cam_bmask.range(7,0)) &
camwr_data.read().range(7,0)).or_reduce() & tmp[0];
    }
  }
}

/* Reset thread */
void n2_stb_cm_64x45_array_proc4()
{
  while(1)
```

Fig. 8.20 (continued)

```
  {
   wait();
    ptag_hit =
"000000000000000000000000000000000000000000000000000000000000000
0";
    byte_match =
"000000000000000000000000000000000000000000000000000000000000000
0";
    byte_overlap =
"000000000000000000000000000000000000000000000000000000000000000
0";
  }
}

/* Operations for data integrity performed at each invocation of CAM object
*/
void n2_stb_cm_64x45_array_proc5()
{
  sc_dt::sc_bv<8> tmp0;
  sc_dt::sc_bv<8> tmp1;
  sc_dt::sc_bv<8> tmp2;
  sc_dt::sc_bv<8> tmp3;
  sc_dt::sc_bv<8> tmp4;
  sc_dt::sc_bv<8> tmp5;
  sc_dt::sc_bv<8> tmp6;
  sc_dt::sc_bv<8> tmp7;
  sc_dt::sc_bv<8> tmp;
  tmp0 = cam_cm_tid.read() == 0 ? "00000001" : "00000000";
  tmp1 = cam_cm_tid.read() == 1 ? "00000001" : "00000000";
  tmp2 = cam_cm_tid.read() == 2 ? "00000001" : "00000000";
  tmp3 = cam_cm_tid.read() == 3 ? "00000001" : "00000000";
  tmp4 = cam_cm_tid.read() == 4 ? "00000001" : "00000000";
  tmp5 = cam_cm_tid.read() == 5 ? "00000001" : "00000000";
  tmp6 = cam_cm_tid.read() == 6 ? "00000001" : "00000000";
  tmp7 = cam_cm_tid.read() == 7 ? "00000001" : "00000000";
  tmp = (tmp0 & byte_overlap.range(7,0)) |
       (tmp1 & byte_overlap.range(15,8)) |
       (tmp2 & byte_overlap.range(23,16)) |
       (tmp3 & byte_overlap.range(31,24)) |
```

**Fig. 8.20** (continued)

```
        (tmp4 & byte_overlap.range(39,32)) |
        (tmp5 & byte_overlap.range(47,40)) |
        (tmp6 & byte_overlap.range(55,48)) |
        (tmp7 & byte_overlap.range(63,56));
   byte_overlap_mx.write(tmp);
  }

  /* Operations for data integrity performed at each invocation of CAM object
  */
  void n2_stb_cm_64x45_array_proc6()
  {
    sc_dt::sc_bv<8> tmp0;
    sc_dt::sc_bv<8> tmp1;
    sc_dt::sc_bv<8> tmp2;
    sc_dt::sc_bv<8> tmp3;
    sc_dt::sc_bv<8> tmp4;
    sc_dt::sc_bv<8> tmp5;
    sc_dt::sc_bv<8> tmp6;
    sc_dt::sc_bv<8> tmp7;
    sc_dt::sc_bv<8> tmp;
  tmp0 = cam_cm_tid.read() == 0 ? "00000001" : "00000000";
  tmp1 = cam_cm_tid.read() == 1 ? "00000001" : "00000000";
  tmp2 = cam_cm_tid.read() == 2 ? "00000001" : "00000000";
  tmp3 = cam_cm_tid.read() == 3 ? "00000001" : "00000000";
  tmp4 = cam_cm_tid.read() == 4 ? "00000001" : "00000000";
  tmp5 = cam_cm_tid.read() == 5 ? "00000001" : "00000000";
  tmp6 = cam_cm_tid.read() == 6 ? "00000001" : "00000000";
  tmp7 = cam_cm_tid.read() == 7 ? "00000001" : "00000000";
  tmp = (tmp0 & byte_match.range(7,0)) |
       (tmp1 & byte_match.range(15,8)) |
       (tmp2 & byte_match.range(23,16)) |
       (tmp3 & byte_match.range(31,24)) |
       (tmp4 & byte_match.range(39,32)) |
       (tmp5 & byte_match.range(47,40)) |
       (tmp6 & byte_match.range(55,48)) |
       (tmp7 & byte_match.range(63,56));
  byte_match_mx.write(tmp);
}
```

**Fig. 8.20** (continued)

```
/* Operations for data integrity performed at each invocation of CAM object
*/
void n2_stb_cm_64x45_array_proc7()
{
  sc_dt::sc_bv<8> tmp0;
  sc_dt::sc_bv<8> tmp1;
  sc_dt::sc_bv<8> tmp2;
  sc_dt::sc_bv<8> tmp3;
  sc_dt::sc_bv<8> tmp4;
  sc_dt::sc_bv<8> tmp5;
  sc_dt::sc_bv<8> tmp6;
  sc_dt::sc_bv<8> tmp7;
  sc_dt::sc_bv<8> tmp;
  tmp0 = cam_cm_tid.read() == 0 ? "00000001" : "00000000";
  tmp1 = cam_cm_tid.read() == 1 ? "00000001" : "00000000";
  tmp2 = cam_cm_tid.read() == 2 ? "00000001" : "00000000";
  tmp3 = cam_cm_tid.read() == 3 ? "00000001" : "00000000";
  tmp4 = cam_cm_tid.read() == 4 ? "00000001" : "00000000";
  tmp5 = cam_cm_tid.read() == 5 ? "00000001" : "00000000";
  tmp6 = cam_cm_tid.read() == 6 ? "00000001" : "00000000";
  tmp7 = cam_cm_tid.read() == 7 ? "00000001" : "00000000";
  tmp = (tmp0 & byte_match.range(7,0)) |
       (tmp1 & ptag_hit.range(15,8)) |
       (tmp2 & ptag_hit.range(23,16)) |
       (tmp3 & ptag_hit.range(31,24)) |
       (tmp4 & ptag_hit.range(39,32)) |
       (tmp5 & ptag_hit.range(47,40)) |
       (tmp6 & ptag_hit.range(55,48)) |
       (tmp7 & ptag_hit.range(63,56));
  ptag_hit_mx.write(tmp);
}

/* Operations for data integrity performed at each invocation of CAM
object */
void n2_stb_cm_64x45_array_proc8()
{
  bool tmp;
  tmp = (ptag_hit_mx.read() &
       byte_match_mx.read() &
```

**Fig. 8.20** (continued)

```
        byte_overlap_mx.read() &
        cam_line_en.read()).or_reduce();
    stb_ld_partial_raw.write(tmp);
  }

  /* Operations for data integrity performed at each invocation of CAM
object */
  void n2_stb_cm_64x45_array_proc9()
  {
    cam_hit.write(ptag_hit_mx.read() &
                  byte_match_mx.read() &
                  cam_line_en.read());
    stb_cam_hit.write(cam_hit.read().or_reduce());
  }

  /* Operations for data integrity performed at each invocation of CAM
object */
  void n2_stb_cm_64x45_array_proc10()
  {
    sc_dt::sc_bv<3> tmp;
    sc_dt::sc_bv<8> tmp0;
    tmp0 = cam_hit.read();
    tmp[0] = tmp0[1] | tmp0[3] | tmp0[5] | tmp0[7] ;
    tmp[1] = tmp0[2] | tmp0[3] | tmp0[6] | tmp0[7] ;
    tmp[2] = tmp0[4] | tmp0[5] | tmp0[6] | tmp0[7] ;
    stb_cam_hit_ptr.write(tmp);
  }

  /* Operations for data integrity performed at each invocation of CAM
object */
  void n2_stb_cm_64x45_array_proc11()
  {
    bool tmp0;
    tmp2 = cam_hit.read();
    nibble0 = tmp2.range(3,0);
    nibble1 = tmp2.range(7,4);
    tmp1[0] = (tmp2[0] & tmp2[1]) |
              (tmp2[2] & tmp2[3]) |
              (tmp2[4] & tmp2[5]) |
```

**Fig. 8.20** (continued)

```
            (tmp2[6] & tmp2[7]) |
            ((tmp2[0] | tmp2[1]) &
             (tmp2[2] | tmp2[3])) |
            ((tmp2[4] | tmp2[5]) &
             (tmp2[6] | tmp2[7])) |
             (nibble0.or_reduce() and nibble1.or_reduce()) ;
    tmp0 = tmp1[0] == "1 " ? true : false;
    stb_cam_mhit.write(tmp0);
  }

  /* Constructor */
  SC_CTOR(n2_stb_cm_64x45_array)
  {
    n2_stb_cm_64x45_array_init();
  /* Initialize CAM array */
    /* Declare/assign threads and specify each thread's sensitivity list */
    SC_THREAD(n2_stb_cm_64x45_array_proc1);
    sensitive << clk << write_vld << rw_addr << camwr_data << cam_vld;
    SC_THREAD(n2_stb_cm_64x45_array_proc2);
    sensitive << clk << read_vld << rw_addr << write_vld;
    /* Declare/assign clocked threads */
    SC_CTHREAD(n2_stb_cm_64x45_array_proc3, clk.pos());
    SC_CTHREAD(n2_stb_cm_64x45_array_proc4, siclk.pos());
    /* Declare/assign methods to be executed each time the CAM object is
 activated */
    SC_METHOD(n2_stb_cm_64x45_array_proc0);
    SC_METHOD(n2_stb_cm_64x45_array_proc5);
    SC_METHOD(n2_stb_cm_64x45_array_proc6);
    SC_METHOD(n2_stb_cm_64x45_array_proc7);
    SC_METHOD(n2_stb_cm_64x45_array_proc8);
    SC_METHOD(n2_stb_cm_64x45_array_proc9);
    SC_METHOD(n2_stb_cm_64x45_array_proc10);
    SC_METHOD(n2_stb_cm_64x45_array_proc11);
  }

  ~n2_stb_cm_64x45_array(){ }
  /*Destructor */
 };

 #endif
```

**Fig. 8.20** (continued)

```
#include "n2_stb_cm_64x45_array.h"
#include <cstdlib>
#include <cstring>

int sc_main(int argc, char **argv)
{
  /* Declare/define channels/signals */
  sc_core::sc_signal<bool> cam_ldq_sig ; /* quad-ld cam */
  sc_core::sc_signal<bool> tcu_array_wr_inhibit_sig;
  sc_core::sc_signal<bool> siclk_sig;
  sc_core::sc_signal< sc_dt::sc_bv<3> > cam_rw_ptr_sig ;
  /* wr pointer for single port. */
  sc_core::sc_signal< sc_dt::sc_bv<3> > cam_rw_tid_sig ;
   /* thread id for rw. */
  sc_core::sc_signal<bool> wptr_vld_sig ; /* write pointer vld */
  sc_core::sc_signal<bool> rptr_vld_sig ; /* read pointer vld */
  sc_core::sc_signal< sc_dt::sc_bv<45> > camwr_data_sig ;
  /* data for compare/write */
  sc_core::sc_signal<bool> cam_vld_sig ; /* cam is required. */
  sc_core::sc_signal< sc_dt::sc_bv<3> > cam_cm_tid_sig ;
  /* thread id for cam operation.*/
  sc_core::sc_signal< sc_dt::sc_bv<8> > cam_line_en_sig;
  /* mask for squashing cam */
  sc_core::sc_signal< sc_dt::sc_bv<45> > stb_rdata_sig;
 /* rd data from CAM RAM. */
 sc_core::sc_signal< bool> stb_ld_partial_raw_sig;
/* ld with partial raw. */
 sc_core::sc_signal< sc_dt::sc_bv<3> > stb_cam_hit_ptr_sig;
 sc_core::sc_signal< bool> stb_cam_hit_sig ;  /* any hit in stb */
 sc_core::sc_signal< bool> stb_cam_mhit_sig ;  /* multiple hits in stb */

 /* Declare/define internal data members */
 sc_dt::sc_bv<45> camwr_data_value;
 sc_dt::sc_bv<45> stb_rdata_value;
 sc_dt::sc_bv<8> cam_line_en_value;
 sc_dt::sc_bv<3> cam_rw_ptr_value;
 sc_dt::sc_bv<3> cam_rw_tid_value;
 sc_dt::sc_bv<3> stb_cam_hit_ptr_value;
 bool cam_ldq_value;
```

**Fig. 8.21** Test harness for T2 64 × 45 content addressable memory

```
    bool tcu_array_wr_inhibit_value;
    bool siclk_value;
    bool wptr_vld_value;
    bool rptr_vld_value;
    bool cam_vld_value;
    bool stb_ld_partial_raw_value;
    bool stb_cam_hit_value;
    bool stb_cam_mhit_value;

    /* Declare/define master clock, CAM array and trace file */
    sc_core::sc_clock clk("clk", 2.0, sc_core::SC_NS, 0.5);
    n2_stb_cm_64x45_array cam_array_64x45("n2_stb_cm_64x45_array_64x45");
    sc_core::sc_trace_file *fp =
  sc_core::sc_create_vcd_trace_file("tr_n2_stb_cm_64x45");
    fp->set_time_unit(1.0, sc_core::SC_NS);

  /* Connect channel/signals and module ports */
    cam_array_64x45.clk(clk);
    cam_array_64x45.cam_ldq(cam_ldq_sig); /* quad-ld cam. */
    cam_array_64x45.tcu_array_wr_inhibit(tcu_array_wr_inhibit_sig);
    cam_array_64x45.siclk(siclk_sig);
    cam_array_64x45.cam_rw_ptr(cam_rw_ptr_sig);
    cam_array_64x45.cam_rw_tid(cam_rw_tid_sig);
    cam_array_64x45.wptr_vld(wptr_vld_sig);
    cam_array_64x45.rptr_vld(rptr_vld_sig);
    cam_array_64x45.camwr_data(camwr_data_sig);
   cam_array_64x45.cam_vld(cam_vld_sig);
   cam_array_64x45.cam_cm_tid(cam_cm_tid_sig);
   cam_array_64x45.cam_line_en(cam_line_en_sig);
   cam_array_64x45.stb_rdata(stb_rdata_sig);
   cam_array_64x45.stb_ld_partial_raw(stb_ld_partial_raw_sig);
   cam_array_64x45.stb_cam_hit_ptr(stb_cam_hit_ptr_sig);
   cam_array_64x45.stb_cam_hit(stb_cam_hit_sig);
   cam_array_64x45.stb_cam_mhit(stb_cam_mhit_sig);

  /* Connect channels/signals and trace file */
   sc_core::sc_trace(fp, clk, "clk");
   sc_core::sc_trace(fp, cam_ldq_sig, "cam_ldq");
   sc_core::sc_trace(fp, tcu_array_wr_inhibit_sig, "tcu_array_wr_inhibit");
```

**Fig. 8.21** (continued)

```
  sc_core::sc_trace(fp, siclk_sig, "siclk");
  sc_core::sc_trace(fp, cam_rw_ptr_sig, "cam_rw_ptr");
  sc_core::sc_trace(fp, cam_rw_tid_sig, "cam_rw_tid");
  sc_core::sc_trace(fp, wptr_vld_sig, "wptr_vld");
  sc_core::sc_trace(fp, rptr_vld_sig, "rptr_vld");
  sc_core::sc_trace(fp, camwr_data_sig, "camwr_data");
  sc_core::sc_trace(fp, cam_vld_sig, "cam_vld");

 /* Assign values to data members and write to channels/signals */
  camwr_data_value="00000001100000001110001001000111100001010101O";
  cam_line_en_value="00000000";
  cam_rw_ptr_value="000";
  cam_ldq_value=true;
  tcu_array_wr_inhibit_value=false;
  siclk_value=false;
  wptr_vld_value=true;
  rptr_vld_value=false;
  cam_vld_value=false;
  cam_ldq_sig.write(cam_ldq_value);

  tcu_array_wr_inhibit_sig.write(tcu_array_wr_inhibit_value);
  siclk_sig.write(siclk_value);
  cam_rw_ptr_sig.write(cam_rw_ptr_value);
  cam_rw_tid_sig.write(cam_rw_tid_value);
  wptr_vld_sig.write(wptr_vld_value);
  rptr_vld_sig.write(rptr_vld_value);
  camwr_data_sig.write(camwr_data_value);
  cam_vld_sig.write(cam_vld_value);

  cam_line_en_sig.write(cam_line_en_value);

  /* Run simulation for pre-defined time interval */
  sc_core::sc_start(2.0, sc_core::SC_NS);

 /* Re-assign values to data members and write to channels/signals */
  camwr_data_value="01010011000110011100011001010010010110110001 1";
  cam_line_en_value="00000000";
  cam_rw_ptr_value="001";
  cam_ldq_value=true;
  tcu_array_wr_inhibit_value=false;
  siclk_value=false;
```

**Fig. 8.21** (continued)

```
  wptr_vld_value=true;
  rptr_vld_value=false;
  cam_vld_value=false;
  cam_ldq_sig.write(cam_ldq_value);
  tcu_array_wr_inhibit_sig.write(tcu_array_wr_inhibit_value);
  siclk_sig.write(siclk_value);
  cam_rw_ptr_sig.write(cam_rw_ptr_value);
  cam_rw_tid_sig.write(cam_rw_tid_value);
  wptr_vld_sig.write(wptr_vld_value);
  rptr_vld_sig.write(rptr_vld_value);
  camwr_data_sig.write(camwr_data_value);
  cam_vld_sig.write(cam_vld_value);
  cam_line_en_sig.write(cam_line_en_value);

  /* Run simulation for pre-defined time interval */
  sc_core::sc_start(2.0, sc_core::SC_NS);

/* Re-assign values to data members and write to channels/signals */
  camwr_data_value="010111110101100111000110110100101101101101011";
  cam_line_en_value="00000000";
  cam_rw_ptr_value="010";
  cam_ldq_value=true;
  tcu_array_wr_inhibit_value=false;
  siclk_value=false;
  wptr_vld_value=true;
  rptr_vld_value=false;
  cam_vld_value=false;
  cam_ldq_sig.write(cam_ldq_value);
  tcu_array_wr_inhibit_sig.write(tcu_array_wr_inhibit_value);
  siclk_sig.write(siclk_value);
  cam_rw_ptr_sig.write(cam_rw_ptr_value);
  cam_rw_tid_sig.write(cam_rw_tid_value);
  wptr_vld_sig.write(wptr_vld_value);
  rptr_vld_sig.write(rptr_vld_value);
  camwr_data_sig.write(camwr_data_value);
  cam_vld_sig.write(cam_vld_value);
  cam_line_en_sig.write(cam_line_en_value);

  /* Run simulation for pre-defined time interval */
  sc_core::sc_start(2.0, sc_core::SC_NS);
```

**Fig. 8.21** (continued)

```
/* Re-assign values to data members and write to channels/signals */
  camwr_data_value="0101100001011001110001101101111011011011101010";
  cam_line_en_value="00000000";
  cam_rw_ptr_value="011";
  cam_ldq_value=true;
  tcu_array_wr_inhibit_value=false;
  siclk_value=false;
  wptr_vld_value=true;
  rptr_vld_value=false;
  cam_vld_value=false;

  cam_ldq_sig.write(cam_ldq_value);
  tcu_array_wr_inhibit_sig.write(tcu_array_wr_inhibit_value);
  siclk_sig.write(siclk_value);
  cam_rw_ptr_sig.write(cam_rw_ptr_value);
  cam_rw_tid_sig.write(cam_rw_tid_value);
  wptr_vld_sig.write(wptr_vld_value);
  rptr_vld_sig.write(rptr_vld_value);
  camwr_data_sig.write(camwr_data_value);
  cam_vld_sig.write(cam_vld_value);
  cam_line_en_sig.write(cam_line_en_value);
  /* Run simulation for pre-defined time interval */
  sc_core::sc_start(1.0, sc_core::SC_NS);

/* Re-assign values to data members and write to channels/signals */
  camwr_data_value="0101101011011001010101101101001011011011101011";
  cam_line_en_value="00000000";
  cam_rw_ptr_value="101";
  cam_ldq_value=true;
  tcu_array_wr_inhibit_value=false;
  siclk_value=false;
  wptr_vld_value=true;
  rptr_vld_value=false;
  cam_vld_value=false;
  cam_ldq_sig.write(cam_ldq_value);
  tcu_array_wr_inhibit_sig.write(tcu_array_wr_inhibit_value);
  siclk_sig.write(siclk_value);
  cam_rw_ptr_sig.write(cam_rw_ptr_value);
  cam_rw_tid_sig.write(cam_rw_tid_value);
  wptr_vld_sig.write(wptr_vld_value);
  rptr_vld_sig.write(rptr_vld_value);
```

**Fig. 8.21** (continued)

```
    camwr_data_sig.write(camwr_data_value);
    cam_vld_sig.write(cam_vld_value);
    cam_line_en_sig.write(cam_line_en_value);

    /* Run simulation for pre-defined time interval */
    sc_core::sc_start(2.0, sc_core::SC_NS);

  /* Re-assign values to data members and write to channels/signals */
    camwr_data_value="0101111101011001110001101101001011011011011011";
    cam_line_en_value="00000000";
    cam_rw_ptr_value="010";
    cam_ldq_value=true;
    tcu_array_wr_inhibit_value=false;
    siclk_value=false;
    wptr_vld_value=false;
    rptr_vld_value=true;
    cam_vld_value=false;

    cam_ldq_sig.write(cam_ldq_value);
    tcu_array_wr_inhibit_sig.write(tcu_array_wr_inhibit_value);
    siclk_sig.write(siclk_value);
    cam_rw_ptr_sig.write(cam_rw_ptr_value);
    cam_rw_tid_sig.write(cam_rw_tid_value);
    wptr_vld_sig.write(wptr_vld_value);
    rptr_vld_sig.write(rptr_vld_value);
    camwr_data_sig.write(camwr_data_value);
    cam_vld_sig.write(cam_vld_value);
    cam_line_en_sig.write(cam_line_en_value);
  /* Run simulation for pre-defined time interval */
  sc_core::sc_start(2.0, sc_core::SC_NS);
  /* Stop simulation and close trace file */
  sc_core::sc_stop();
  sc_core::sc_close_vcd_trace_file(fp);
  return 0;
}
```

**Fig. 8.21** (Continued)

boost converters, to name a few. Conceptually, it is very simple—a very high frequency carrier signal is *modulated* by a much lower frequency modulation signal. Depending on where the two signals overlap, the pulse width of the output signal varies—hence *pulse width modulation*. Figures 8.24, 8.25, 8.26, and 8.27 show the source code, test harness, and input/output traces, respectively, for a PWM scheme

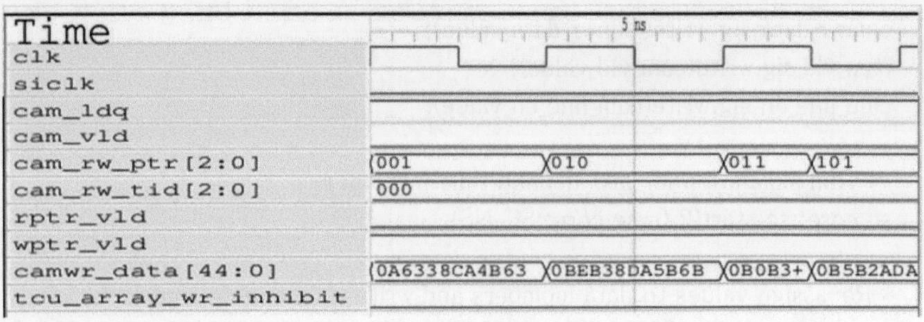


**Fig. 8.22** Partial input/output traces for T2 54 × 45 content addressable memory SystemC-2.2.0

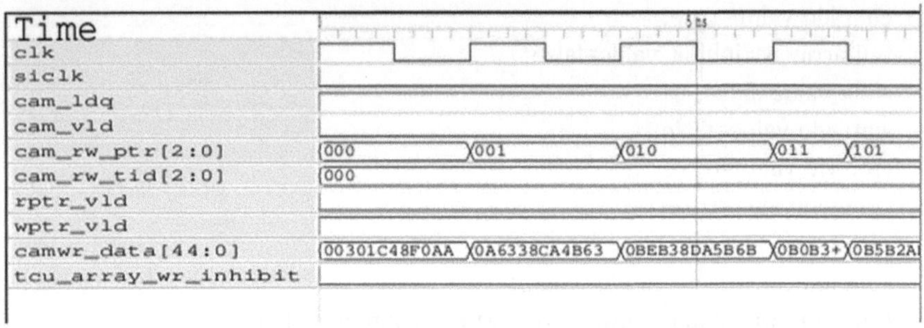


**Fig. 8.23** Partial input/output traces for T2 54 × 45 content addressable memory SystemC-2.3.0

using high frequency triangle wave carrier and DC modulation signal. Other choices of carrier/modulation signals are used as the situation demands, e.g., triangle/sine (for DC–AC inverters). Although at first sight the reader might ask where combinational and sequential logic blocks fit into this design, careful review shows that *the high frequency carrier and low frequency modulation signal generation is sequential, as both need to be synchronized with some clock, while the actual modulation process is 'combinational' in nature, as no clock synchronization is required.*

When executed from the command prompt as:./sim 5 5000 2.5, the output is as in Fig. 8 26.

When executed from the command prompt as:./sim 5 10000 2.5, the output is as in Fig. 8 27. Note the new DC modulation value. Another popular variation of PWM (used in DC–AC inverters) uses high frequency triangle carrier and low frequency sine wave modulation.

## 8.6 T2 Flip-Flop Bank

A bank of flip-flops connected in series is a very common sub-circuit in digital hardware. Here, a simple bank of 16 series connected flip-flops is examined in detail. Figures 8.28, 8.29 contain the source code and test harness, respectively,

**Fig. 8.24** Triangle wave carrier, DC signal modulator pulse width modulation (*PWM*)

```
#ifndef PWM_H
#define PWM_H

#include <systemc>
SC_MODULE(pwmtrstl)
{
  /* Declare/define input/output ports */
  sc_core::sc_in<double> intriangle;
  sc_core::sc_in<double> refsigin;
  sc_core::sc_out<double> pwmout;
  /* Declare/define internal members */
  double refval;
  double trival;
  double pwmval;

  /* Pulse width modulation main operation thread */
  void pwmtrstl_proc0()
  {
    while(1)
    {
     wait();
     refval = refsigin.read();
     trival = intriangle.read();
     pwmval = trival > refval ? 1.0 : 0.0;
     pwmout.write(pwmval);
    }
  }

 /* Constructor */
 SC_CTOR(pwmtrstl):pwmval(0.0),
                   refval(0.0),
                   trival(0.0)
 {
    /*Declare/assign main thread */
   SC_THREAD(pwmtrstl_proc0);
    /*Sensitivity list for thread */
   sensitive << refsigin << intriangle;
 }
 ~pwmtrstl(){ }
/* Destructor */
};
#endif
```

```
#include "pwmsysc.h"
#include "syscsrcs.h"
#include "sysctrace.h"
#include <cstdlib>

int sc_main(int argc, char* argv[])
{
  /* Read in triangle carrier signal
     amplitude, frequency, DC signal value */
  if(argc < 4)
  {
   std::cout<<"Insufficient parameters ..."<<std::endl;
   std::cout<<"usage ./sim <amplitude> <frequency> <offset>"<<std::endl;
   exit(0);
  }
  /* Declare/define channels/signals */
  sc_core::sc_signal<double> pwm;
  sc_core::sc_signal<double> triangle;
  sc_core::sc_signal<double> refsig;

  double amplitude = strtod(argv[1], NULL);
  double frequency = strtod(argv[2], NULL);
  double offset = strtod(argv[3], NULL);

  /* Declare/define clock for reference and high frequency
     triangle carrier wave generation */
  sc_core::sc_clock clk("clk", 25.0, sc_core::SC_US, 0.5);
  /* Declare/define PWM module, signal sources and trace file */

  pwmtrstl pwm_trstl("pwm_tr_stl");
  refsrc ref_src("ref_src");
  trianglesrc triangle_src("triangle_src");
  ssctracedbldbl ssctrdbldbl("tr_pwmstl");

  /* Connect module ports and channels/signals */
  pwm_trstl.refsigin(refsig);
  pwm_trstl.intriangle(triangle);
  pwm_trstl.pwmout(pwm);
```

**Fig. 8.25** Test harness for triangle wave carrier, DC signal modulator pulse width modulation (*PWM*)

```
    ref_src.clk(clk);
    ref_src.refout(refsig);
    ref_src.refval = offset;

    triangle_src.clk(clk);
    triangle_src.sigout(triangle);
    triangle_src.amplitude = amplitude;
    triangle_src.frequency = frequency;

    ssctrdbldbl.din0(triangle);
    ssctrdbldbl.din1(pwm);
    /* Run simulation for pre-defind time-interval and then stop */
    sc_core::sc_start(5000.0, sc_core::SC_US);
    sc_core::sc_stop();
    return 0;
}
```

**Fig. 8.25** (continued)

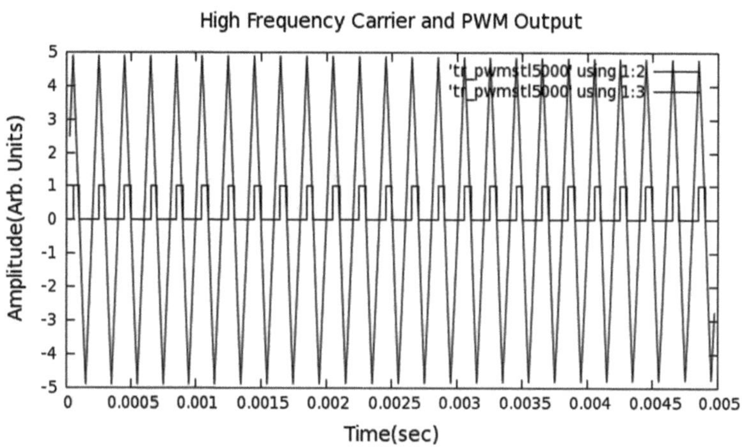


**Fig. 8.26** 6.0 kHz triangle wave carrier, DC modulation input, and pulse width modulation output

for the flip-flop bank, and Figs. 8.30 (SystemC-2.2.0) and 8.31 (SystemC-2.3.0) shows partial input/output traces. A flip-flop bank is a hierarchical combination of both combinational and sequential logic sub-circuits—intermediate buffers are purely combinational, but flip-flops are purely sequential. There are 16 flip-flops, in master–slave configuration.

The T2 [3] flip-flop bank also uses a set of extra control signals for scan, etc.

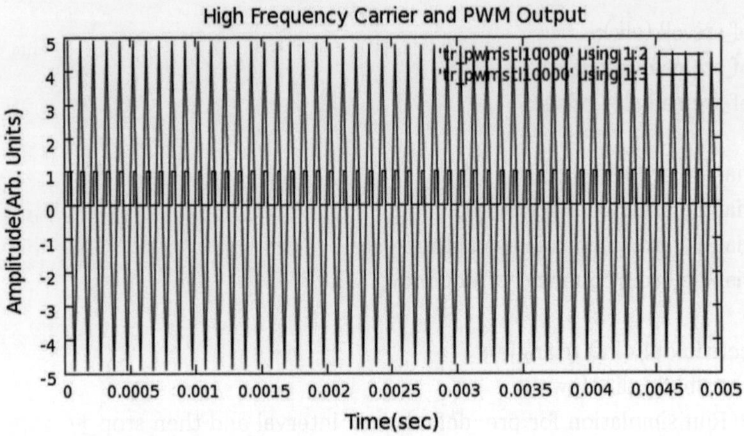


**Fig. 8.27** 6.0 kHz triangle wave carrier, DC modulation input, and pulse width modulation output

## 8.7 Simple Custom Blocking Signal Interface and Channel

While built-in SystemC [4] channels/signals are excellent for almost all data transfer between SystemC modules, sometimes custom signal channels are required for to address the needs of specific purposes, e.g., while analyzing a network protocol. A custom channel may be created by specifying a SystemC interface whose pure virtual methods are implemented by a SystemC module. The SystemC module overrides the pure virtual functions and provides the signal channel implementation. Sender and receiver objects must be passed a reference to a signal channel instance for them to exchange data over that channel. A custom channel may be blocking/non-blocking and half/full duplex. Although this example cannot be classified strictly as combinational or sequential, it illustrates how a customized channel may be implemented, as the situation demands. Here, a simple half duplex signal channel is analyzed. Figures 8.32, 8.33, and 8.34 contain the signal channel interface/implementation source code, the sender/receiver source code, and the test harness, respectively.

From the receiver and sender classes, it is seen that the statement *typedef < class name > SC_CURRENT_USER_MODULE* has the same effect as *SC_HAS_PROCESS(< class name >).*

The console output appears as follows:
SystemC 2.2.0—Aug 30 2011 19:39:07
Copyright (c) 1996–2006 by all Contributors
ALL RIGHTS RESERVED

A read:
B wrote: Test character string for customized SystemC channel
A read: Test character string for customized SystemC channel
B wrote: Test character string for customized SystemC channel
A read: Test character string for customized SystemC channel

```
#ifndef N2_FLOP_BANK_CUST_H
#define N2_FLOP_BANK_CUST_H

#include <systemc>

/* Buffer class */
SC_MODULE(cl_u1_buf_32x)
{
  sc_core::sc_in<bool> in;
  sc_core::sc_out<bool> out;

  /* Main buffer thread */
  void u1_buf_32x_proc0()
  {
    out.write(in.read());
  }

  SC_CTOR(cl_u1_buf_32x)  /*Constructor */
  {
    SC_THREAD(u1_buf_32x_proc0);
 /* Declare/assign thread */
    sensitive << in;
 /* Sensitivity list */
  }

  ~cl_u1_buf_32x(){ }
};

/* L1 Header class */
SC_MODULE(cl_sc1_l1hdr_32x)
{
   /* Declare/define input/output ports */
   sc_core::sc_in < bool > l2clk;
   sc_core::sc_in < bool > se;
   sc_core::sc_in < bool > pce;
   sc_core::sc_in < bool > pce_ov;
   sc_core::sc_in < bool > stop;
   sc_core::sc_out < bool > l1clk;
   bool FORMAL_TOOL;
   bool SCAN_MODE;
   bool l1en;
```

**Fig. 8.28** Series connected flip-flop bank in T2 processor

```
/* Header class method */
void sc1_l1hdr_32x_proc0()
{
  if(FORMAL_TOOL == true)
  {
    l1en = (!(stop.read()) & ( pce.read() | pce_ov.read() ));
    l1clk.write((l2clk.read() & l1en) | se.read());
  }
}

/* Header class main thread */
void sc1_l1hdr_32x_proc1()
{
  while(1)
  {
    wait();
    if(SCAN_MODE == true)
    {
      if (l2clk.read() == false)
       l1en = (~(stop.read()) & (pce.read() | pce_ov.read()));
    }
  }
}
/* Header class method */
void sc1_l1hdr_32x_proc2()
{
  if(SCAN_MODE == false)
  {
    l1en = (!(stop.read()) & ( pce.read() | pce_ov.read() ));
  }
}

void sc1_l1hdr_32x_proc3()
{
  l1clk.write((l2clk.read() & l1en) | se.read());
}

/* Consructor */
SC_CTOR(cl_sc1_l1hdr_32x)
{
```

**Fig. 8.28** (continued)

```
    /* Declare/assign threads, methods and sensitivity lists for each */
    SC_METHOD(sc1_l1hdr_32x_proc0);
    SC_METHOD(sc1_l1hdr_32x_proc3);
    SC_THREAD(sc1_l1hdr_32x_proc1);
    sensitive << l2clk << stop << pce << pce_ov;
    SC_CTHREAD(sc1_l1hdr_32x_proc2, l2clk.neg());
  }

  ~cl_sc1_l1hdr_32x(){ }
 /* Destructor */
};


/* Master-slave flip-flop class */
SC_MODULE(cl_sc1_msff_8x)
{
  /* Declare/define input/output ports */
  sc_core::sc_in<bool> d;
  sc_core::sc_in<bool> l1clk;
  sc_core::sc_in<bool> si;
  sc_core::sc_in<bool> siclk;
  sc_core::sc_in<bool> soclk;
  sc_core::sc_out<bool> q;
  sc_core::sc_out<bool> so;
/* Declare members */
  bool l1;
  bool si_unused;
  bool siclk_unused;
  bool soclk_unused;
  bool FAST_FLUSH;
  bool SCAN_MODE;
  bool INITLATZERO;

  /* Flip-flop method */
  void cl_sc1_msff_8x_proc0()
  {
    if(SCAN_MODE == true && INITLATZERO == true)
    {
      l1 = false;
    }
```

**Fig. 8.28** (continued)

```
}

/* Flip-flop threadsand methods */
void cl_sc1_msff_8x_proc1()
{
 while(1)
 {
  wait();
  if(SCAN_MODE == true && FAST_FLUSH == true)
  {
   if(siclk.read() == true) q.write(false);
   else q.write(d.read());
  }
 }
}

void cl_sc1_msff_8x_proc2()
{
 while(1)
 {
  wait();
  if(SCAN_MODE == true && FAST_FLUSH == false)
  {
   if(l1clk.read() == false &&
     siclk.read() == false) l1 = d.read();
  else if(l1clk.read() == true &&
        siclk.read() == true) l1 = si.read();
  else if(l1clk.read() == false &&
        siclk.read() == true) l1 = false;
  else if(l1clk.read() == true &&
        siclk.read() == false &&
        soclk.read() == false)  q.write(l1);
  if(l1clk.read() == true &&
    siclk.read() == true &&
    soclk.read() == true) q.write(si.read());
 }
 }
}

void cl_sc1_msff_8x_proc3()
```

**Fig. 8.28** (continued)

```
    {
      if(SCAN_MODE == false && FAST_FLUSH == false)
       {
         si_unused = si.read();
         siclk_unused = siclk.read();
         soclk_unused = soclk.read();
       }
    }

    void cl_sc1_msff_8x_proc4()
    {
      while(1)
      {
        wait();
         if(siclk.read() == false &&
            soclk.read() == false) q.write(d.read());
         else q.write(false);
      }
    }

    /* Constructor */
    SC_CTOR(cl_sc1_msff_8x)
    {
      /* Declare/assign methods/threads and sensitivity lists */
      SC_METHOD(cl_sc1_msff_8x_proc0);
      SC_METHOD(cl_sc1_msff_8x_proc3);
      SC_THREAD(cl_sc1_msff_8x_proc1);
      sensitive << l1clk.pos() << siclk.pos();
      SC_THREAD(cl_sc1_msff_8x_proc2);
      sensitive << l1clk << siclk << soclk << d << si;
      SC_THREAD(cl_sc1_msff_8x_proc4);
      sensitive << l1clk.pos();
    }
    ~cl_sc1_msff_8x(){ }
  /* destructor */
 };

 SC_MODULE(n2_flop_bank_cust)
 {
   /* Declare/deine input/output ports */
   sc_core::sc_in<bool> l2clk;
```

**Fig. 8.28** (continued)

```
  sc_core::sc_in<bool> scan_in;
  sc_core::sc_in<bool> tcu_aclk;
  sc_core::sc_in<bool> tcu_bclk;
  sc_core::sc_in<bool> tcu_scan_en;
  sc_core::sc_in<bool> tcu_pce_ov;
  sc_core::sc_in<bool> local_stop;
  sc_core::sc_in< sc_dt::sc_bv<16> > data_in;
  sc_core::sc_out<bool> scan_out;
  sc_core::sc_out< sc_dt::sc_bv<16> > data_out;
 /* Declare/define channels/signals */
  sc_core::sc_signal<bool> siclk;
  sc_core::sc_signal<bool> soclk;
  sc_core::sc_signal<bool> se;
  sc_core::sc_signal<bool> pce_ov;
  sc_core::sc_signal<bool> l1clk;
  sc_core::sc_signal<bool> datain_ff_scanin;
  sc_core::sc_signal<bool> datain_ff_scanout;
  sc_core::sc_signal<bool> true_sig;

  /* Declare/define members and intermediate channels/signals  */
  sc_dt::sc_bv<16> data_in_tmp;
  sc_dt::sc_bv<16> data_out_tmp;
  sc_core::sc_signal<bool> fdin0;
  sc_core::sc_signal<bool> fdin1;
  sc_core::sc_signal<bool> fdin2;
  sc_core::sc_signal<bool> fdin3;
  sc_core::sc_signal<bool> fdin4;
  sc_core::sc_signal<bool> fdin5;
  sc_core::sc_signal<bool> fdin6;
  sc_core::sc_signal<bool> fdin7;
  sc_core::sc_signal<bool> fdin8;
  sc_core::sc_signal<bool> fdin9;
  sc_core::sc_signal<bool> fdin10;
  sc_core::sc_signal<bool> fdin11;
  sc_core::sc_signal<bool> fdin12;
  sc_core::sc_signal<bool> fdin13;
  sc_core::sc_signal<bool> fdin14;

  sc_core::sc_signal<bool> fo0;
  sc_core::sc_signal<bool> fo1;
  sc_core::sc_signal<bool> fo2;
```

**Fig. 8.28** (continued)

```
sc_core::sc_signal<bool> fo3;
sc_core::sc_signal<bool> fo4;
sc_core::sc_signal<bool> fo5;
sc_core::sc_signal<bool> fo6;
sc_core::sc_signal<bool> fo7;
sc_core::sc_signal<bool> fo8;
sc_core::sc_signal<bool> fo9;
sc_core::sc_signal<bool> fo10;

sc_core::sc_signal<bool> fo11;
sc_core::sc_signal<bool> fo12;
sc_core::sc_signal<bool> fo13;
sc_core::sc_signal<bool> fo14;
sc_core::sc_signal<bool> fo15;
sc_core::sc_signal<bool> dins0;
sc_core::sc_signal<bool> dins1;
sc_core::sc_signal<bool> dins2;
sc_core::sc_signal<bool> dins3;
sc_core::sc_signal<bool> dins4;
sc_core::sc_signal<bool> dins5;
sc_core::sc_signal<bool> dins6;
sc_core::sc_signal<bool> dins7;
sc_core::sc_signal<bool> dins8;
sc_core::sc_signal<bool> dins9;
sc_core::sc_signal<bool> dins10;
sc_core::sc_signal<bool> dins11;
sc_core::sc_signal<bool> dins12;
sc_core::sc_signal<bool> dins13;
sc_core::sc_signal<bool> dins14;
sc_core::sc_signal<bool> dins15;

sc_core::sc_signal<bool> dos0;
sc_core::sc_signal<bool> dos1;
sc_core::sc_signal<bool> dos2;
sc_core::sc_signal<bool> dos3;
sc_core::sc_signal<bool> dos4;
sc_core::sc_signal<bool> dos5;
sc_core::sc_signal<bool> dos6;
sc_core::sc_signal<bool> dos7;
sc_core::sc_signal<bool> dos8;
```

**Fig. 8.28** (continued)

```
sc_core::sc_signal<bool> dos9;
sc_core::sc_signal<bool> dos10;
sc_core::sc_signal<bool> dos11;
sc_core::sc_signal<bool> dos12;
sc_core::sc_signal<bool> dos13;
sc_core::sc_signal<bool> dos14;
sc_core::sc_signal<bool> dos15;

bool din0;
bool din1;
bool din2;
bool din3;
bool din4;
bool din5;
bool din6;
bool din7;
bool din8;
bool din9;
bool din10;
bool din11;
bool din12;
bool din13;
bool din14;
bool din15;

bool do0;
bool do1;
bool do2;
bool do3;
bool do4;
bool do5;
bool do6;
bool do7;
bool do8;
bool do9;
bool do10;
bool do11;
bool do12;
bool do13;
bool do14;
bool do15;
```

**Fig. 8.28** (continued)

```
/* Method to format input data for buffersand flip-flops and output data
   for output signals/ports */
void n2_flop_bank_cust_proc0()
{
   se.write(tcu_scan_en.read());
   pce_ov.write(tcu_pce_ov.read());
   data_in_tmp = data_in.read();
   din0 = data_in_tmp[0] == "1" ? true : false;
   din1 = data_in_tmp[1] == "1" ? true : false;
   din2 = data_in_tmp[2] == "1" ? true : false;
   din3 = data_in_tmp[3] == "1" ? true : false;
   din4 = data_in_tmp[4] == "1" ? true : false;
   din5 = data_in_tmp[5] == "1" ? true : false;
   din6 = data_in_tmp[6] == "1" ? true : false;
   din7 = data_in_tmp[7] == "1" ? true : false;
   din8 = data_in_tmp[8] == "1" ? true : false;
   din9 = data_in_tmp[9] == "1" ? true : false;
   din10 = data_in_tmp[10] == "1" ? true : false;
   din11 = data_in_tmp[11] == "1" ? true : false;
   din12 = data_in_tmp[12] == "1" ? true : false;
   din13 = data_in_tmp[13] == "1" ? true : false;
   din14 = data_in_tmp[14] == "1" ? true : false;
   din15 = data_in_tmp[15] == "1" ? true : false;

   dins0.write(din0);
   dins1.write(din1);
   dins2.write(din2);
   dins3.write(din3);
   dins4.write(din4);
   dins5.write(din5);
   dins6.write(din6);
   dins7.write(din7);
   dins8.write(din8);
   dins9.write(din9);
   dins10.write(din10);
   dins11.write(din11);
   dins12.write(din12);
   dins13.write(din13);
   dins14.write(din14);
   dins15.write(din15);
```

**Fig. 8.28** (continued)

```
        do0 = dos0.read();
        do1 = dos1.read();
        do2 = dos2.read();
        do3 = dos3.read();
        do4 = dos4.read();
        do5 = dos5.read();
        do6 = dos6.read();
        do7 = dos7.read();
        do8 = dos8.read();
        do9 = dos9.read();
        do10 = dos10.read();
        do11 = dos11.read();
        do12 = dos12.read();
        do13 = dos13.read();
        do14 = dos14.read();
        do15 = dos15.read();

        data_out_tmp[0] = do0 == true ? "1" : "0";
        data_out_tmp[1] = do1 == true ? "1" : "0";
        data_out_tmp[2] = do2 == true ? "1" : "0";
        data_out_tmp[3] = do3 == true ? "1" : "0";
        data_out_tmp[4] = do4 == true ? "1" : "0";
        data_out_tmp[5] = do5 == true ? "1" : "0";
        data_out_tmp[6] = do6 == true ? "1" : "0";
        data_out_tmp[7] = do7 == true ? "1" : "0";
        data_out_tmp[8] = do8 == true ? "1" : "0";
        data_out_tmp[9] = do9 == true ? "1" : "0";
        data_out_tmp[10] = do10 == true ? "1" : "0";
        data_out_tmp[11] = do11 == true ? "1" : "0";
        data_out_tmp[12] = do13 == true ? "1" : "0";
        data_out_tmp[14] = do14 == true ? "1" : "0";
        data_out_tmp[15] = do15 == true ? "1" : "0";
        data_out.write(data_out_tmp);
    }

  /* Declare all internal buffers and flip-flops */
    cl_u1_buf_32x buf_0_;
    cl_u1_buf_32x buf_1_;
    cl_u1_buf_32x buf_2_;
    cl_u1_buf_32x buf_3_;
```

**Fig. 8.28** (continued)

```
  cl_u1_buf_32x buf_4_;
  cl_u1_buf_32x buf_5_;
  cl_u1_buf_32x buf_6_;
  cl_u1_buf_32x buf_7_;
  cl_u1_buf_32x buf_8_;
  cl_u1_buf_32x buf_9_;
  cl_u1_buf_32x buf_10_;
  cl_u1_buf_32x buf_11_;
  cl_u1_buf_32x buf_12_;
  cl_u1_buf_32x buf_13_;
  cl_u1_buf_32x buf_14_;
  cl_u1_buf_32x buf_15_;

  cl_sc1_l1hdr_32x c_0;
  cl_sc1_msff_8x d0_0;
  cl_sc1_msff_8x d0_1;
  cl_sc1_msff_8x d0_2;
  cl_sc1_msff_8x d0_3;
  cl_sc1_msff_8x d0_4;
  cl_sc1_msff_8x d0_5;
  cl_sc1_msff_8x d0_6;
  cl_sc1_msff_8x d0_7;
  cl_sc1_msff_8x d0_8;
  cl_sc1_msff_8x d0_9;
  cl_sc1_msff_8x d0_10;
  cl_sc1_msff_8x d0_11;
  cl_sc1_msff_8x d0_12;
  cl_sc1_msff_8x d0_13;
  cl_sc1_msff_8x d0_14;
  cl_sc1_msff_8x d0_15;

/* Constructor with member initialization */
  SC_CTOR(n2_flop_bank_cust):buf_0_("buf_0_"), buf_1_("buf_1_"),
                      buf_2_("buf_2_"), buf_3_("buf_3_"),
                      buf_4_("buf_4_"), buf_5_("buf_5_"),
                      buf_6_("buf_6_"), buf_7_("buf_7_"),
                      buf_8_("buf_8_"), buf_9_("buf_9_"),
                      buf_10_("buf_10_"), buf_11_("buf_11_"),
                      buf_12_("buf_12_"), buf_13_("buf_13_"),
                      buf_14_("buf_14_"), buf_15_("buf_15_"),
```

**Fig. 8.28** (continued)

```
                c_0("c_0"), d0_0("d0_0"), d0_1("d0_1"),
                d0_2("d0_2"), d0_3("d0_3"), d0_4("d0_4"),
                d0_5("d0_5"), d0_6("d0_6"), d0_7("d0_7"),
                d0_8("d0_8"), d0_9("d0_9"), d0_10("d0_10"),
                d0_11("d0_11"), d0_12("d0_12"),
                d0_13("d0_13"), d0_14("d0_14"),
                d0_15("d0_15")
{
  /* Connect module(buffer/flip-flop) ports and channels */
  buf_15_.in(fo15);
  buf_15_.out(dos15);
  buf_14_.in(fo14);
  buf_14_.out(dos14);
  buf_13_.in(fo13);
  buf_13_.out(dos13);
  buf_12_.in(fo12);
  buf_12_.out(dos12);
  buf_11_.in(fo11);
  buf_11_.out(dos11);
  buf_10_.in(fo10);
  buf_10_.out(dos10);
  buf_9_.in(fo9);
  buf_9_.out(dos9);
  buf_8_.in(fo8);
  buf_8_.out(dos8);
  buf_7_.in(fo7);
  buf_7_.out(dos7);
  buf_6_.in(fo6);
  buf_6_.out(dos6);
  buf_5_.in(fo5);
  buf_5_.out(dos5);
  buf_4_.in(fo4);
  buf_4_.out(dos4);
  buf_3_.in(fo3);
  buf_3_.out(dos3);
  buf_2_.in(fo2);
  buf_2_.out(dos2);
  buf_1_.in(fo1);
  buf_1_.out(dos1);
  buf_0_.in(fo0);
```

**Fig. 8.28** (continued)

```
buf_0_.out(dos0);

c_0.l2clk(l2clk);
c_0.pce(true_sig);
c_0.l1clk(l1clk);
c_0.se(se);
c_0.pce_ov(pce_ov);
c_0.stop(local_stop);

d0_0.l1clk(l1clk);
d0_0.siclk(tcu_aclk);
d0_0.soclk(tcu_bclk);
d0_0.d(dins0);
d0_0.q(fo0);
d0_0.si(scan_in);
d0_0.so(fdin0);

d0_1.l1clk(l1clk);
d0_1.siclk(tcu_aclk);
d0_1.soclk(tcu_bclk);
d0_1.d(dins1);
d0_1.q(fo1);
d0_1.si(fdin0);
d0_1.so(fdin1);

d0_2.l1clk(l1clk);
d0_2.siclk(tcu_aclk);
d0_2.soclk(tcu_bclk);
d0_2.d(dins2);
d0_2.q(fo2);
d0_2.si(fdin1);
d0_2.so(fdin2);

d0_3.l1clk(l1clk);
d0_3.siclk(tcu_aclk);
d0_3.soclk(tcu_bclk);
d0_3.d(dins3);
d0_3.q(fo3);
d0_3.si(fdin2);
d0_3.so(fdin3);
```

**Fig. 8.28** (continued)

```
d0_4.l1clk(l1clk);
d0_4.siclk(tcu_aclk);
d0_4.soclk(tcu_bclk);
d0_4.d(dins4);
d0_4.q(fo4);
d0_4.si(fdin3);
d0_4.so(fdin4);

d0_5.l1clk(l1clk);
d0_5.siclk(tcu_aclk);
d0_5.soclk(tcu_bclk);
d0_5.d(dins5);
d0_5.q(fo5);
d0_5.si(fdin4);
d0_5.so(fdin5);

d0_6.l1clk(l1clk);
d0_6.siclk(tcu_aclk);
d0_6.soclk(tcu_bclk);
d0_6.d(dins6);
d0_6.q(fo6);
d0_6.si(fdin5);
d0_6.so(fdin6);

d0_7.l1clk(l1clk);
d0_7.siclk(tcu_aclk);
d0_7.soclk(tcu_bclk);
d0_7.d(dins7);
d0_7.q(fo7);
d0_7.si(fdin6);
d0_7.so(fdin7);

d0_8.l1clk(l1clk);
d0_8.siclk(tcu_aclk);
d0_8.soclk(tcu_bclk);
d0_8.d(dins8);
d0_8.q(fo8);
d0_8.si(fdin7);
d0_8.so(fdin8);
```

**Fig. 8.28** (continued)

```
d0_9.l1clk(l1clk);
d0_9.siclk(tcu_aclk);
d0_9.soclk(tcu_bclk);
d0_9.d(dins9);
d0_9.q(fo9);
d0_9.si(fdin8);
d0_9.so(fdin9);

d0_10.l1clk(l1clk);
d0_10.siclk(tcu_aclk);
d0_10.soclk(tcu_bclk);
d0_10.d(dins10);
d0_10.q(fo10);
d0_10.si(fdin9);
d0_10.so(fdin10);

d0_11.l1clk(l1clk);
d0_11.siclk(tcu_aclk);
d0_11.soclk(tcu_bclk);
d0_11.d(dins11);
d0_11.q(fo11);
d0_11.si(fdin10);
d0_11.so(fdin11);

d0_12.l1clk(l1clk);
d0_12.siclk(tcu_aclk);
d0_12.soclk(tcu_bclk);
d0_12.d(dins12);
d0_12.q(fo12);
d0_12.si(fdin11);
d0_12.so(fdin12);

d0_13.l1clk(l1clk);
d0_13.siclk(tcu_aclk);
d0_13.soclk(tcu_bclk);
d0_13.d(dins13);
d0_13.q(fo13);
d0_13.si(fdin12);
d0_13.so(fdin13);
```

**Fig. 8.28** (continued)

```
    d0_14.l1clk(l1clk);
    d0_14.siclk(tcu_aclk);
    d0_14.soclk(tcu_bclk);
    d0_14.d(dins14);
    d0_14.q(fo14);
    d0_14.si(fdin13);
    d0_14.so(fdin14);

    d0_15.l1clk(l1clk);
    d0_15.siclk(tcu_aclk);
    d0_15.soclk(tcu_bclk);
    d0_15.d(dins15);
    d0_15.q(fo15);
    d0_15.si(fdin14);
    d0_15.so(scan_out);

    SC_METHOD(n2_flop_bank_cust_proc0);
  }

  ~n2_flop_bank_cust() { }
  /*Destructor */
};
#endif
```

**Fig. 8.28** (continued)

```
#include "n2_flop_bank_cust.h"
#include "syscsrcs.h"

int sc_main(int argc, char **argv)
{
  /* Declare/define channels/signals */
  sc_core::sc_signal<bool> scan_in_sig;
  sc_core::sc_signal<bool> tcu_scan_en_sig;
  sc_core::sc_signal<bool> tcu_pce_ov_sig;
  sc_core::sc_signal<bool> local_stop_sig;
  sc_core::sc_signal< sc_dt::sc_bv<16> > data_in_sig;
  sc_core::sc_signal<bool> scan_out_sig;
  sc_core::sc_signal< sc_dt::sc_bv<16> > data_out_sig;
```

**Fig. 8.29** Test harness for T2 series connected flip-flops

```
  sc_core::sc_signal<bool> clkA;
  sc_core::sc_signal<bool> clkB;
  sc_core::sc_signal<bool> clkC;

  /* Declare/define three clocks */
  /* First the master clock */
  sc_core::sc_clock masterclock("masterclock",
                                2.0,
                                sc_core::SC_NS, 0.5);

  /* Then three derived clocks */
  clkdiv_2_4_8 clk_div_2_4_8("clk_div_2_4_8");

  /* Declare/define intermediate variables */
  sc_dt::sc_bv<16> data_in_value;
  sc_dt::sc_bv<16> data_out_value;
  bool scan_in_value;
  bool tcu_scan_en_value;
  bool tcu_pce_ov_value;
  bool local_stop_value;

/* Declare/define flip-flop bank and trace file object */
  n2_flop_bank_cust n2_flop_bank_cust_OBJ("n2_flop_bank_cust_OBJ");
  sc_core::sc_trace_file *fp =
sc_core::sc_create_vcd_trace_file("tr_flop_bank");

  fp->set_time_unit(500.0, sc_core::SC_PS);

  /* Connect master and derived clocks */
  clk_div_2_4_8.clk(masterclock);
  clk_div_2_4_8.clk2(clkA);
  clk_div_2_4_8.clk4(clkB);
  clk_div_2_4_8.clk8(clkC);

  /* Connect flip-flop bank ports and channels/signals */
  n2_flop_bank_cust_OBJ.l2clk(clkA);
  n2_flop_bank_cust_OBJ.tcu_aclk(clkB);
  n2_flop_bank_cust_OBJ.tcu_bclk(clkC);
  n2_flop_bank_cust_OBJ.scan_in(scan_in_sig);
```

**Fig. 8.29** (continued)

```
n2_flop_bank_cust_OBJ.tcu_scan_en(tcu_scan_en_sig);
n2_flop_bank_cust_OBJ.tcu_pce_ov(tcu_pce_ov_sig);
n2_flop_bank_cust_OBJ.local_stop(local_stop_sig);
n2_flop_bank_cust_OBJ.data_in(data_in_sig);
n2_flop_bank_cust_OBJ.scan_out(scan_out_sig);
n2_flop_bank_cust_OBJ.data_out(data_out_sig);

/* Connect data channels/signals with trace file */
sc_core::sc_trace(fp, clkA, "clk");
sc_core::sc_trace(fp, clkB, "aclk");
sc_core::sc_trace(fp, clkC, "bclk");
sc_core::sc_trace(fp, scan_in_sig, "scan_in");
sc_core::sc_trace(fp, tcu_scan_en_sig, "scan_en");
sc_core::sc_trace(fp, tcu_pce_ov_sig, "pce_ov");
sc_core::sc_trace(fp, data_in_sig, "data_in");
sc_core::sc_trace(fp, data_out_sig, "data_out");

/* Generate initial input data and write to channels */
data_in_value = "0000000000000000";
scan_in_value = false;
tcu_scan_en_value = false;
tcu_pce_ov_value = false;
local_stop_value = true;
data_in_sig.write(data_in_value);
scan_in_sig.write(scan_in_value);
tcu_scan_en_sig.write(tcu_scan_en_value);
tcu_pce_ov_sig.write(tcu_pce_ov_value);
local_stop_sig.write(local_stop_value);
std::cout<<"Start"<<std::endl;

/* Run simulation for pre-defined time-interval */
sc_core::sc_start(25.0, sc_core::SC_NS);

/* Change input data and write to channels */
data_in_value = "0010101010011110";
scan_in_value = true;
tcu_scan_en_value = true;
tcu_pce_ov_value = false;
```

**Fig. 8.29** (continued)

```
local_stop_value = false;
data_in_sig.write(data_in_value);
scan_in_sig.write(scan_in_value);
tcu_scan_en_sig.write(tcu_scan_en_value);
tcu_pce_ov_sig.write(tcu_pce_ov_value);
local_stop_sig.write(local_stop_value);

/* Run simulation for pre-defined time-interval */
sc_core::sc_start(500.0, sc_core::SC_PS);

/* Change input data and write to channels */
data_in_value = "0110101010011000";
scan_in_value = true;
tcu_scan_en_value = true;
tcu_pce_ov_value = true;
local_stop_value = false;
data_in_sig.write(data_in_value);
scan_in_sig.write(scan_in_value);
tcu_scan_en_sig.write(tcu_scan_en_value);
tcu_pce_ov_sig.write(tcu_pce_ov_value);
local_stop_sig.write(local_stop_value);

/* Run simulation for pre-defined time-interval */
sc_core::sc_start(25.0, sc_core::SC_NS);

/* Change input data and write to channels */
data_in_value = "1010101010000110";
scan_in_value = true;
tcu_scan_en_value = true;
tcu_pce_ov_value = true;
local_stop_value = true;
data_in_sig.write(data_in_value);
scan_in_sig.write(scan_in_value);
tcu_scan_en_sig.write(tcu_scan_en_value);
tcu_pce_ov_sig.write(tcu_pce_ov_value);
local_stop_sig.write(local_stop_value);

/* Run simulation for pre-defined time-interval */
sc_core::sc_start(25.0, sc_core::SC_NS);
```

**Fig. 8.29** (continued)

```
  /* Change input data and write to channels */
  data_in_value = "0010101011111110";
  scan_in_value = true;
  tcu_scan_en_value = true;
  tcu_pce_ov_value = false;
  local_stop_value = true;
  data_in_sig.write(data_in_value);
  scan_in_sig.write(scan_in_value);
  tcu_scan_en_sig.write(tcu_scan_en_value);
  tcu_pce_ov_sig.write(tcu_pce_ov_value);
  local_stop_sig.write(local_stop_value);

  /* Run simulation for pre-defined time-interval */
  sc_core::sc_start(25.0, sc_core::SC_NS);

  /* Change input data and write to channels */
  data_in_value = "0010101010011110";
  scan_in_value = true;
  tcu_scan_en_value = true;
  tcu_pce_ov_value = false;
  local_stop_value = false;
  data_in_sig.write(data_in_value);
  scan_in_sig.write(scan_in_value);
  tcu_scan_en_sig.write(tcu_scan_en_value);
  tcu_pce_ov_sig.write(tcu_pce_ov_value);
  local_stop_sig.write(local_stop_value);

  /* Run simulation for pre-defined time-interval */
  sc_core::sc_start(25.0, sc_core::SC_NS);

  /* Stop simulation and close trace file */
  sc_core::sc_stop();
  sc_core::sc_close_vcd_trace_file(fp);
  return 0;
}
```

**Fig. 8.29** (continued)

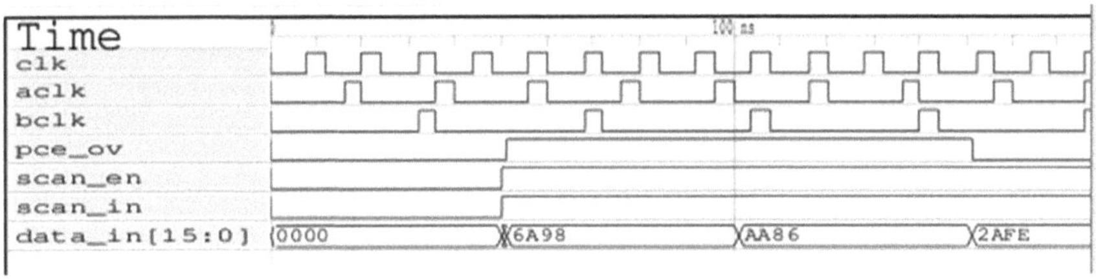


**Fig. 8.30** Partial input/output trace for T2 series connected flip-flops SystemC-2.2.0

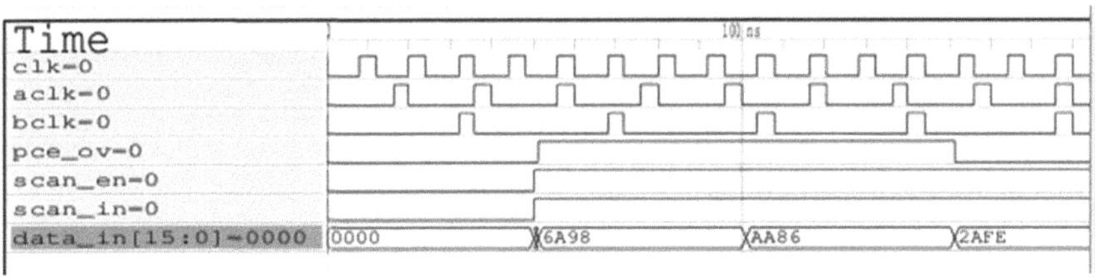


**Fig. 8.31** Partial input/output trace for T2 series connected flip-flops SystemC-2.3.0

B wrote: Test character string for customized SystemC channel
A read: Test character string for customized SystemC channel
B wrote: Test character string for customized SystemC channel
A read: Test character string for customized SystemC channel
B wrote: Test character string for customized SystemC channel
SystemC: simulation stopped by user.

SystemC 2.3.0-ASI—Jul 19 2012 18:53:11
Copyright (c) 1996-2012 by all Contributors,
ALL RIGHTS RESERVED

A read:
B wrote: Test character string for customized SystemC channel
A read: Test character string for customized SystemC channel
B wrote: Test character string for customized SystemC channel
A read: Test character string for customized SystemC channel
B wrote: Test character string for customized SystemC channel
A read: Test character string for customized SystemC channel
B wrote: Test character string for customized SystemC channel
A read: Test character string for customized SystemC channel
B wrote: Test character string for customized SystemC channel
Info: /OSCI/SystemC: Simulation stopped by user.

It is worth noting that the very first time the receiver attempts to read data from the channel, it is empty. Then, as the sender writing data to the channel, it read data from it.

```
#ifndef CHNL_H
#define CHNL_H

#include <systemc>
#include <iostream>
#include <cstring>
#include <cstdlib>

/* Custom signal channel interface declaration/definition */
class chnlIf : public sc_core::sc_interface
{
  public:
  virtual void write_proc(char *) = 0;
  virtual void read_proc(char *) = 0;
  /* Pure virtual interface methods to be overridden by channel
      implementation class */
};

/* Signal channel implementation class */
class chnl : public chnlIf, public sc_core::sc_module
{
  private:
   char msg[64];
   enum states{READY = 0, BUSY };
   unsigned int chnlstatus;
   unsigned int waitperiod;

   /* Private data members, including array for channel data storage */
  public:
   SC_HAS_PROCESS(chnl);
   sc_core::sc_in<bool> clk;

   void write_proc(char cp)
 /* Method for writing data to signal channel */
   {
     if(cp != NULL)
     {
```

**Fig. 8.32** Custom signal interface and channel

```
      wait();
      /* Channel is busy -- wait for a random period of time */
      while(chnlstatus == BUSY)
      {
        std::cout<<"waiting to write ..."<<std::endl;
        waitperiod = std::rand() % 10;
        wait(waitperiod);
      }

      /* Channel is ready to be written to - set channel
         status to BUSY and copy data to buffer.
         When data writing is complete, set channel
         status to READY */
      if(chnlstatus == READY)
      {
        chnlstatus = BUSY;
        strncpy(msg, cp, strlen(cp));
        chnlstatus = READY;
      }
    }
  }

  void read_proc(char cp)
/* Method to read data from channel */
  {
    if(cp != NULL)
    {
      wait();
      /* Channel is busy -- wait for a random period of time */
      while(chnlstatus == BUSY)
      {
        std::cout<<"Channel busy wait - read ..."<<std::endl;
        waitperiod = std::rand() % 10;
        wait(waitperiod);
      }
       /* Channel is ready for data read  - set channel
         status to BUSY and copy data from channel buffer.
         When data reading is complete, set channel
         status to READY */
      if(chnlstatus == READY)
      {
```

**Fig. 8.32** (continued)

```
        chnlstatus = BUSY;
        strncpy(cp, msg, strlen(msg));
        chnlstatus = READY;
      }
    }
  }

  /* Consrructor */
  chnl (sc_core::sc_module_name nm) : sc_core::sc_module(nm),
                          chnlstatus(READY),
                          waitperiod(0)
  {
    bzero(&msg[0], 64*sizeof(char));
    sensitive<<clk.pos();
 /* Assign sensitivity list */
  }

  ~chnl() { }
};

#endif
```

**Fig. 8.32** (continued)

```
#ifndef CHNLENDS_H
#define CHNLENDS_H

#include "chnl.h"

const unsigned int MAX = 64;
const char *str = "Test character string for customized SystemC channel";

/* Receiver class */
class chnlendA : public sc_core::sc_module
{
  private:
   chnl &chnlPtr;
      /* Reference to signal/channel */
   char tmpA[MAX];
 /* Buffer to read data into */
 public:
  sc_core::sc_in<bool> clk;
```

**Fig. 8.33** Receiver and sender classes for custom signal interface and class

```
  typedef chnlendA SC_CURRENT_USER_MODULE;
  /* Thread for reading data from channel */
  void readtest()
  {
    while(1)
    {
      wait();
      /* Invoke signal/channel's 'read' method to read data from it */
      chnlPtr.read_proc(&tmpA[0]);
      std::cout<<name()<<" read : "<<tmpA<<std::endl;
      bzero(&tmpA[0], MAX*sizeof(char));
 /* Clear buffer for next read */
    }
  }

  /* Constructor */
  chnlendA(sc_core::sc_module_name nm,
          chnl &chnlPtr) : sc_core::sc_module(nm), chnlPtr(chnlPtr)
  {
    SC_THREAD(readtest);
 /* Declare/assign thread and sensitivity list */
    sensitive<<clk.pos();
    bzero(&tmpA[0], MAX*sizeof(char));
  }
  ~chnlendA(){ }
 /* Destructor */
};

/* Sender class */
class chnlendB : public sc_core::sc_module
{
  private:
   chnl &chnlPtr;
 /* Reference to signal/channel */
   char tmpB[MAX];
 /* Temporary data storage before writing to channel */
  public:
   sc_core::sc_in<bool> clk;
  SC_HAS_PROCESS(chnlendB);
  /* Data write thread */
  void writetest()
  {
    while(1)
```

**Fig. 8.33** (continued)

```
    {
      wait();
      sprintf(&tmpB[0], "%s", str);
      /* Invoke signal/channel's 'write' method to  write data to it */
      chnlPtr.write_proc(&tmpB[0]);
      std::cout<<name()<<" wrote : "<<tmpB<<std::endl;
      bzero(&tmpB[0], MAX*sizeof(char));
    }
  }

  /* Consrructor */
  chnlendB(sc_core::sc_module_name nm,
        chnl &chnlPtr) : sc_core::sc_module(nm), chnlPtr(chnlPtr)
  {
    SC_THREAD(writetest);
 /* Declare/assign thread and sensitivity list */
    sensitive<<clk.pos();
    bzero(&tmpB[0], MAX*sizeof(char));
  }

  ~chnlendB(){ }
 /*Destructor */
};

#endif
```

**Fig. 8.33** (continued)

```
#include "chnlends.h"
int sc_main (int argc, char **argv)
{
  sc_core::sc_clock clock("Clock", 10.0, sc_core::SC_US, 0.5);
  chnl chnltest("chnl_test");
  chnltest.clk(clock);
  chnlendA chendA("A", chnltest);
  chendA.clk(clock);
  chnlendB chendB("B", chnltest);
  chendB.clk(clock);
  sc_core::sc_start(100.0, sc_core::SC_US);
  sc_core::sc_stop();
  return 0;
}
```

**Fig. 8.34** Test harness for custom signal/channel, sender, and receiver

## 8.8 Simple Moving Average Filter

A moving average is a finite impulse response filter to analyze a data set by creating a series of averages of subsets of the full data set to smooth out short-term seemingly random variations and extract log-term trends, for example number of packets per unit time arriving at a backbone router. Given a large set of numbers and a fixed subset size, the first element of the moving average is obtained by taking the average of the initial fixed subset of the large set. Then, the subset is modified by '*shifting forward*,' by excluding the first number of the series and including the next number following the original subset in the series, resulting in a new average value. This process is repeated over the entire data set. There are several variations of this scheme, often involving using weights for different elements of the subset and so on. The raw data is hard to comprehend, but correct analysis yields a wealth of hidden information. There are several moving average filter algorithms, and the choice of one scheme for a given data set depends entirely on the domain knowledge of the user. A simple moving average filter and test harness are in Figs. 8.35, 8.36. The input and filter output (SystemC 2.3.0) are in Figs. 8.37, 8.38, respectively.

The smoothing action of the filter is clear. While the input values range between 3.068E + 07 and 2.13E + 09, the output values range between 9.39E + 08 and 1.25E + 09.

## 8.9 Level-Sensitive Scan: Clock Generator

While increasing the number of MOSFETs/transistors on an integrated circuit is essential for performance enhancements, fast, accurate testing techniques are essential to verify that the final design satisfies all specifications. A popular technique is *Level-Sensitive Scan* (LSS) [1]. Scan design increases the controllability and observability of a digital logic circuit by having special 'scan' registers to create a scan path. Level-sensitive scan design (LSSD) uses separate scan and system clocks to separate the normal and scan modes. Latches are used in pairs, each with a normal data input, data output and clock for system operation. In the scan mode, two latches form a master–slave pair with one scan input, one scan output, and *non-overlapping* scan clocks. The scan clocks are held low during normal circuit operation and are high in scan mode, allowing scan data to be latched. In a single-latch LSS configuration, the second latch is only used in scan configuration, allowing to be used as a system latch during normal operation and reducing silicon overhead.

Here, a simple clock generator for LSS is analyzed. It is hierarchical combination of 3-input, 2-input NOR gates, and two inverter types. The source code and test harness are in Figs. 8.39, 8.40, respectively. The input/output traces are in Figs. 8.41 (SystemC-2.2.0) and 8.42 (SystemC-2.3.0).

```
#ifndef MOVAVG_H
#define MOVAVG_H

#include <systemc>

const unsigned int MAXFILTELEM = 55;
/* Define fixed subset size */

SC_MODULE(movavg)
{
 /* Input/output ports and members */
 public:
  sc_core::sc_in<bool> clk;
  sc_core::sc_in<double> insig;
  sc_core::sc_out<double> outsig;

   bool first;
   double lastsample;
   double sum;
   double sum_copy;
   double tmp;
   double tmp0;
   unsigned int count;

 /* Moving average filter calculation thread */
   void movavg_proc0()
   {
     while(1)
     {
       wait();
       tmp0 = insig.read();
       if(count < MAXFILTELEM)
       {
         /* Check for start of large data set */
         if(first == false && count == 0)
         {
           sum += tmp0;
           first = true;
         }
 else if(first == true && count == 0)
 {
```

**Fig. 8.35** Simple moving average filter algorithm

```
        /* Use empirical weights for data elements at
           boundary of data subset */
        sum += 0.25*tmp0;
        sum += 0.75*lastsample;
      }
      else if(first == true && count > 0)
      {
        sum += tmp0;
      }
      count += 1;
      if(count == MAXFILTELEM - 1)
      {
        lastsample = tmp0;
        sum_copy = sum;
        sum_copy /= MAXFILTELEM;
      }
      }
      else
      {
        count = 0;
        sum = 0.0;
      }
      outsig.write(sum_copy);
    }
  }

  SC_CTOR(movavg): count(0),
                   first(false),
                   lastsample(0.0),
                   sum(0.0),
                   sum_copy(0.0),
                   tmp0(0.0)
  {
    SC_THREAD(movavg_proc0);
    sensitive << clk.pos();
  }

  ~movavg(){ }
};

#endif
```

**Fig. 8.35** (continued)

```
#include "movavg.h"
#include "syscsrcs.h"
#include "sysctrace.h"

int sc_main(int argc, char **argv)
{
  /* Declare/define channels */
  sc_core::sc_signal<double> dblsig;
  sc_core::sc_signal<double> avgsig;

  /* Define clock for synchronization */
  sc_core::sc_clock clk("clk", 10.0, sc_core::SC_US, 0.5);
  /* Declare/define moving average filter */
  movavg mavg("mavg");
  /* Declare/define input data source and tracing object */
  syncdatasrcdbl sdtsrc("sdtsrc");
  ssctracedbldbl sscmvg("tr_movavg");
  /* Connect module ports and channels */
  mavg.clk(clk);
  mavg.insig(dblsig);
  mavg.outsig(avgsig);
  sdtsrc.clk(clk);
  sdtsrc.dout(dblsig);
  sscmvg.din0(dblsig);
  sscmvg.din1(avgsig);
  /* Run simulation for pre-defind time period and stop */
sc_core::sc_start(20000.0, sc_core::SC_US);
  sc_core::sc_stop();
  return 0;
}
```

**Fig. 8.36** Test harness for moving average filter

Here, 'clk1' is high when scan mode is high, and 'clk2' is high when scan mode is low, and 'clk1' is low. 'clk3' is a delayed copy of the master clock.

## 8.10 Level-Sensitive Scan: Reconfigurable D Flip-Flop

The clock generator discussed previously can be used with a reconfigurable master slave D flip-flop whose master and slave sections are clocked with different clock outputs of the previous clock generator. The D flip-flop input is provided by

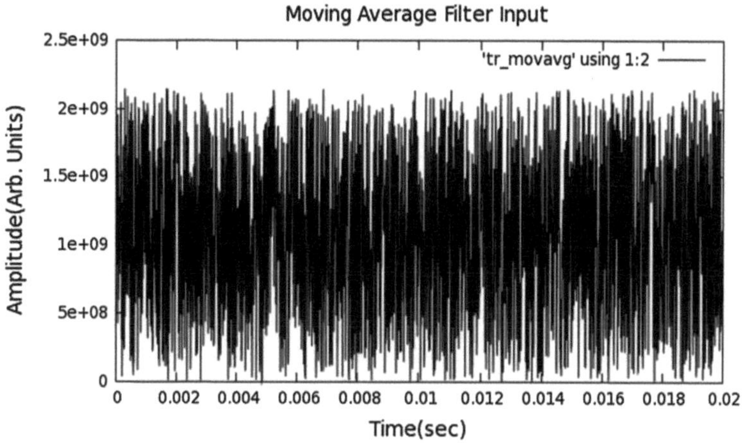


**Fig. 8.37** Raw data input to moving average filter

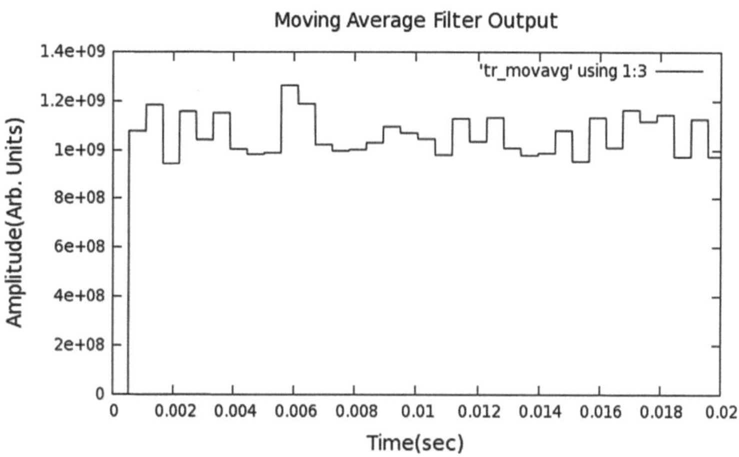


**Fig. 8.38** Moving average filter output for input of Fig. 8.31

a random bit generator synchronized with the master clock that feeds the clock generator. The three output clocks of the clock generator drive the reconfigurable master–slave D flip-flop [1]. This design uses a novel two-output NOR gate. The source code and input/output traces are in Figs. 8.43, 8.44 (SystemC 2.2.0), and 8.45 (SystemC-2.3.0), respectively. The test harness consists of minor modifications to the clock generator test harness and is not provided here.

```
#ifndef CLKGEN_H
#define CLKGEN_H

#include <systemc>

/* Inverter with one output port */
SC_MODULE(inv)
{
 sc_core::sc_in<bool> in0;
 sc_core::sc_out<bool> out0;

 void inv_proc0()
 {
  while(1)
  {
   wait();
   out0.write(!(in0.read()));
  }
 }

 SC_CTOR(inv)
 {
  SC_THREAD(inv_proc0);
  sensitive << in0;
 }

 ~inv(){ }
};

/* Inverter with two output ports */
SC_MODULE(inv_out)
{
 sc_core::sc_in<bool> in0;
 sc_core::sc_out<bool> out0;
 sc_core::sc_out<bool> out1;
 bool b0;

 void inv_out_proc0()
 {
```

**Fig. 8.39** Level-sensitive scan (*LSS*) clock generator

```
    while(1)
    {
      wait();
      b0 = in0.read();
      out0.write(!b0);
      out1.write(!b0);
    }
  }

  SC_CTOR(inv_out)
  {
    SC_THREAD(inv_out_proc0);
    sensitive << in0;
  }

  ~inv_out(){ }
};

/* Two input NOR gate */
SC_MODULE(nor2)
{
  sc_core::sc_in<bool> in0;
  sc_core::sc_in<bool> in1;
  sc_core::sc_out<bool> out0;

  void nor2_proc0()
  {
    while(1)
    {
      wait();
      out0.write(!(in0.read() | in1.read()));
    }
  }

  SC_CTOR(nor2)
  {
    SC_THREAD(nor2_proc0);
    sensitive << in0 << in1;
  }
  ~nor2(){ }
};
```

**Fig. 8.39** (continued)

```
/* 3 input NOR gate */
SC_MODULE(nor3)
{
  sc_core::sc_in<bool> in0;
  sc_core::sc_in<bool> in1;
  sc_core::sc_in<bool> in2;
  sc_core::sc_out<bool> out0;

  void nor3_proc0()
  {
    while(1)
    {
      wait();
      out0.write(!(in0.read() | in1.read() | in2.read()));
    }
  }

  SC_CTOR(nor3)
  {
    SC_THREAD(nor3_proc0);
    sensitive << in0 << in1 << in2;
  }

  ~nor3(){ }
};

/* Clock generatior */
SC_MODULE(clkgen)
{
  sc_core::sc_in<bool> slct;
     /* Scan mode select */
  sc_core::sc_in<bool> clk_in;
  /* Master input clock */
  sc_core::sc_out<bool> clk1;
   /* Output clocks */
  sc_core::sc_out<bool> clk2;
  sc_core::sc_out<bool> clk3;
  /* Signal channels */
  sc_core::sc_signal<bool> sig0;
  sc_core::sc_signal<bool> sig1;
```

**Fig. 8.39** (continued)

```
sc_core::sc_signal<bool> sig2;
sc_core::sc_signal<bool> sig3;
sc_core::sc_signal<bool> sig4;
sc_core::sc_signal<bool> sig5;
sc_core::sc_signal<bool> sig6;
sc_core::sc_signal<bool> sig7;
sc_core::sc_signal<bool> sig8;
sc_core::sc_signal<bool> sig9;
sc_core::sc_signal<bool> sig10;
sc_core::sc_signal<bool> sig11;
sc_core::sc_signal<bool> sig12;

/* Component NOR gates and inverters */
nor3 nor3_0;
nor3 nor3_1;
nor3 nor3_2;

nor2 nor2_0;
nor2 nor2_1;

inv inv_0;
inv inv_1;
inv inv_2;
inv inv_3;
inv inv_4;

inv_out inv_out_0;
inv_out inv_out_1;
inv_out inv_out_2;


/* Constructor - initialize members and connect channels and ports */
SC_CTOR(clkgen):nor3_0("nor3_0"),
        nor3_1("nor3_1"),
        nor3_2("nor3_2"),
        nor2_0("nor2_0"),
        nor2_1("nor2_1"),
        inv_0("inv_0"),
        inv_1("inv_1"),
        inv_2("inv_2"),
        inv_3("inv_3"),
```

**Fig. 8.39** (continued)

```
              inv_4("inv_4"),
              inv_out_0("inv_out_0"),
              inv_out_1("inv_out_1"),
              inv_out_2("inv_out_2"),
              sig0("sig0"), sig1("sig1"),
              sig2("sig2"), sig3("sig3"),
              sig4("sig4"), sig5("sig5"),
              sig6("sig6"), sig7("sig7"),
              sig8("sig8"), sig9("sig9"),
              sig10("sig10")
  {
    nor3_0.in0(sig0);
    nor3_0.in1(sig3);
    nor3_0.in2(sig7);
    nor3_0.out0(sig6);

    nor3_1.in0(sig0);
    nor3_1.in1(sig3);
    nor3_1.in2(sig8);
    nor3_1.out0(sig5);

    nor3_2.in0(sig1);
    nor3_2.in1(sig2);
    nor3_2.in2(clk_in);
    nor3_2.out0(sig11);

    nor2_0.in0(sig6);
    nor2_0.in1(slct);
    nor2_0.out0(sig7);

    nor2_1.in0(sig5);
    nor2_1.in1(sig4);
    nor2_1.out0(sig8);

    inv_0.in0(sig6);
    inv_0.out0(sig9);

    inv_1.in0(sig5);
    inv_1.out0(sig10);

    inv_2.in0(sig11);
```

**Fig. 8.39** (continued)

```
    inv_2.out0(sig12);

    inv_3.in0(slct);
    inv_3.out0(sig4);

    inv_4.in0(clk_in);
    inv_4.out0(sig3);

    inv_out_0.in0(sig9);
    inv_out_0.out0(sig1);
    inv_out_0.out1(clk1);

    inv_out_1.in0(sig10);
    inv_out_1.out0(sig2);
    inv_out_1.out1(clk2);

    inv_out_2.in0(sig12);
    inv_out_2.out0(sig0);
    inv_out_2.out1(clk3);
  }

  ~clkgen(){ }
};

#endif
```

**Fig. 8.39** (continued)

```
#include "clkgen.h"
#include "systemc.h"

int sc_main(int argc, char **argv)
{
  /* Declare/define signal channels */
  sc_core::sc_signal<bool> slctsig;
  sc_core::sc_signal<bool> clk_1;
  sc_core::sc_signal<bool> clk_2;
  sc_core::sc_signal<bool> clk_3;
```

**Fig. 8.40** Test harness for level-sensitive scan (*LSS*) clock generator

```
/* Declare/define master clock, clock generator and trace file */
sc_core::sc_clock clk("clk", 10.0, sc_core::SC_NS, 0.5);
clkgen clk_gen("clk_gen");
sc_core::sc_trace_file *fp;
fp = sc_create_vcd_trace_file("tr_clkgen");
fp->set_time_unit(1.0, sc_core::SC_NS);
bool slctval;

/* Connect module ports and channels and trace file */
clk_gen.clk_in(clk);
clk_gen.slct(slctsig);
clk_gen.clk1(clk_1);
clk_gen.clk2(clk_2);
clk_gen.clk3(clk_3);

sc_core::sc_trace(fp, clk, "clk_in");
sc_core::sc_trace(fp, slctsig, "slct");
sc_core::sc_trace(fp, clk_1, "clk1");
sc_core::sc_trace(fp, clk_2, "clk2");
sc_core::sc_trace(fp, clk_3, "clk3");
/* Set scan mode select and run simulation for pre-defined time period */
slctval = false;
slctsig.write(slctval);
sc_core::sc_start(50.0, sc_core::SC_NS);

/* Reset scan mode select and run simulation for pre-defined time period */

slctval = true;
slctsig.write(slctval);
sc_core::sc_start(50.0, sc_core::SC_NS);

/* Reset scan mode select and run simulation for pre-defined time period */

slctval = false;
slctsig.write(slctval);
sc_core::sc_start(50.0, sc_core::SC_NS);

/* Reset scan mode select and run simulation for pre-defined time period */

slctval = true;
```

**Fig. 8.40** (continued)

```
    slctsig.write(slctval);
    sc_core::sc_start(50.0, sc_core::SC_NS);

    /* Reset scan mode select and run simulation for pre-defined time period */

    slctval = false;
    slctsig.write(slctval);
    sc_core::sc_start(50.0, sc_core::SC_NS);

    ?* Stop simulation and close trace file */
    sc_core::sc_stop();
    sc_core::sc_close_vcd_trace_file(fp);
    return 0;
}
```

**Fig. 8.40** (continued)

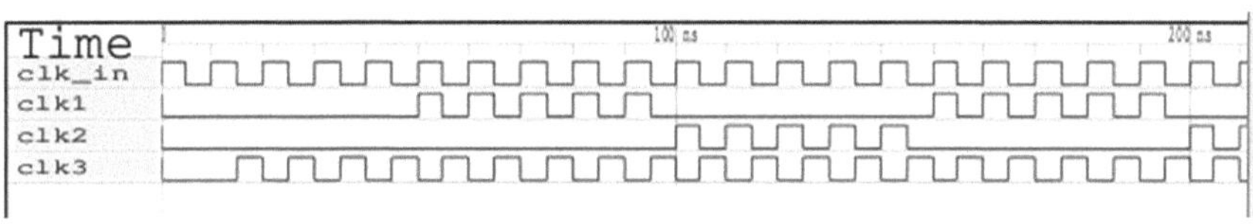


**Fig. 8.41** Input/output traces for level-sensitive scan (*LSS*) clock generator SystemC-2.2.0

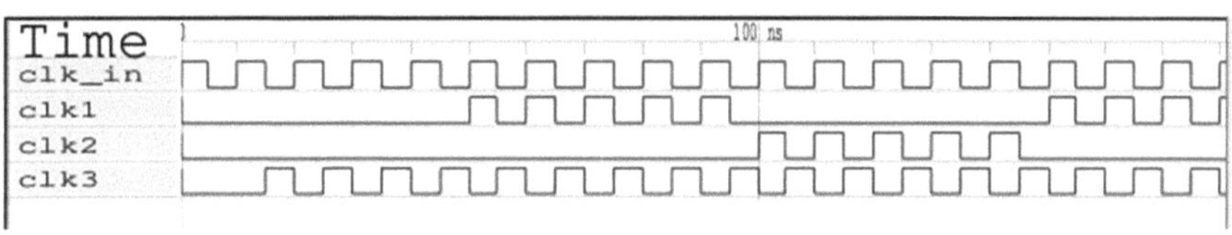


**Fig. 8.42** Input/output traces for level-sensitive scan (*LSS*) clock generator SystemC-2.3.0

## 8.11 Built-In Self-Test: Signature Analysis

Built-in self-test (BIST) [1] is an effective scheme for evaluating/testing a digital circuit, and one of its components is signature analysis (SA). Signature analysis was originally designed to monitor the output of a node on a circuit board using a large number of clock cycles. The original circuit used for this was a maximal length LFSR with an exclusive OR gate (XOR) to modify the input. An additional XOR gate is used in a feedback loop in front of the input XOR gate, fed by the final LFSR output, and the output of first stage of the LFSR. Ideally, if the test inputs are incorrect, the LFSR's final state will not be the same as in the case when testing a functionally correct circuit. This basic technique is often extended to produce the signature for N parallel inputs, using N flip-flops (in the LFSR) with a XOR gate at the input of each flip-flop. In this case, the input of a flip-flop depends on both the output of the previous flip-flop and one of the N parallel inputs. A six-stage, six-input signature analysis circuit is examined here.

```
SC_MODULE(nor2_out2)  /* Two output NOR gate */
{
  /* Declare/define input/output ports */
  sc_core::sc_in<bool> in0;
  sc_core::sc_in<bool> in1;
  sc_core::sc_out<bool> out0;
  sc_core::sc_out<bool> out1;
  bool b0;

  /* Two ouput NOR gate main thread */
  void nor2_out2_proc0()
  {
    while(1)
    {
      wait();
      b0 = !(in0.read() | in1.read());
      out0.write(b0);
      out1.write(b0);
    }
  }

  /* Constructor */
  SC_CTOR(nor2_out2)
  {
    /* Declare/assign thread */
    SC_THREAD(nor2_out2_proc0);
    sensitive << in0 << in1;
  }

  ~nor2_out2(){ }
};

/* Reconfigurable master-slave D flip-flop */
SC_MODULE(reconDff)
```

**Fig. 8.43** Reconfigurable D flip-flop driven by three output clocks of the clock generator

```
{
  /* Declare/define input/output ports */
  sc_core::sc_in<bool> clk1;
  sc_core::sc_in<bool> clk2;
  sc_core::sc_in<bool> clk3;
  sc_core::sc_in<bool> din;
  sc_core::sc_in<bool> slct;
  sc_core::sc_out<bool> dout0;
  sc_core::sc_out<bool> dout1;
  /* Declare/define channels */
  sc_core::sc_signal<bool> sig0;
  sc_core::sc_signal<bool> sig1;
  sc_core::sc_signal<bool> sig2;
  sc_core::sc_signal<bool> sig3;
  sc_core::sc_signal<bool> sig4;
  sc_core::sc_signal<bool> sig5;
  sc_core::sc_signal<bool> sig6;
  sc_core::sc_signal<bool> sig7;
  sc_core::sc_signal<bool> sig8;
  sc_core::sc_signal<bool> sig9;
  sc_core::sc_signal<bool> sig10;
  sc_core::sc_signal<bool> sig11;

  /* Declare/define members */
  and2 and2_0;
  and2 and2_1;
  and2 and2_2;
  and2 and2_3;
  and2 and2_4;
  and2 and2_5;

  inv inv_0;
  inv inv_1;

  nor2_out2 nor2_out2_0;
  nor2_out2 nor2_out2_1;

  nor3 nor3_0;
  nor3 nor3_1;
```

**Fig. 8.43** (continued)

```
/* Construcyor - initialuize members */
SC_CTOR(reconDff): and2_0("and2_0"),
               and2_1("and2_1"),
               and2_2("and2_2"),
               and2_3("and2_3"),
               and2_4("and2_4"),
               and2_5("and2_5"),
               inv_0("inv_0"),
               inv_1("inv_1"),
               nor3_0("nor3_0"),
               nor3_1("nor3_1"),
               nor2_out2_0("nor2_out2_0"),
               nor2_out2_1("nor2_out2_1")
 {
   /* Connect module ports and channels */
   and2_0.in0(din);
   and2_0.in1(clk1);
   and2_0.out0(sig2);
   and2_1.in0(clk1);
   and2_1.in1(sig1);
   and2_1.out0(sig4);
   and2_2.in0(slct);
   and2_2.in1(clk2);
   and2_2.out0(sig3);
   and2_3.in0(clk2);
   and2_3.in1(sig0);
   and2_3.out0(sig5);
   and2_4.in0(sig7);
   and2_4.in1(clk3);
   and2_4.out0(sig9);
   and2_5.in0(sig6);
   and2_5.in1(clk3);
   and2_5.out0(sig8);
   inv_0.in0(din);
   inv_0.out0(sig1);
   inv_1.in0(slct);
   inv_1.out0(sig0);
   nor3_0.in0(sig2);
   nor3_0.in1(sig3);
```

**Fig. 8.43** (continued)

```
        nor3_0.in2(sig6);
        nor3_0.out0(sig7);
        nor3_1.in0(sig4);
        nor3_1.in1(sig5);
        nor3_1.in2(sig7);
        nor3_1.out0(sig6);
        nor2_out2_0.in0(sig9);
        nor2_out2_0.in1(sig11);
        nor2_out2_0.out0(sig10);
        nor2_out2_0.out1(dout0);
        nor2_out2_1.in0(sig8);
        nor2_out2_1.in1(sig10);
        nor2_out2_1.out0(sig11);
        nor2_out2_1.out1(dout1);
    }

    ~reconDff(){ } /* Destructor */
};
```

**Fig. 8.43** (continued)

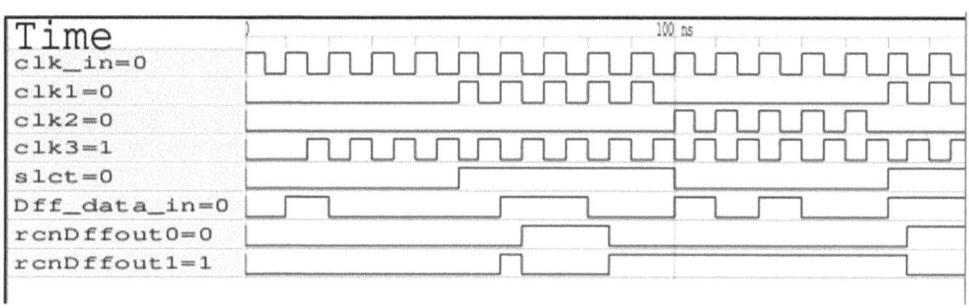


**Fig. 8.44** Reconfigurable master–slave D flip-flop input/output traces SystemC-2.2.0

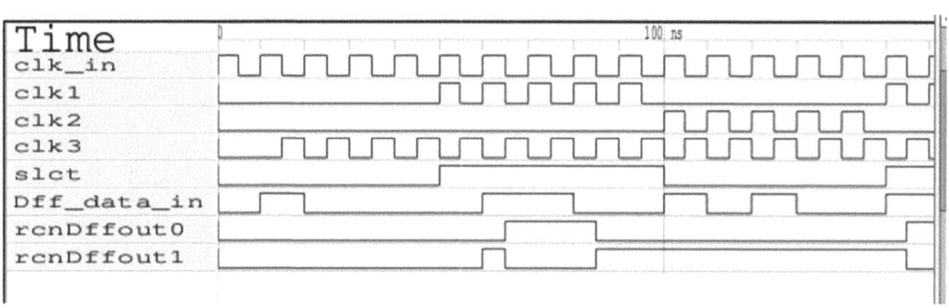


**Fig. 8.45** Reconfigurable master–slave D flip-flop input/output traces SystemC-2.3.0

Figures 8.46, 8.47, 8.48 (SystemC-2.2.0), and 8.49 (SystemC-2.3.0) contain, respectively, the circuit diagram, SystemC source code, the test harness, and input/output traces.

## 8.12 Simplified Built-In Logic Block Observation

Built-in logic block observation (BILBO) [1] is a digital logic testing technique that combines level-sensitive scan design (LSSD—discussed earlier) with maximal length LFSR input test vector generator with multiple-level signature analysis. A simplified BILBO scheme [1], with four register cells, is examined here, although in reality sixteen or thirty-two cell registers are used. Its power and versatility arises from the control signals (in this case three—C1, C2, and C3) which allow the BILBO register to be used as a maximal length LFSR, a parallel input signature analysis register, a shift register for shifting in test inputs, shifting out compressed outputs, or as latches to be used in regular operation of a state machine. SystemC source code, test harness, and sample input/output traces are in Figs. 8.50, 8.51, 8.52 (SystemC-2.2.0), and 8.53 (SystemC-2.3.0), respectively.

```
#ifndef SIGANAL_H
#define SIGANAL_H

#include <systemc>

SC_MODULE(dff)
{
  /* D flip-flop for each stage of LFSR */
  sc_core::sc_in<bool> clk;
  sc_core::sc_in<bool> in0;
  sc_core::sc_out<bool> out0;

  void dff_proc0()
  {
    while(1)
    {
      wait();
      out0.write(in0.read());
    }
  }
```

**Fig. 8.46** Source code for 6-input 6-stage signature analysis circuit

```
  SC_CTOR(dff)
  {
    SC_THREAD(dff_proc0);
    sensitive << clk.pos();
  }

  ~dff(){ }
};

SC_MODULE(xorgate)
{
  /* XOR gate with two outputs */
  sc_core::sc_in<bool> in0;
  sc_core::sc_in<bool> in1;
  sc_core::sc_out<bool> out0;
  sc_core::sc_out<bool> out1;
  bool b0;
  bool b1;
  bool b2;

  void xorgate_proc0()
  {
    while(1)
    {
      wait();
      b0 = in0.read();
      b1 = in1.read();
      b2 = ((b0 & !b1) | (!b0 & b1));
      out0.write(b2);
      out1.write(b2);
    }
  }

  SC_CTOR(xorgate)
  {
    SC_THREAD(xorgate_proc0);
    sensitive << in0 << in1;
  }

  ~xorgate(){ }
};
```

**Fig. 8.46** (continued)

```
/* 6 input 6 stage signature analysis circuit */
SC_MODULE(siganal6x6)
{
  /* Declare/define input ports */
  sc_core::sc_in<bool> clk;
  sc_core::sc_in<bool> in0;
  sc_core::sc_in<bool> in1;
  sc_core::sc_in<bool> in2;
  sc_core::sc_in<bool> in3;
  sc_core::sc_in<bool> in4;
  sc_core::sc_in<bool> in5;
  sc_core::sc_in<bool> in6;

  /* Declare/define output ports */
  sc_core::sc_out<bool> out0;
  sc_core::sc_out<bool> out1;
  sc_core::sc_out<bool> out2;
  sc_core::sc_out<bool> out3;
  sc_core::sc_out<bool> out4;
  sc_core::sc_out<bool> out5;
  sc_core::sc_out<bool> out6;
  sc_core::sc_out<bool> out7;

  /* Channels */
  sc_core::sc_signal<bool> sig0;
  sc_core::sc_signal<bool> sig1;
  sc_core::sc_signal<bool> sig2;
  sc_core::sc_signal<bool> sig3;
  sc_core::sc_signal<bool> sig4;
  sc_core::sc_signal<bool> sig5;
  sc_core::sc_signal<bool> sig6;
  sc_core::sc_signal<bool> sig7;
  sc_core::sc_signal<bool> sig8;
  sc_core::sc_signal<bool> sig9;
  sc_core::sc_signal<bool> sig10;
  sc_core::sc_signal<bool> sig11;
  sc_core::sc_signal<bool> sig12;
  sc_core::sc_signal<bool> sig13;
  sc_core::sc_signal<bool> sig14;
```

**Fig. 8.46** (continued)

```
/* Declare/define member D flip-flops and XOR gates */
 dff df0;
 dff df1;
 dff df2;
 dff df3;
 dff df4;
 dff df5;
 dff df6;
 xorgate xog0;
 xorgate xog1;
 xorgate xog2;
 xorgate xog3;
 xorgate xog4;
 xorgate xog5;
 xorgate xog6;
 xorgate xog7;

 /* Constructor - initialize members */
 SC_CTOR(siganal6x6):df0("df0"), df1("df1"), df2("df2"), df3("df3"),
                 df4("df4"), df5("df5"), df6("df6"), xog0("xog0"),
                 xog1("xog1"), xog2("xog2"), xog3("xog3"),
                 xog4("xog4"), xog5("xog5"), xog6("xog6"),
                 xog7("xog7"), sig0("sig0"), sig1("sig1"),
                 sig2("sig2"), sig3("sig3"), sig4("sig4"),
                 sig5("sig5"), sig6("sig6"), sig7("sig7"),
                 sig8("sig8"), sig9("sig9"), sig10("sig10"),
                 sig11("sig11"),sig12("sig12"), sig13("sig13"),
                 sig14("sig14")
 {
   /* Connect modules with ports and channels */
   df0.clk(clk);
   df0.in0(sig0);
   df0.out0(sig1);

   df1.clk(clk);
   df1.in0(sig2);
   df1.out0(sig3);

   df2.clk(clk);
   df2.in0(sig4);
   df2.out0(sig5);
```

**Fig. 8.46** (continued)

```
df3.clk(clk);
df3.in0(sig6);
df3.out0(sig7);

df4.clk(clk);
df4.in0(sig8);
df4.out0(sig9);

df5.clk(clk);
df5.in0(sig10);
df5.out0(sig11);

df6.clk(clk);
df6.in0(sig12);
df6.out0(sig13);

xog0.in0(in0);
xog0.in1(sig14);
xog0.out0(out0);
xog0.out1(sig0);

xog1.in0(in1);
xog1.in1(sig1);
xog1.out0(out1);
xog1.out1(sig2);

xog2.in0(in2);
xog2.in1(sig3);
xog2.out0(out2);
xog2.out1(sig4);

xog3.in0(in3);
xog3.in1(sig5);
xog3.out0(out3);
xog3.out1(sig6);

xog4.in0(in4);
xog4.in1(sig7);
xog4.out0(out4);
xog4.out1(sig8);
```

**Fig. 8.46** (continued)

```
    xog5.in0(in5);
    xog5.in1(sig9);
    xog5.out0(out5);
    xog5.out1(sig10);

    xog6.in0(in6);
    xog6.in1(sig11);
    xog6.out0(out6);
    xog6.out1(sig12);

    xog7.in0(sig1);
    xog7.in1(sig13);
    xog7.out0(out7);
    xog7.out1(sig14);
  }

  ~siganal6x6(){ }
};

#endif
```

**Fig. 8.46** (continued)

```
#include "siganal.h"
#include "systemc.h"
int sc_main(int argc, char **argv)
{
  /* Declare/define channels */
  sc_core::sc_signal<bool> sig0;
  sc_core::sc_signal<bool> sig1;
  sc_core::sc_signal<bool> sig2;
  sc_core::sc_signal<bool> sig3;
  sc_core::sc_signal<bool> sig4;
  sc_core::sc_signal<bool> sig5;
  sc_core::sc_signal<bool> sig6;
  sc_core::sc_signal<bool> sig7;
  sc_core::sc_signal<bool> sig8;
  sc_core::sc_signal<bool> sig9;
  sc_core::sc_signal<bool> sig10;
  sc_core::sc_signal<bool> sig11;
```

**Fig. 8.47** Test harness for 6-input 6-stage signature analysis circuit

```
sc_core::sc_signal<bool> sig12;
sc_core::sc_signal<bool> sig13;
sc_core::sc_signal<bool> sig14;

/* Declare/define master clock and local members */
sc_core::sc_clock clk("clk", 2.0, sc_core::SC_NS, 0.5);
bool b0;
bool b1;
bool b2;
bool b3;
bool b4;
bool b5;
bool b6;

/* Declare/define signature analysis circuit and trace file object */
siganal6x6 siganal("siganal");
sc_core::sc_trace_file *fp;
fp = sc_create_vcd_trace_file("tr_siganal");
fp->set_time_unit(1.0, sc_core::SC_NS);

/* Connect module with ports and channels */
siganal.clk(clk);
siganal.in0(sig0);
siganal.in1(sig1);
siganal.in2(sig2);
siganal.in3(sig3);
siganal.in4(sig4);
siganal.in5(sig5);
siganal.in6(sig6);
siganal.out0(sig7);
siganal.out1(sig8);
siganal.out2(sig9);
siganal.out3(sig10);
siganal.out4(sig11);
siganal.out5(sig12);
siganal.out6(sig13);
siganal.out7(sig14);
```

**Fig. 8.47** (continued)

```
  sc_core::sc_trace(fp, clk, "clk");
  sc_core::sc_trace(fp, sig0, "d0");
  sc_core::sc_trace(fp, sig1, "d1");
  sc_core::sc_trace(fp, sig2, "d2");
  sc_core::sc_trace(fp, sig3, "d3");
  sc_core::sc_trace(fp, sig4, "d4");
  sc_core::sc_trace(fp, sig5, "d5");
  sc_core::sc_trace(fp, sig6, "d6");
  sc_core::sc_trace(fp, sig7, "o0");
  sc_core::sc_trace(fp, sig8, "o1");
  sc_core::sc_trace(fp, sig9,  "o2");
  sc_core::sc_trace(fp, sig10, "o3");
  sc_core::sc_trace(fp, sig11, "o4");
  sc_core::sc_trace(fp, sig12, "o5");
  sc_core::sc_trace(fp, sig13, "o6");
  sc_core::sc_trace(fp, sig14, "o7");

/* Set inputs and write to input channels */
  b0 = true;
  b1 = false;
  b2 = true;
  b3 = false;
  b4 = true;
  b5 = false;
  b6 = true;

  sig0.write(b0);
  sig1.write(b1);
  sig2.write(b2);
  sig3.write(b3);
  sig4.write(b4);
  sig5.write(b5);
  sig6.write(b6);

   /* Run simulation for pre-defined time period */
  sc_core::sc_start(5.0, sc_core::SC_NS);

  /* Reset inputs and write to input channels */
  b0 = true;
  b1 = true;
  b2 = true;
```

**Fig. 8.47** (continued)

```
b3 = false;
b4 = false;
b5 = false;
b6 = true;

sig0.write(b0);
sig1.write(b1);
sig2.write(b2);
sig3.write(b3);
sig4.write(b4);
sig5.write(b5);
sig6.write(b6);

/* Run simulation for pre-defined time period */
sc_core::sc_start(5.0, sc_core::SC_NS);

/* Reset inputs and write to input channels */
b0 = false;
b1 = true;
b2 = false;
b3 = false;
b4 = false;
b5 = false;
b6 = true;

sig0.write(b0);
sig1.write(b1);
sig2.write(b2);
sig3.write(b3);
sig4.write(b4);
sig5.write(b5);
sig6.write(b6);

/* Run simulation for pre-defined time period */
 sc_core::sc_start(5.0, sc_core::SC_NS);

/* Reset inputs and write to input channels */
b0 = true;
b1 = true;
b2 = true;
```

**Fig. 8.47** (continued)

```
b3 = false;
b4 = false;
b5 = false;
b6 = false;

sig0.write(b0);
sig1.write(b1);
sig2.write(b2);
sig3.write(b3);
sig4.write(b4);
sig5.write(b5);
sig6.write(b6);

/* Run simulation for pre-defined time period */
 sc_core::sc_start(5.0, sc_core::SC_NS);

/* Reset inputs and write to input channels */
b0 = true;
b1 = false;
b2 = true;
b3 = false;
b4 = true;
b5 = false;
b6 = true;

sig0.write(b0);
sig1.write(b1);
sig2.write(b2);
sig3.write(b3);
sig4.write(b4);
sig5.write(b5);
sig6.write(b6);

/* Run simulation for pre-defined time period */
sc_core::sc_start(5.0, sc_core::SC_NS);

/* Reset inputs and write to input channels */
b0 = false;
b1 = false;
b2 = true;
```

**Fig. 8.47** (continued)

```
b3 = true;
b4 = true;
b5 = false;
b6 = false;

sig0.write(b0);
sig1.write(b1);
sig2.write(b2);
sig3.write(b3);
sig4.write(b4);
sig5.write(b5);
sig6.write(b6);

/* Run simulation for pre-defined time period */
sc_core::sc_start(5.0, sc_core::SC_NS);

/* Reset inputs and write to input channels */
b0 = false;
b1 = false;
b2 = false;
b3 = false;
b4 = true;
b5 = true;
b6 = true;

sig0.write(b0);
sig1.write(b1);
sig2.write(b2);
sig3.write(b3);
sig4.write(b4);
sig5.write(b5);
sig6.write(b6);

/* Run simulation for pre-defined time period */
sc_core::sc_start(5.0, sc_core::SC_NS);

/* Reset inputs and write to input channels */
b0 = true;
b1 = true;
```

**Fig. 8.47** (continued)

```
b2 = true;
b3 = true;
b4 = true;
b5 = false;
b6 = true;

sig0.write(b0);
sig1.write(b1);
sig2.write(b2);
sig3.write(b3);
sig4.write(b4);
sig5.write(b5);
sig6.write(b6);

/* Run simulation for pre-defined time period */
sc_core::sc_start(5.0, sc_core::SC_NS);

/* Stop simulation and close trace file */
sc_core::sc_stop();
sc_core::sc_close_vcd_trace_file(fp);

return 0;

}
```

**Fig. 8.47** (continued)

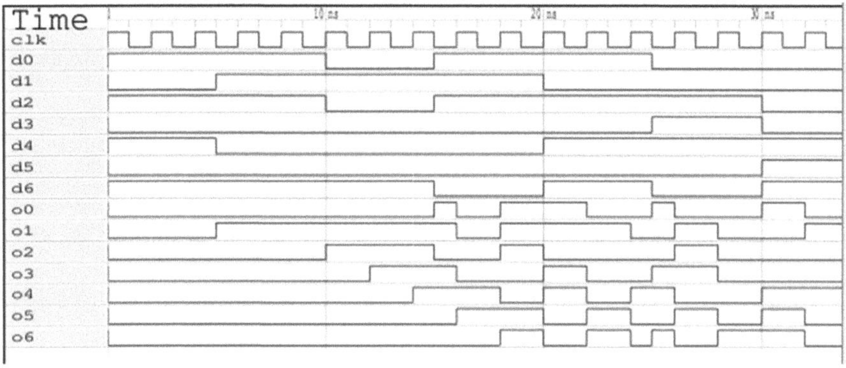


**Fig. 8.48** Input/output traces for 6-input 6-stage signature analysis circuit: SystemC-2.2.0

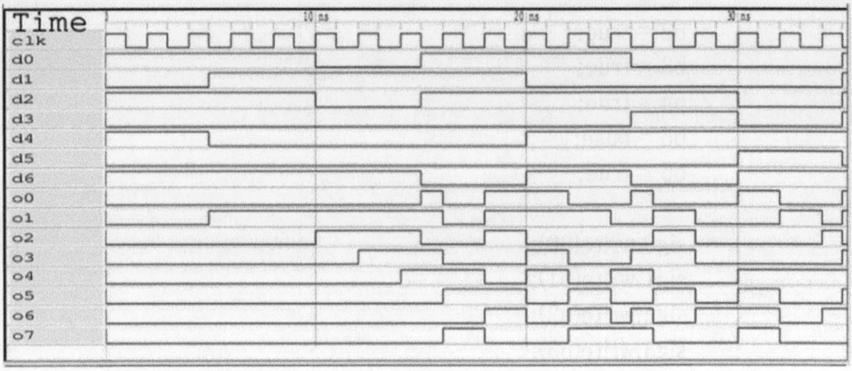


**Fig. 8.49** Input/output traces for 6-input 6-stage signature analysis circuit: SystemC-2.3.0

```
#ifndef BILBO_H
#define BILBO_H

#include <systemc>

SC_MODULE(dff)
{
  /* D flip-flop with single output */
  sc_core::sc_in<bool> clk;
  sc_core::sc_in<bool> in0;
  sc_core::sc_out<bool> out0;

  void dff_proc0()
  {
    while(1)
    {
      wait();
      out0.write(in0.read());
    }
  }

  SC_CTOR(dff)
  {
    SC_THREAD(dff_proc0);
    sensitive << clk.pos();
  }

  ~dff(){ }
};
```

**Fig. 8.50** SystemC source file simplified 4-cell BILBO register

```
SC_MODULE(dffout2)
{
  /* D flip-flop with both normal and inverted outputs */
  sc_core::sc_in<bool> clk;
  sc_core::sc_in<bool> in0;
  sc_core::sc_out<bool> out0;
  sc_core::sc_out<bool> out1;

  void dffout2_proc0()
  {
    while(1)
    {
      wait();
      out0.write(in0.read());
      out1.write(!(in0.read()));
    }
  }

  SC_CTOR(dffout2)
  {
    SC_THREAD(dffout2_proc0);
    sensitive << clk.pos();
  }

  ~dffout2(){ }
};

/* 2 x 1 multiplexer */
SC_MODULE(mux2)
{
  sc_core::sc_in<bool> in0;
  sc_core::sc_in<bool> in1;
  sc_core::sc_in<bool> ctrl;
  sc_core::sc_out<bool> out0;

  void mux2_proc0()
  {
    while(1)
    {
      wait();
      if(ctrl.read() == true) out0.write(in0.read());
```

**Fig. 8.50** (continued)

```
      else if(ctrl.read() == false) out0.write(in1.read());
    }
  }

  SC_CTOR(mux2)
  {
    SC_THREAD(mux2_proc0);
    sensitive << ctrl;
  }

  ~mux2(){ }
};

/* Two input NAND */
SC_MODULE(nand2)
{
  sc_core::sc_in<bool> in0;
  sc_core::sc_in<bool> in1;
  sc_core::sc_out<bool> out0;

  void nand2_proc0()
  {
    while(1)
    {
      wait();
      out0.write(!(in0.read()) & !(in1.read()));
    }
  }

  SC_CTOR(nand2)
  {
    SC_THREAD(nand2_proc0);
    sensitive << in0 << in1;
  }

  ~nand2(){ }
};

/* Two input NOR */
SC_MODULE(nor2)
{
```

**Fig. 8.50** (continued)

```
    sc_core::sc_in<bool> in0;
    sc_core::sc_in<bool> in1;
    sc_core::sc_out<bool> out0;

    void nor2_proc0()
    {
      while(1)
      {
        wait();
        out0.write(!(in0.read()) | !(in1.read()));
      }
    }

    SC_CTOR(nor2)
    {
      SC_THREAD(nor2_proc0);
      sensitive << in0 << in1;
    }

    ~nor2(){ }
};

/* XOR gate */
SC_MODULE(xorgate)
{
    sc_core::sc_in<bool> in0;
    sc_core::sc_in<bool> in1;
    sc_core::sc_out<bool> out0;
    bool b0;
    bool b1;
    bool b2;

    void xorgate_proc0()
    {
      while(1)
      {
        wait();
        b0 = in0.read();
        b1 = in1.read();
        b2 = ((b0 & !b1) | (!b0 & b1));
        out0.write(b2);
```

**Fig. 8.50** (continued)

```
    }
  }

  SC_CTOR(xorgate):b0(false), b1(false), b2(false)
  {
    SC_THREAD(xorgate_proc0);
    sensitive << in0 <<in1;
  }

  ~xorgate() { }
};

/* XOR gate with both normal and negated outputs */
SC_MODULE(xorgateneg)
{
  sc_core::sc_in<bool> in0;
  sc_core::sc_in<bool> in1;
  sc_core::sc_out<bool> out0;
  bool b0;
  bool b1;
  bool b2;
  void xorgateneg_proc0()
  {
    while(1)
    {
      wait();
      b0 = in0.read();
      b1 = in1.read();
      b2 = !((b0 & !b1) | (!b0 & b1));
      out0.write(b2);
    }
  }

  SC_CTOR(xorgateneg):b0(false), b1(false), b2(false)
  {
    SC_THREAD(xorgateneg_proc0);
    sensitive << in0 <<in1;
  }

  ~xorgateneg() { }
};
```

**Fig. 8.50** (continued)

```
/* BILBO register */
SC_MODULE(bilbo)
{
  /* Declare/define input/output ports */
  sc_core::sc_in<bool> clk;
  sc_core::sc_in<bool> shftin;
  sc_core::sc_in<bool> c1;
  sc_core::sc_in<bool> c2;
  sc_core::sc_in<bool> c3;
  sc_core::sc_in<bool> t1;
  sc_core::sc_in<bool> t2;
  sc_core::sc_in<bool> t3;
  sc_core::sc_in<bool> t4;
  sc_core::sc_out<bool> shftout;

  /* Declare/define internal signal channels */
  sc_core::sc_signal<bool> sig0;
  sc_core::sc_signal<bool> sig1;
  sc_core::sc_signal<bool> sig2;
  sc_core::sc_signal<bool> sig3;
  sc_core::sc_signal<bool> sig4;
  sc_core::sc_signal<bool> sig5;
  sc_core::sc_signal<bool> sig6;
  sc_core::sc_signal<bool> sig6a;
  sc_core::sc_signal<bool> sig7;
  sc_core::sc_signal<bool> sig8;
  sc_core::sc_signal<bool> sig9;
  sc_core::sc_signal<bool> sig10;
  sc_core::sc_signal<bool> sig11;
  sc_core::sc_signal<bool> sig12;
  sc_core::sc_signal<bool> sig13;
  sc_core::sc_signal<bool> sig14;
  sc_core::sc_signal<bool> sig15;
  sc_core::sc_signal<bool> sig16;
  sc_core::sc_signal<bool> sig17;

  ?* Declare component modules */
  dff df0;
  dff df1;
  dff df2;
```

**Fig. 8.50** (continued)

```
dffout2 dffo;

mux2 mux_2;

nand2 nand_0;
nand2 nand_1;
nand2 nand_2;
nand2 nand_3;

nor2 nor_0;
nor2 nor_1;
nor2 nor_2;
nor2 nor_3;

xorgateneg xorn_0;
xorgateneg xorn_1;
xorgateneg xorn_2;
xorgateneg xorn_3;

xorgate xor_0;

/* Constructor - initialize member modules and signals */
SC_CTOR(bilbo):df0("df0"), df1("df1"), df2("df2"),
          dffo("dffo"), mux_2("mux_2"),
          nand_0("nand_0"), nand_1("nand_1"),
          nand_2("nand_2"), nand_3("nand_3"),
          nor_0("nor_0"), nor_1("nor_1"),
          nor_2("nor_2"), nor_3("nor_3"),
          xorn_0("xorn_0"), xorn_1("xorn_1"),
          xorn_2("xorn_2"), xorn_3("xorn_3"),
          sig0("sig0"), sig1("sig1"), sig2("sig2"),
          sig3("sig3"), sig4("sig4"), sig5("sig5"),
          sig6("sig6"), sig6a("sig6a"), sig7("sig7"),
          sig8("sig8"), sig9("sig9"), sig10("sig10"),
          sig11("sig11"), sig12("sig12"), sig13("sig13"),
          sig14("sig14"), sig15("sig15"), sig16("sig16"),
          sig17("sig17"), xor_0("xor_0")
{
  /* Connect internal mdules with ports and channels */
  mux_2.in0(shftin);
```

**Fig. 8.50** (continued)

```
mux_2.in1(sig0);
mux_2.ctrl(c3);
mux_2.out0(sig17);

nor_0.in0(c2);
nor_0.in1(sig17);
nor_0.out0(sig15);

nand_0.in0(c1);
nand_0.in1(t1);
nand_0.out0(sig16);

xorn_0.in0(sig16);
xorn_0.in1(sig15);
xorn_0.out0(sig14);

df0.clk(clk);
df0.in0(sig14);
df0.out0(sig13);

nor_1.in0(c2);
nor_1.in1(sig13);
nor_1.out0(sig11);

nand_1.in0(c1);
nand_1.in1(t2);
nand_1.out0(sig12);

xorn_1.in0(sig11);
xorn_1.in1(sig12);
xorn_1.out0(sig10);

df1.clk(clk);
df1.in0(sig10);
df1.out0(sig9);

nor_2.in0(c2);
nor_2.in1(sig9);
nor_2.out0(sig6);

nand_2.in0(c1);
```

**Fig. 8.50** (continued)

```
        nand_2.in1(t3);
        nand_2.out0(sig7);

        xorn_2.in0(sig7);
        xorn_2.in1(sig6);
        xorn_2.out0(sig6a);

        df2.clk(clk);
        df2.in0(sig6a);
        df2.out0(sig5);

        nor_3.in0(c2);
        nor_3.in1(sig5);
        nor_3.out0(sig4);

        nand_3.in0(c1);
        nand_3.in1(t4);
        nand_3.out0(sig3);

        xorn_3.in0(sig3);
        xorn_3.in1(sig4);
        xorn_3.out0(sig2);

        dffo.clk(clk);
        dffo.in0(sig2);
        dffo.out0(shftout);
        dffo.out1(sig1);

        xor_0.in0(sig1);
        xor_0.in1(sig13);
        xor_0.out0(sig0);
    }

    ~bilbo(){ }
    /* Destructor */
};

#endif
```

**Fig. 8.50** (continued)

```
#include "bilbo.h"
#include "systemc.h"

int sc_main(int argc, char **argv)
{

 /* Declare/define channels */
  sc_core::sc_signal<bool> shftinsig;
  sc_core::sc_signal<bool> c1sig;
  sc_core::sc_signal<bool> c2sig;
  sc_core::sc_signal<bool> c3sig;
  sc_core::sc_signal<bool> t1sig;
  sc_core::sc_signal<bool> t2sig;
  sc_core::sc_signal<bool> t3sig;
  sc_core::sc_signal<bool> t4sig;
  sc_core::sc_signal<bool> outsig;

  /* Declare/define Boolean local variables */
  bool shftinval;
  bool c1val;
  bool c2val;
  bool c3val;
  bool t1val;
  bool t2val;
  bool t3val;
  bool t4val;
  /* Define master clock */

  sc_core::sc_clock clk("clk", 2.0, sc_core::SC_NS, 0.5);

  /* Declare/define BILBO object and trace file object */
  bilbo bil_bo("bil_bo");
  sc_core::sc_trace_file *fp;
  fp = sc_create_vcd_trace_file("tr_bilbo");
  fp->set_time_unit(1.0, sc_core::SC_NS);

  /* Connect BILBO object ports and channels */
  bil_bo.clk(clk);
  bil_bo.shftin(shftinsig);
  bil_bo.c1(c1sig);
  bil_bo.c2(c2sig);
```

**Fig. 8.51** Test harness for simplified 4-cell BILBO register

```
bil_bo.c3(c3sig);
bil_bo.t1(t1sig);
bil_bo.t2(t2sig);
bil_bo.t3(t3sig);
bil_bo.t4(t4sig);
bil_bo.shftout(outsig);

/* Connect trace file and data channels */
sc_core::sc_trace(fp, clk, "clk");
sc_core::sc_trace(fp, shftinsig, "shiftin");
sc_core::sc_trace(fp, c1sig, "c1");
sc_core::sc_trace(fp, c2sig, "c2");
sc_core::sc_trace(fp, c3sig, "c3");
sc_core::sc_trace(fp, t1sig, "t1");
sc_core::sc_trace(fp, t2sig, "t2");
sc_core::sc_trace(fp, t3sig, "t3");
sc_core::sc_trace(fp, t4sig, "t4");
sc_core::sc_trace(fp, outsig, "out");

/* Assign values to variables and write to channels */
shftinval = false;
c1val = false;
c2val = false;
c3val = false;
t1val = false;
t2val = false;
t3val = false;
t4val = false;
shftinsig.write(shftinval);
c1sig.write(c1val);
c2sig.write(c2val);
c3sig.write(c3val);
t1sig.write(t1val);
t2sig.write(t2val);
t3sig.write(t3val);
t4sig.write(t4val);

/* Run simulation fro pre-defined time interval */
sc_core::sc_start(10.0, sc_core::SC_NS);

/* Re-assign values to variables and write to channels */
```

**Fig. 8.51** (continued)

```
shftinval = false;
shftinval = true;
c1val = true;
c2val = false;
c3val = true;
t1val = true;
t2val = false;
t3val = true;
t4val = false;
shftinsig.write(shftinval);
c1sig.write(c1val);
c2sig.write(c2val);
c3sig.write(c3val);
t1sig.write(t1val);
t2sig.write(t2val);
t3sig.write(t3val);
t4sig.write(t4val);

/* Run simulation fro pre-defined time interval */
 /* Run simulation fro pre-defined time interval */
sc_core::sc_start(10.0, sc_core::SC_NS);

/* Re-assign values to variables and write to channels */
shftinval = false;
c1val = false;
c2val = true;
c3val = false;
t1val = false;
t2val = true;
t3val = true;
t4val = false;
shftinsig.write(shftinval);
c1sig.write(c1val);
c2sig.write(c2val);
c3sig.write(c3val);
t1sig.write(t1val);
t2sig.write(t2val);
t3sig.write(t3val);
t4sig.write(t4val);

 /* Run simulation fro pre-defined time interval */
```

**Fig. 8.51** (continued)

```
sc_core::sc_start(10.0, sc_core::SC_NS);

/* Re-assign values to variables and write to channels */
shftinval = true;
c1val = true;
c2val = true;
c3val = true;
t1val = false;
t2val = true;
t3val = true;
t4val = false;
shftinsig.write(shftinval);
c1sig.write(c1val);
c2sig.write(c2val);
c3sig.write(c3val);
t1sig.write(t1val);
t2sig.write(t2val);
t3sig.write(t3val);
t4sig.write(t4val);

 /* Run simulation fro pre-defined time interval */
sc_core::sc_start(10.0, sc_core::SC_NS);


/* Re-assign values to variables and write to channels */
shftinval = false;
c1val = true;
c2val = true;
c3val = false;
t1val = false;
t2val = true;
t3val = true;
t4val = true;
shftinsig.write(shftinval);
c1sig.write(c1val);
c2sig.write(c2val);
c3sig.write(c3val);
t1sig.write(t1val);
t2sig.write(t2val);
t3sig.write(t3val);
t4sig.write(t4val);
```

**Fig. 8.51** (continued)

```
 /* Run simulation fro pre-defined time interval */
sc_core::sc_start(10.0, sc_core::SC_NS);

/* Re-assign values to variables and write to channels */
shftinval = true;
c1val = false;
c2val = false;
c3val = true;
t1val = false;
t2val = true;
t3val = true;
t4val = false;
shftinsig.write(shftinval);
c1sig.write(c1val);
c2sig.write(c2val);
c3sig.write(c3val);
t1sig.write(t1val);
t2sig.write(t2val);
t3sig.write(t3val);
t4sig.write(t4val);

 /* Run simulation fro pre-defined time interval */
sc_core::sc_start(10.0, sc_core::SC_NS);

/* Re-assign values to variables and write to channels */
shftinval = false;
c1val = false;
c2val = true;
c3val = true;
t1val = false;
t2val = false;
t3val = true;
t4val = false;
shftinsig.write(shftinval);
c1sig.write(c1val);
c2sig.write(c2val);
c3sig.write(c3val);
t1sig.write(t1val);
t2sig.write(t2val);
t3sig.write(t3val);
```

**Fig. 8.51** (continued)

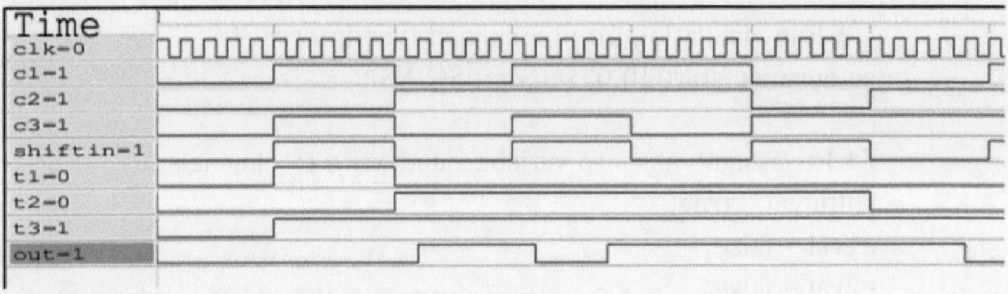


**Fig. 8.52** Sample input/output traces for simplified 4-cell BILBO register SystemC-2.2.0

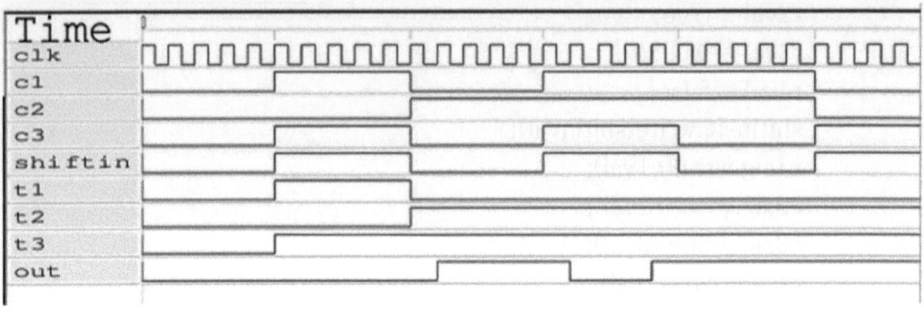


**Fig. 8.53** Sample input/output traces for simplified 4-cell BILBO register SystemC-2.3.0

# References


1. Kenneth, Martin. *Digital Integrated Circuit Design*. Oxford University Press, 2000. Print
2. Mhambrey, S. S., Clark, L. T., Maurya, S. K., Berezowski, K (2010) *Out- of_order Issue Logic Using Sorting Networks – GLSVLSI - '10*
3. Sun MicroSystems - Austin. IEEE Xplore. IEEE, 12 Nov. 2007 *UltraSparc T2 A highly-threaded, power-efficient SPARC SOC* http://ieeexplore.ieee.org/xpl/login.jsp?tp=&arnumber=4425786&url=http%3A%2F%2Fieeexplore.ieee.org%2Fxpls%2Fabs_all.jsp%3Farnumber%3D4425786#/xpl/articleDetails.jsp?tp=&arnumber=4425786&url=http%3A%2F
4. IEEE Standards Board, IEEE Standards Association Standards Board IEEE-SA—IEEE Get Program. IEEE, 2011 *IEEE Standard 1666 Open SystemC Language Reference Manual (LRM)* http://standards.ieee.org/getieee/1666/download/1666-2011.pdf

# Chapter 9
# Introduction to SystemC-AMS

**Abstract** Before using the SystemC-AMS library, the analog and mixed-signal system (AMS) designer must understand why it was created, specifically which problem does it address. This brief chapter examines this issue and introduces SystemC-AMS [1].

## 9.1 Introduction

Despite digital circuits being a special type of analog circuits, methods or techniques to analyze and understand analog and mixed-signal systems (AMS) have so far been rare, or too complicated/time-consuming to use, for the analog system designer. The SystemC-AMS [1] extension to ANSI C++ is a crucial starting point in addressing this issue and builds upon existing SystemC (IEEE 1666–2005 specifications) to address the special requirements of analog systems and digital hardware/software systems coupled to/interacting with their physical analog environment. For example, when digital hardware/software interacts with RF systems, sensors and actuators and power electronics, the analysis must not only tackle the issues specific to pure analog and pure digital systems, and in addition, their interaction in real time. SystemC-AMS addresses these special requirements.

Building upon existing SystemC infrastructure, the new AMS language standard allows designers to build system-level executable specifications to model and analyze AMS and refine it to a virtual prototype, allowing design space exploration and integration validation. It exploits both discrete-time static non-linear (non-conservative behavior) and continuous-time dynamic linear (conservative and non-conservative behavior) model abstractions to provide three modeling formalisms of timed data flow (TDF), linear signal flow (LSF), and electrical linear networks (ELN). The SystemC-AMS architecture builds upon and interacts with the existing pure digital SystemC environment or infrastructure as summarized in Fig. 9.1. *SystemC-AMS infrastructure does not use SystemC's clocking/timing environment, but can interact with it.* This allows a mixed-signal design

A. Banerjee and B. Sur, *SystemC and SystemC-AMS in Practice*,
DOI: 10.1007/978-3-319-01147-9_9,

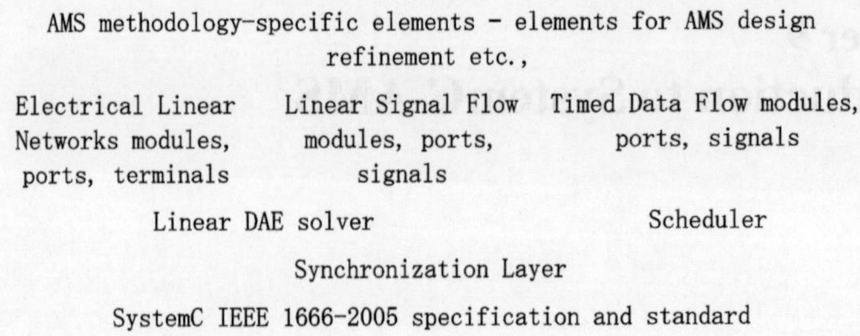


**Fig. 9.1** SystemC-AMS architecture

in which, e.g., a pure digital control signal can control a pure analog circuit, using appropriately conditioned feedback signals from the analog circuit. More importantly, SystemC-AMS exploits the same TLM framework as pure SystemC, allowing the user to focus on how data are processed inside a processing element, rather than how data were read in or how data have to be sent to the next processing element.

Thus, SystemC-AMS [1] abstracts away low-level circuit/component-specific details, freeing the user to focus on understanding system behavior, interaction of its major components, and adjust system parameters so that simulated system behavior matches desired system behavior, i.e., *explore design space*. While traditional circuit design tools as SPICE allow the engineer to estimate circuit performance characteristics very accurately, they are inappropriate when trying to understand the behavior of a new design/system, as the engineer has to spend a lot of effort/time to create an accurate SPICE representation of the same, taking into account properties of non-linear electronic devices (e.g., BJT or MOSFET), impedance matching of transmission lines, drive current at an output stage and so on. With any change to a sub-component in the SPICE representation of a circuit, the designer must ensure that the effects of that change are accurately taken into account in the connected sub-components. SystemC-AMS makes this task easy, by abstracting out low-level details unnecessary to analyzing/understanding/optimizing the behavior of a new design/system. This feature, coupled with the fact that SystemC-AMS, like SystemC is an ANSI C++ library, makes SystemC-AMS so powerful. Once the designer understands the behavior of new complex design/circuit/system, he/she may revert to traditional tools as SPICE to estimate/optimize performance characteristics at finer levels of granularity (e.g., sub-micron MOSFET characteristics). Our approach to writing this book is from the viewpoint of a practicing engineer/system designer who is too hard-pressed to read through a detailed language reference manual (LRM), digest the knowledge, and apply that knowledge to tackle the problem at hand. That person would find it extremely helpful to examine a previously worked out (with SystemC-AMS library classes and components) example of a real-world system, draw the necessary inferences, and apply those inferences to solve the problem at hand.

# Reference

1. SystemC-AMS Working Group. SystemC-AMS (Analog Mixed Signal) — Accellera Systems Initiative. http://www.accellera.org/downloads/standards/systemc/ams and http://www.accellera.org/activities/committees/systemc-ams/

# Chapter 10
# Downloading, Installing, and Getting Started with SystemC-AMS

**Abstract** Before the analog and mixed-signal system designer can use SystemC-AMS, it must be downloaded from the official Accellera Web site (http://www.accellera.org) and then compiled and installed correctly, bearing in mind that SystemC has to be preinstalled on the same computer, but at the same time, the latest SystemC-AMS (1.0) library is fully compatible with the latest stable SystemC (2.3.0). These download, compilation, and final installation steps are examined here.

## 10.1 How to Download and Install SystemC-AMS

To successfully install and run SystemC-AMS [1] on a given computer, *SystemC must already be installed and running on it.* Like SystemC, SystemC-AMS has been developed and runs best on computers with the Linux and Unix-like operating systems. There are ways to install and run it on Microsoft Windows operating system. For example, one can create a Linux/Unix-like environment on a Windows machine by installing the versatile Cygwin/MingW package on it, followed by SystemC/SystemC-AMS. The more complicated way is to compile the libraries directly with Microsoft Visual Studio. All code examples in this book were created and tested on a Red Hat Enterprise Linux 4.0 (with GCC compiler version 3.2.6) and a Red Hat Fedora 14 machine (with GCC compiler version 4.5.1) running SystemC 2.3.0 and SystemC-AMS 1.0. The SystemC-AMS 1.0 stable release is fully compatible with SystemC 2.3.0. In fact, the initial compilation process automatically detects the directory path to SystemC 2.3.0.

The SystemC-AMS source tarball may be downloaded easily from http://www.accellera.org. On a Linux machine with GCC installed, the executable libraries may be created/installed easily by decompressing the tarball and following simple instructions ('configure,' 'make,' 'make check/test,' and 'make install') in the README or INSTALL files inside the source file directory. These sequence of commands ('configure,' 'make,' 'make check/test,' and 'make install') are executed at the command line of a standard Linux shell window, with appropriate

A. Banerjee and B. Sur, *SystemC and SystemC-AMS in Practice*,
DOI: 10.1007/978-3-319-01147-9_10,

parameters. As mentioned earlier, SystemC must be preinstalled and running on the same machine, as its library path has to be provided, as a parameter for the first 'configure' command (from the command chain). Queries regarding issues at installation time may be directed to the relevant OSCI user forums and so far from our experience, the responses are very prompt. The following is part of contents of the INSTALL file in the System-AMS 1.0 Beta2 directory:

Installation Instructions

Copyright (C) 1994, 1995, 1996, 1999, 2000, 2001, 2002, 2004 Free Software Foundation, Inc.

This file is free documentation; the Free Software Foundation gives unlimited permission to copy, distribute, and modify it.

Contents:

1. Installation Notes for Linux
2. Installation Notes for Windows

1. Installation Notes for Linux

Basic Installation

These are generic installation instructions.

The 'configure' shell script attempts to guess correct values for various system-dependent variables used during compilation. It uses those values to create a 'Makefile' in each directory of the package. It may also create one or more '.h' files containing system-dependent definitions. Finally, it creates a shell script 'config.status' that you can run in the future to recreate the current configuration, and a file 'config.log' containing compiler output (useful mainly for debugging 'configure').

It can also use an optional file (typically called 'config.cache' and enabled with '–cache-file = config.cache' or simply '-C') that saves the results of its tests to speed up reconfiguring. (Caching is disabled by default to prevent problems with accidental use of stale cache files.) If you need to do unusual things to compile the package, please try to figure out how 'configure' could check whether to do them and mail diffs or instructions to the address given in the 'README' so they can be considered for the next release. If you are using the cache, and at some point, 'config.cache' contains results you do not want to keep, you may remove or edit it.

The file 'configure.ac' (or 'configure.in') is used to create 'configure' by a program called 'autoconf.' You only need 'configure.ac' if you want to change it or regenerate 'configure' using a newer version of 'autoconf.'

The simplest way to compile this package is

1. 'cd' to the directory containing the package's source code and type './configure' to configure the package for your system. If you are using 'csh' on an old version of System V, you might need to type 'sh./configure' instead to prevent 'csh' from trying to execute 'configure' itself.

Running ‘configure’ takes awhile. While running, it prints some messages telling which features it is checking for.

2. Type ‘make’ to compile the package.
3. Optionally, type ‘make check’ to run any self-tests that come with the package.
4. Type ‘make install’ to install the programs and any data files and documentation.

Compiling and executing any SystemC-AMS executable is identical to that of executing a pure SystemC 2.3.0 executable, except that the directory path to the pure SystemC 2.3.0 library must be specified in the compile path along with the directory path to the SystemC-AMS 1.0 library. Consider a SystemC-AMS executable source file named test.cc. To compile and run it in the Linux bash shell, one must execute the following steps in the order listed, *provided SELinux is NOT enabled.*

<user_system_prompt> g++ -I. -I/<directory_path_to_SystemC_2.3.0>/include
–I<directory_path_to_SystemC_AMS 1.0>/include -L.
–L/<directory_path_to_SystemC_2.3.0>/lib-linux -L/<directory_path_to_SystemC_AMS 1.0>/lib-linux -o sim text.cc -lsystemc-ams -lsystemc -lm
<user_system_prompt> export
LD_LIBRARY_PATH=/<directory_path_to_SystemC_2.3.0>/lib-linux
<user_system_prompt> echo $LD_LIBRARY_PATH
<directory_path_to_SystemC_2.3.0>/lib-linux
<user_system_prompt> ./sim.

*If however, SELinux is enabled*, it must be temporarily disabled as root. The following sequence of steps need to be executed in that order.

<user_system_prompt> su root
Password:
[root@localhost <directory_name>]# setenforce 0

*The ‘setenforce’ command, executed in root mode with parameter 0, temporarily disables SELinux.*

[root@localhost <directory_name>]# g++ -I.
-I/<directory_path_to_SystemC_2.3.0>/include -I/<directory_path_to_SystemC_AMS 1.0>/include -L. -L/<directory_path_to_SystemC_2.3.0>/lib-linux
-L/<directory_path_to_SystemC_AMS 1.0>/lib-linux -o sim test.cc -lsystemc-ams –lsystemc -lm
[root@localhost <directory_name>]# export
LD_LIBRARY_PATH=<directory_path_to_SystemC_2.3.0>/lib-linux
[root@localhost <directory_name>]# echo $LD_LIBRARY_PATH
<directory_path_to_SystemC_2.3.0>/lib-linux
[root@localhost <directory_name>]# ./sim

After the program has been executed successfully, the user (in root mode) types:

[root@localhost <directory_name>] setenforce 1

This activates SELinux temporarily suspended features once more.

Finally, the user exits from root mode as:

[root@localhost <directory_name>] exit

Note that for both cases above, the directory path to the SystemC 2.3.0 or SystemC-AMS 1.0 libraries could be either absolute or real. In Addition, SELinux might give some runtime warnings. These may be deleted at the users' discretion.

## Reference

1. SystemC-AMS Working Group. SystemC-AMS (Analog Mixed Signal)—Accellera Systems Initiative. http://www.accellera.org/downloads/standards/systemc/ams and http://www.accellera.org/activities/committees/systemc-ams/

# Chapter 11
# SystemC-AMS Formalisms, Data Types, and Main Language Constructs

**Abstract** To use SystemC-AMS [1] effectively, the core concepts, formalisms, and infrastructure that underpin its design must be examined carefully and understood. We start by briefly reviewing the concepts underlying the three components of SystemC-AMS—timed data flow (TDF), linear signal flow (LSF), and electrical linear networks (ELN). We have liberally used available documentation from http://www.accellera.org.

## 11.1 Timed Data Flow

Timed data flow (TDF) [1] is a *model of computation* (MOC) based on the familiar synchronous data flow MOC. *Data is conceptualized as signals sampled in time, and carry discrete or continuous information, as signal amplitude, despite being tagged discrete in time.* A TDF model is a set of connected TDF *modules* to form a directed graph (called *TDF cluster*). The nodes in the directed graph are the TDF modules, while the edges are the TDF channels (*signals*). Mathematical functions in a TDF module are executed using direct inputs and internal states. A given function is *processed* only if required a number of input data values are available, and the results written to the output ports. The number of input data elements for a single invocation of a function may not equal the number of data elements produced from that invocation. However, the number of input data elements required for and produced by a single invocation is fixed. A time tag, called *time step*, is associated with each data element—hence the name *TDF*.

TDF modeling formalism is versatile, arising from the user's ability to specify the properties of a TDF module and each of its ports, for example,

- Specify time step of a module and each of its ports.
- Specify port rate of a module's port—i.e., the number of data elements read in or written to a port per each read/write operation.
- Specify delays and time offsets for each port.

A. Banerjee and B. Sur, *SystemC and SystemC-AMS in Practice*,
DOI: 10.1007/978-3-319-01147-9_11,

TDF imposes three strict constraints on the data flow.

1. Attributes assigned to ports and modules must be compatible—(a Boolean value cannot be written to a port assigned for an unsigned integer).
2. Rate of data flow (samples/time) must match at sending and receiving ports.
3. All feedback loops must have port delays.

Figures 11.1, 11.2, 11.3, and 11.4 show typical TDF module topologies where a 'D' indicates delay, and 'R' indicates rate. A TDF module might interact with non-TDF modules via converter ports—Figures 11.5, 11.6. In all these diagrams, small filled-in black boxes are TDF-specific ports on TDF modules. Borders only small boxes are converter ports. *A converter port, as the name suggests, allows a TDF module to read in/write data to (using appropriate channels), linear signal flow (LSF), electrical linear networks (ELN), or even discrete-event (pure SystemC) modules.* This versatility allows SystemC-AMS modules to interact directly with pure SystemC modules, allowing for mixed signal analysis, for example, an analog-to-digital converter (ADC).

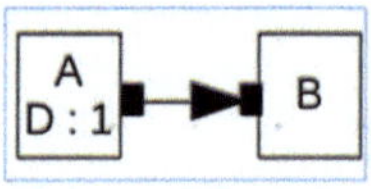


**Fig. 11.1** TDF model port delay; *D* delay

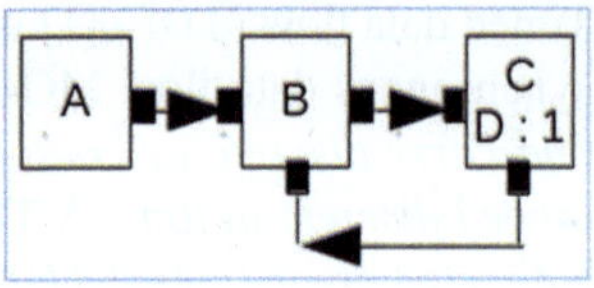


**Fig. 11.2** TDF module with loop and port delay (*D*)

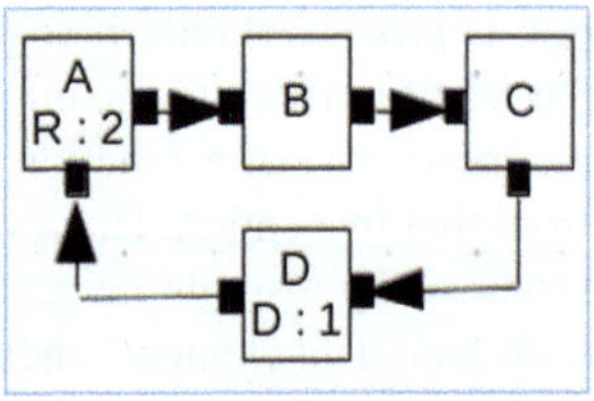


**Fig. 11.3** Multi-rate TDF model with delay *D* and rate *R*

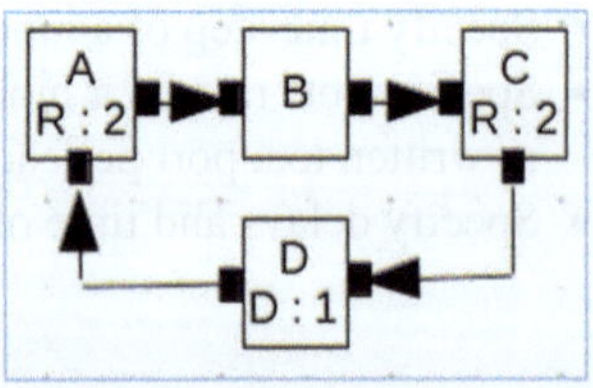


**Fig. 11.4** Multi-rate TDF model with loop and incompatible rates—samples accumulate in loop with delay *D* and rate *R*

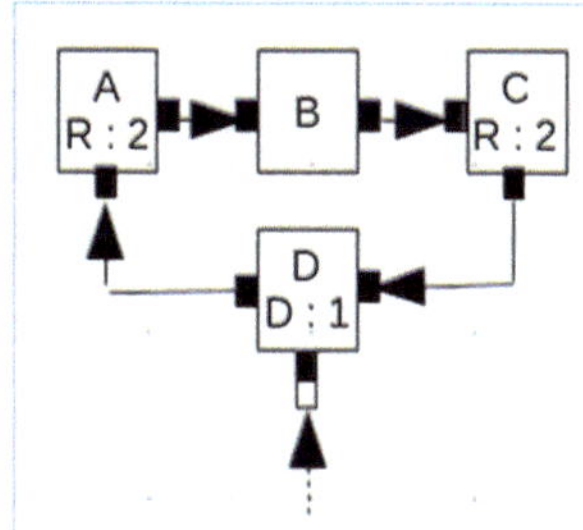


**Fig. 11.5** TDF model with discrete-event domain interface

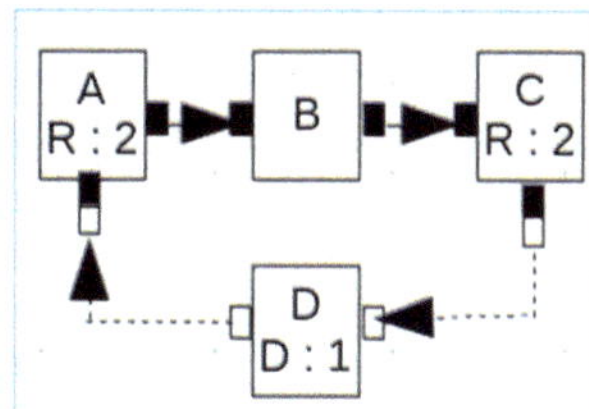


**Fig. 11.6** TDF module with loop in discrete-event domain

| Step | Action | Phase |
|---|---|---|
| 1 | Set TDF module attributes : execute *void set_attributes* for each module in cluster | Elaboration |
| 2 | Define/compute TDF time step and propogate it through all modules in cluster – check time step consistency | |
| 3 | Define/compute/check cluster schedule – computability check | |
| 4 | Initialize all TDF modules in cluster by executing *void initialize*(optional method) method for each | |
| 5 | Activate all TDF modules and start processing : execute *void processing* method of each module | Simulation |
| 6 | TDF post processing : execute all *end of simulation* member functions | |

**Fig. 11.7** TDF elaboration (first 3 steps) and simulation

Just like SystemC, SystemC-AMS uses the *elaboration–simulation* approach, Fig. 11.7. Also, recent announcements from the OSCI SystemC-AMS group indicate that TDF will be modified in future releases to allow dynamic rates—dynamic TDF (DTDF).

All TDF modules have a regular structure, i.e., the user, while creating his/her own modules, overrides (SystemC-AMS is based on C++ and thus supports method overriding) a set of methods of the base TDF class, to provide the required functionality. First, TDF modules may be declared/defined inside a macro *SCA_TDF_MODULE*, with the class constructor declared via another macro *SCA_CTOR*. The overridden methods are as follows:

| | |
|---|---|
| void initialize() | Initialize class specific data structures |
| void set_attributes() | Initialize input/output delay, rate and time step |
| void processing() | Performs class specific computation |

While 'void initialixe()' may not be overridden in any particular case (that TDF module might not have internal data structures), the other two have to be overridden to make the instantiated object do the tasks it is designed to do. Each of 'void initialize(),' 'void set_attributes(),' and 'void processing()' is invoked implicitly.

## 11.2 Linear Signal Flow

The Linear Signal Flow (LSF) [1] computational model allows modeling and analysis of an AMS system defined in terms of relations between variables of a system of linear algebraic equations—i.e., a non-conservative system with continuous-time, directed real-valued signals. Each real-valued quantity represents a signal. Diagrammatically, a LSF model consists of a set of blocks (LSF modules) interconnected by arrows (LSF signals) with LSF input and output ports for the model to interact with, e.g., a TDF module. *Unlike TDF, the user cannot write customized code for the model, but rather use members of a ore-defined set of LSF models (addition, subtraction, multiplication, derivative,* etc,), *as required.* A few of these primitive LSF models are as in Figs. 11.8, 11.9, and 11.10.

Similar to the TDF model, the time step of a LSF module might be assigned directly, or propagated. In the hybrid case, with an LSF module connected to a

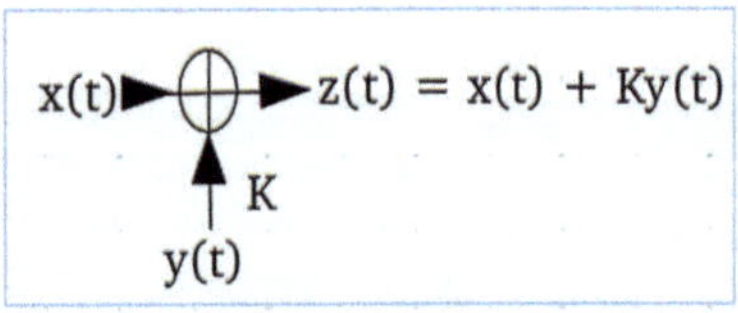


**Fig. 11.8** Weighted LSF addition

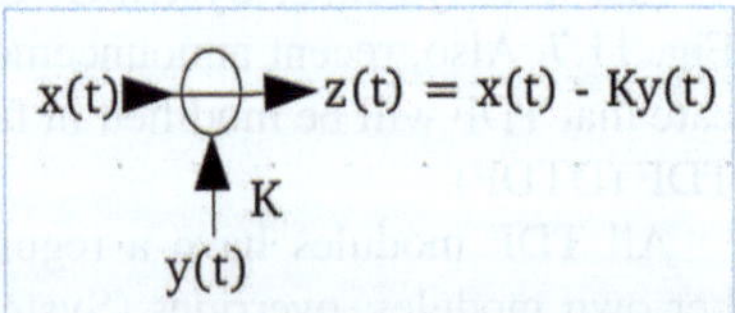


**Fig. 11.9** Weighted LSF subtraction

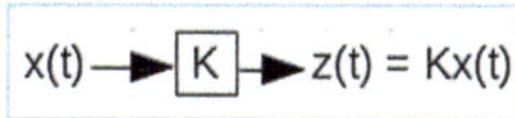


Fig. 11.10 Weighted LSF multiplication

| Step | Action | Phase |
|---|---|---|
| 1 | Define LSF simulation time step and verify consistency | Elaboration |
| 2 | Define LSF equation system for all modules and verify that they can be solved | |
| 3 | Set LSF module initial conditions | |
| 4 | Execute LSF simulation and generate results at pre-defined time points | Simulation |

Fig. 11.11 Linear signal flow (LSF) elaboration and simulation

TDF module, the time step from the TDF ports is propagated to the LSF model. Consistency between locally defined LSF module time step and propagated time step is crucial for correct data communication between the modules. During simulation, the LSF equation system is solved numerically with appropriate time steps. The LSF model follows the familiar elaboration–simulation sequence, as in Fig. 11.11.

## 11.3 Electrical Linear Networks

The Electrical Linear Networks [1] (ELN) MOC introduces electrical primitives and their connections to model and analyze continuous-time, conservative electrical circuits. A ELN model consists of electrical primitives (e.g., capacitor) connected to nodes, to form an electrical network. The mathematical relationships between the primitives, obeying Kirchoff's current/voltage laws (KCL/KVL), are represented by a set of differential algebraic equations and are resolved during simulation. A ELN model consists of a set of ELN primitives connected via terminals, to form a ELN cluster or equation system. Just like the LSF formalism, the user has to use members of a pre-defined set of ELN primitives to form the ELN network to be analyzed. The available ELN primitive modules and their description are in Appendix D, consisting of dependent and independent sources (current, voltage), lumped elements (capacitor, inductor, resistor), linear distributed element (transmission line), ideal amplifier (ideal operational amplifier), linear gyrator, and ideal switch. *It is very important to note that unlike the electrical/electronics engineer's standard circuit simulation tool SPICE, SystemC-AMS attempts to achieve the same results at a higher level of abstraction.* Thus, a nonlinear electronic switch (BJT or MOSFET) may be modeled as an ideal switch, as embodied in the ELN primitive sca_eln::sca_rswiitch.

Similar to the TDF model, the time step of a ELN module might be assigned directly, or propagated. In the hybrid case, with an ELN module connected to a TDF module, the time step from the TDF ports is propagated to the LSF model. Consistency between locally defined ELN module time step and propagated time step is crucial for correct data communication between the modules. During simulation, the ELN equation system is solved numerically with appropriate time steps.

As a TDF signal may represent both a continuous-time or discrete-time signal, in a lot of cases, both ELN and LSF modules interact with TDF modules. Then, it is essential that special converter ports be used to translate between the data formats of the three formalisms.

## Reference

1. SystemC-AMS Working Group. SystemC-AMS (Analog Mixed Signal)—Accellera Systems Initiative. http://www.accellera.org/downloads/standards/systemc/ams and http://www.accellera.org/activities/committees/systemc-ams/

# Chapter 12
# Small Signal, Linear Domain, and Hybrid Models

**Abstract** As per standard circuit analysis techniques, a small alternating current (AC) signal, at various frequencies, is used to analyze the frequency domain characteristics of a given circuit. The AC signal is either sinusoidal or a noise source, and the test circuit is linearized around a direct current (DC) operating point, eliminating all large signal behavior, e.g., distortion. Small signal analysis is built into both LSF and ELN formalisms.

## 12.1 Small Signal Analysis

SystemC-AMS [1] allows both small signal and small signal noise analysis for hybrid test systems consisting of all of TDF, LSF, and ELN modules or some subset of these. Unlike time domain analysis, a hybrid system under small signal analysis is described by a *linear complex equation* system, where the transformation from time domain to frequency domain is enabled via Laplace transforms. Specifically,

- A time derivative d/dt is replaced with $jw$.
- An time integration is replaced by $1/jw$.
- A delay of the form $f(t - delay)$ is replaced by $\exp(-(jw - delay))$.

The small signal analysis of TDF modules can be realized with built-in utility classes as *sca_util::sca_complex*. In addition, the time domain TDF module function *void processing(){...}* is replaced by *void ac_processing(){...}*. However, all time and frequency domain consistency checks are user defined. The complex value of all non-converter TDF ports may be accessed using the utility function *sca_util::sca_complex& sca_ac_analysis::sca_ac(<TDF port instance>)*. The returned sca_util::sca_complex references constant or non-constant depending on whether the argument TDF port is input or output, respectively. The values returned from *sca_ac_analysis::sca_ac* and *sca_ac_analysis::sca_ac_noise* are solely implementation defined and do not have any physical meaning. The utility

A. Banerjee and B. Sur, *SystemC and SystemC-AMS in Practice*,
DOI: 10.1007/978-3-319-01147-9_12,

function *sca_ac_analysis::sca_ac_delay* allows delays to be added, as appropriate. In addition, analysis in the z-domain is supported by a related set of utility functions as *sca_ac_analysis::sca_ac_z*.

## Reference

1. SystemC-AMS Working Group. SystemC-AMS (Analog Mixed Signal)—Accellera Systems Initiative. http://www.accellera.org/downloads/standards/systemc/ams and http://www.accellera.org/activities/committees/systemc-ams/

# Chapter 13
# Timed Data Flow in Practice and Theory

**Abstract** Of the three formalisms underlying SystemC-AMS [1], timed data flow (TDF) is the most versatile. It allows the system designer the complete freedom to set up the analysis/simulation with custom SystemC-AMS modulus, incorporating all possible detailed steps that are necessary to successfully analyze/model the system. In contrast, as will be clear in the subsequent chapters, both the LSF and ELN formalisms offer a fixed set of built-in modules, which must be connected appropriately to analyze/model the system at hand; the designer does not have the freedom to add custom modules and built up from scratch.

## 13.1 Fifth-Order Low-Pass Butterworths Filter

The story has it that each forest has a king, and the king of the analog design world is the analog filter designer. A digital filter is designed by starting from its analog counterpart, and using Z transforms to translate to the digital version.

A fifth-order low-pass Butterworths filter example is presented here. An analog filter may be designed/represented either as a pure electrical network (e.g., RC low-pass filter) or as the frequency domain counterpart (pole–zero form). Each of these representations may be derived from the other. The pole–zero form allows detailed analysis of the design and exploits the fact that tables of coefficients of the pole–zero equations for different common filter types (Butterworths. Bessel, Chebyshev, etc.) are readily available. The electrical network (corresponding to our design) is presented in Fig. 13.1a, b, and c. A fifth-order low-pass filter [2] (odd order) consists of a first-order low-pass filter [2] followed by two second-order low-pass filters [2]. The two second-order low-pass filters may be of any type—we use the popular Sallen-Key [2] design. For low frequencies, operational amplifiers may be used as a filter component, but for high frequencies (e.g., RF) only discrete components may be used.

In terms of an electrical network, the transfer function of a unity-gain first-order low-pass filter is $H(s) = \frac{1}{1+w_c R_1 C_1 s}$ where $s = jw$ for sinusoidal input, $w_c$ is the angular cut-off frequency, and $R_1$ and $C_1$ are the resistor and capacitor.

A. Banerjee and B. Sur, *SystemC and SystemC-AMS in Practice*,
DOI: 10.1007/978-3-319-01147-9_13,

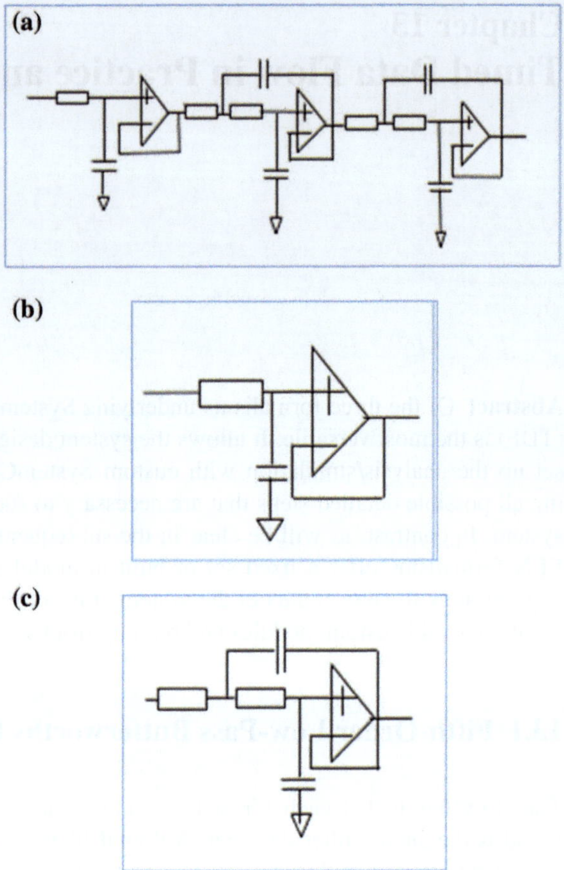

**Fig. 13.1** **a** Fifth-order unity-gain low-pass Butterworths filter electrical network. **b** Unity-gain first-order low-pass filter. **c** Second-order unity-gain Sallen-Key low-pass filter

Similarly, for a second-order low-pass unity-gain Sallen-Key filter, the transfer function is:

$$H(s) = \frac{1}{1 + w_c C_1(R_1 + R_1)s + w_c^2 R_1 R_2 C_1 C_2 s^2}$$

In the standard canonical notation, the transfer functions for an unity-gain first-order low-pass filter and an unity-gain second-order low-pass Sallen-Key filters are:

$H(s) = \frac{1}{1+a_1 s^2}$ and $H(s) = \frac{1}{1+a_2 s+b_2 s^2}$ respectively, where the coefficients $a_1$,$a_2$, $b_2$ are obtained from standard look-up tables, and their numerical values are different depending on filter type (Bessel, Butterworths, or Chebyschev etc.). The canonical form for the fifth-order low-pass filter is then:

$$H(s) = \frac{1}{(1 + a_1 s)(1 + a_2 s + b_2 s^2)(1 + a_3 s + b_3 s^2)}$$

Comparing the two forms of the transfer function, and some simple calculations using standard coefficient values for the first-order unity-gain low-pass filter and the two unity-gain second-order low-pass Sallen-Key filters, reveal that for a cut-off frequency of 60.0 Hz:

$$a_1 = w_c R_1 C_1 \quad a_2 = w_c C_1 (R_1 + R_2) \quad b_2 = w_c^2 C_1 C_2 R_1 R_2 \quad a_3 = w_c C_1 (R_1 + R_2) \quad b_3 = w_c^2 C_1 C_2 R_1 R_2$$

where $a_3$ $b_3$ are for the second unity-gain second-order low-pass Sallen-Key filter. From filter coefficient look-up tables for fifth-order low-pass Butterworths filter, we get:

$$a_1 = 1 \; a_2 = 1.618 \; b_2 = 1 \; a_3 = 0.618 \; b_3 = 1$$

so that for $w_c = 60\text{Hz}$ $R_1 C_1 = 0.0026$ $C_1(R_1 + R_2) = 0.0043$ $C_1 C_2 R_1 R_2 = 0.000007$ (for first Sallen-Key filter) and $C_1(R_1 + R_2) = 0.0016$ $C_1 C_2 R_1 R_2 = 0.000007$ (for second Sallen-Key filter).

With the required filter coefficients, the source code is in Figs. 13.2, 13.3, 13.4, and 13.5.

```
#ifndef AC_ANAL_L_P_FO_SDOM_H
#define AC_ANAL_L_P_FO_SDOM_H

#include <systemc-ams>

SCA_TDF_MODULE(lpo1sdom)
{
  private:
  /* Scaled Laplace transfer function in the time domain
     in the numerator - denominator form */
  sca_tdf::sca_ltf_nd ltf_nd;
  /* Vecior to store denominator coefficients */
  sca_util::sca_vector<double> denom;
  /* Vector to store numerator coefficients */
  sca_util::sca_vector<double> numer;

  public:
   /* Input/output ports */
   sca_tdf::sca_in<double> sigin;
   sca_tdf::sca_out<double> sigout;

   /* Initialize first-order low-pass filter coefficients */
   void initialize()
   {
```

**Fig. 13.2** First-stage first-order low-pass filter of fifth-order Butterworths filter

```
      numer(0) = 1.0;
      denom(0) = 1.0;
      denom(1) = 2.65E-3;
    }

    /* Initialize port properties */
    void set_attributes()
    {
      /* Time step for output port */
      sigout.set_timestep(1.0, sc_core::SC_US);
      /* Specify the number of data samples that pass through
         the input/output ports per time step */
      sigout.set_rate(1);
      sigin.set_rate(1);
      /* Feedback loop necessitates a delay in the output port */
      sigout.set_delay(1);
    }

    void processing()
    {
      /* Main module computation */
      sigout.write(ltf_nd(numer, denom, sigin.read(),1.0));
    }

    SCA_CTOR(lpo1sdom){ }  /* Constructor */
    ~lpo1sdom() { } /* Destructor */
};
```

**Fig. 13.2** (continued)

```
SCA_TDF_MODULE(lpo2sdom)
{
  private:
   /*  Scaled Laplace transfer function in the time domain
      in the numerator - denominator form */
  sca_tdf::sca_ltf_nd ltf_nd;
  /* Vecior to store denominator coefficients */
  sca_util::sca_vector<double> denom;
   /* Vector to store numerator coefficients */
  sca_util::sca_vector<double> numer;
```

**Fig. 13.3** Second-stage second-order low-pass Sallen-Key filter of fifth-order Butterworths filter

```
public:
  /* Input/output ports */
  sca_tdf::sca_in<double> sigin;
  sca_tdf::sca_out<double> sigout;

  /* Initialize second order low-pass Sallen-Key filter coefficients */
  void initialize()
  {
    numer(0) = 1.0;
    denom(0) = 1.0;
    denom(1) = 4.3E-3;
    denom(2) = 7.0E-6;
  }

  /* Initialize port properties */
  void set_attributes()
  {
    sigout.set_timestep(1.0, sc_core::SC_US);
    sigout.set_rate(1);
    sigout.set_delay(1);
    sigin.set_rate(1);
  }

  void processing()
  {
    /* Main module computation */
    sigout.write(ltf_nd(numer, denom, sigin.read(),1.0));
  }

  SCA_CTOR(lpo2sdom){ } /* Constructor */
  ~lpo2sdom() { } /* Destructor */
};
```

**Fig. 13.3** (continued)

```
SCA_TDF_MODULE(lpo3sdom)
{
  private:
   /*  Scaled Laplace transfer function in the time domain
      in the numerator - denominator form */
  sca_tdf::sca_ltf_nd ltf_nd;
  /* Vecior to store denominator coefficients */
  sca_util::sca_vector<double> denom;
   /* Vector to store numerator coefficients */
  sca_util::sca_vector<double> numer;

  public:
   /* Input/output ports */
   sca_tdf::sca_in<double> sigin;
   sca_tdf::sca_out<double> sigout;

   /* Initialize third order low-pass filter coefficients */
   void initialize()
   {
     numer(0) = 1.0;
     denom(0) = 1.0;
     denom(1) = 1.6E-3;
     denom(2) = 7.0E-6;
   }

    /* Initialize port properties */
   void set_attributes()
   {
     sigout.set_timestep(1.0, sc_core::SC_US);
     sigout.set_rate(1);
     sigout.set_delay(1);
     sigin.set_rate(1);
   }

   void processing()
   {
      /* Main module computation */
     sigout.write(ltf_nd(numer, denom, sigin.read(),1.0));
   }
```

**Fig. 13.4** Third (final)-stage second-order low-pass Sallen-Key filter of fifth-order Butterworths filter

```
    SCA_CTOR(lpo3sdom){ }  /* Constructor */
    ~lpo3sdom() { } /* Destructor */
  };
```

Fig. 13.4 (continued)

```
#include "ac_anal_l_p_fo_sdom.h"
#include "commonsrcs.h"
#include "gendatatrace.h"
#include <cstdlib>
#include <cstring>

int sc_main(int argc, char **argv)
{
  double amplitude;
  double frequency;

  /* Check for input parameters */
  if(argc < 3)
  {
    cout<<"Insufficient input parameters ..."<<endl;
    cout<<"usage ./sim <amplitude> <frequency>"<<endl;
    exit(0);
  }

/* Declare/define inter-module channels */
sca_tdf::sca_signal<double> sigin0;
sca_tdf::sca_signal<double> sigout0;
sca_tdf::sca_signal<double> sigin1;
sca_tdf::sca_signal<double> sigin1;
sca_tdf::sca_signal<double> sigout1;
sca_tdf::sca_signal<double> sigout2;

amplitude = strtod(argv[1], NULL);
frequency = strtod(argv[2], NULL);

/* Declare/define filter modules */
lpo1sdom f1("f1");
lpo2sdom f2("f2");
lpo3sdom f3("f3");
```

Fig. 13.5 Test harness for fifth-order low-pass Butterworths filter with a cut-off frequency of 60 Hz

```
    /* Declare/define filter signal input source and trace file */
    sinsrc sin_src("sin_src");
    tracedoublecombo trdbl("tr_nd");

    /* Connect module ports and signal
       channels and set input parameters */
    sin_src.sigout(sigin0);
    sin_src.amplitude = amplitude;
    sin_src.frequency = frequency;

    f1.sigin(sigin0);
    f1.sigout(sigout0);
    f2.sigin(sigout0);
    f2.sigout(sigout1);

    f3.sigin(sigout1);
    f3.sigout(sigout2);

    trdbl.in1(sigin0);
    trdbl.in2(sigout2);

    /* Run simulation for pre-defined time period and then stop */
    sc_core::sc_start(100.0, sc_core::SC_MS);
    sc_core::sc_stop();
    return 0;
}
```

**Fig. 13.5** (continued)

Careful attention must be paid to the structure of a TDF module. The '*void set_attributes ()*' and '*void processing()*' methods have to be overwritten, and '*void initialize()*' may or may not be overridden. In this example, 'void initialixe()' is overridden to initialize the internal data structures—vectors for storing the denominator and numerator coefficients for the low-pass filter in pole–zero form. Most importantly, the SystemC-AMS built-in *sca_tdf::sca_ltf_nd* module represents the scaled Laplace transfer function in the time domain in the numerator–denominator form.

The above source code example may be compiled to an executable using the following command:

*g++-I. -I<path_to_SystemC_directory>/include -I<path_to_SystemC-AMS_directory>/include -L. -L<path_to_SystemC_directory>/lib-linux -L<path_to_SystemC-AMS_directory>/lib-linux -o sim<test_bench_file_name>.cc -lsystemc-ams -lsystemc -lm.*

A standard Linux/Unix Makefile or shell script may be used.

As mentioned at start, we have compiled all our examples on Red Hat Enterprise Linux 4.0 (with GCC compiler version 3.2.6) and a Red Hat Fedora 14 machine (with GCC compiler version 4.5.1) running SystemC 2.3.0 and SystemC-AMS 1.0. Traditionally, the final executable is named *sim* and may be invoked from the command line as. */sim<optional parameter list>*.

The SystemC-AMS class structure shown here is based on the standard TLM infrastructure with processing elements communicating via channels (*sca_tdf::sca_signal<T><channel_name>*), with input/output ports (*sca_tdf::sca_in<T><input_port_name>, sca_tdf::sca_out<T><output_port_name>*) interfacing between the channel and processing element. A SystemC-AMS timed data flow (TDF) class is typically declared using a macro as *SCA_TDF_MODULE*(*<module_name>*) and, like any other C++ class, can have private/public members and methods. The macro *SCA_CTOR* specifies the module constructor. The TDF formalism provides the user a set of standard methods to initialize (*void initialize()*) data members and set module and port properties (*void set_attributes()*) specifically:

- Specify time step of a module and each of its ports
- Specify port rate of a module's port—i.e., the number of data elements read in or written to a port per each read/write operation
- Specify delays and time offsets for each port.

The computation associated with a module is performed in the `void processing()` method. It is very important to note that System-AMS TDF modules can interact with SystemC's clocking/timing mechanism, but do not depend on it in any way; time step of any TDF module and each of its ports can be set explicitly from within the module, along with port rate and port delays/offsets. However, the main simulation uses the core SystemC method sc_core::sc_start.

For the pole–zero analog filter description formalism, the coefficients of the transfer function $H(s) = \frac{X(s)}{Y(s)}$ are declared and/or specified using the two vectors *sca_util::sca_vector<T>numer* and *sca_util::sca_vector<T>denom*, respectively. In the test bench file, the channels are defined first, followed by the individual modules.

sca_tdf::sca_signal<double>sigin0;

sca_tdf::sca_signal<double>sigout0;
sca_tdf::sca_signal<double>sigin1;
sca_tdf::sca_signal<double>sigout1;
sca_tdf::sca_signal<double>sigout2;

lpo1sdom f1("f1");

lpo2sdom f2("f2");
lpo3sdom f3("f3");
sinsrc sin_src("sin_src");
tracedoublecombo trdbl("tr_nd");

This is followed by the crucial task of connecting the modules with the appropriate channels.

```
sin_src.sigout(sigin0);

sin_src.amplitude = amplitude;
sin_src.frequency = frequency;
f1.sigin(sigin0);
f1.sigout(sigout0);
f2.sigin(sigout0);
f2.sigout(sigout1);
f3.sigin(sigout1);
f3.sigout(sigout2);
trdbl.in1(sigin0);
trdbl.in2(sigout2);
```

Finally, the simulation is started and the total simulation time is specified.

```
sc_core::sc_start(100.0, sc_core::SC_MS);
```

We use custom modules to format the output numerical data so that it can be displayed with Gnuplot (Fig. 13.6).

```
tracedoublecombo trdbl("tr_nd");
trdbl.in1(sigin0);
trdbl.in2(sigout2);
```

The frequency response and phase response for this filter are shown in Figs. 13.6 and 13.7. Figures 13.8, 13.9, 13.10, and 13.11 show some sample input and output waveforms for the filter. As expected, the amplitude of the output waveform

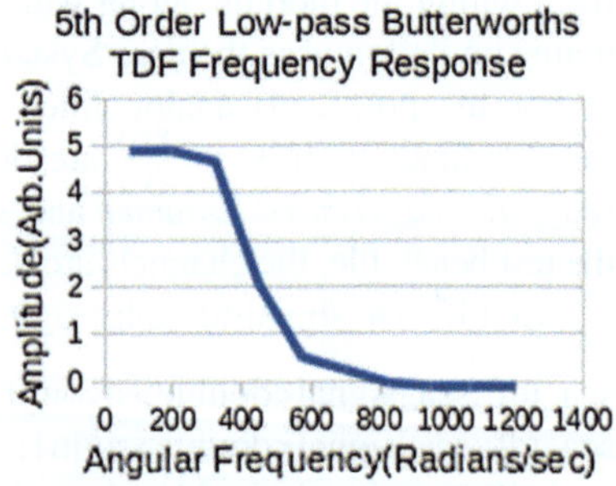


**Fig. 13.6** Fifth-order low-pass unity-gain Butterworths frequency response, cut-off frequency of 60 Hz

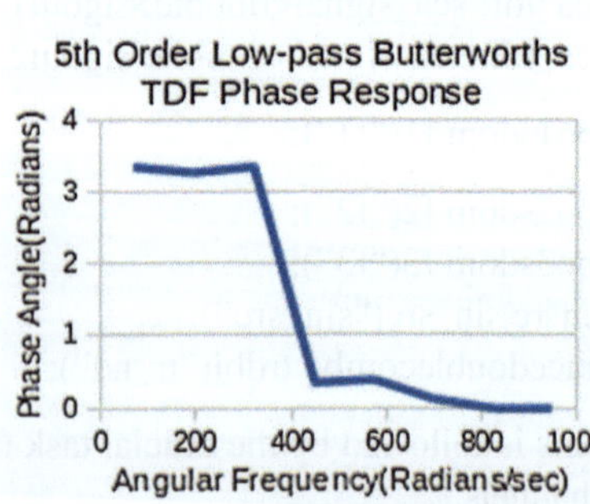


**Fig. 13.7** Fifth-order low-pass unity-gain Butterworths phase response, cut-off frequency of 60 Hz

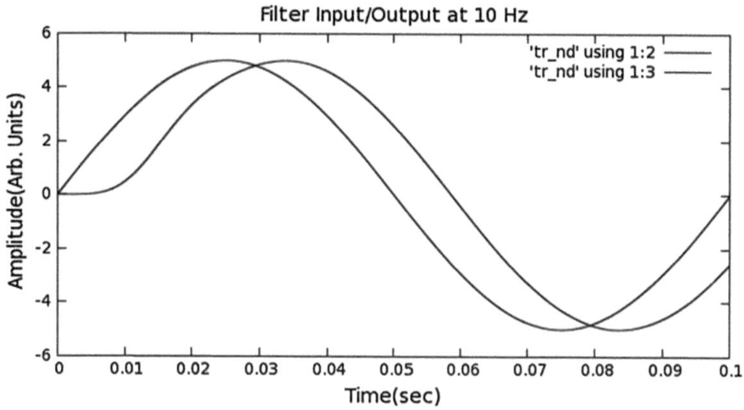


**Fig. 13.8** Fifth-order low-pass unity-gain Butterworths filter output for 10 Hz input, cut-off frequency of 60 Hz

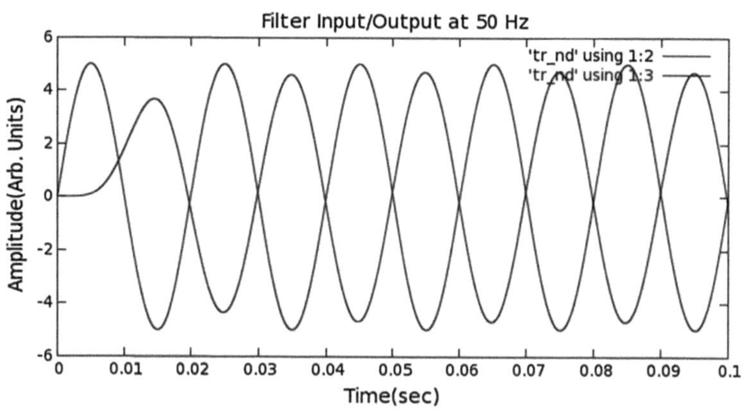


**Fig. 13.9** Fifth-order low-pass unity-gain Butterworths filter output for 50 Hz input, cut-off frequency of 60 Hz

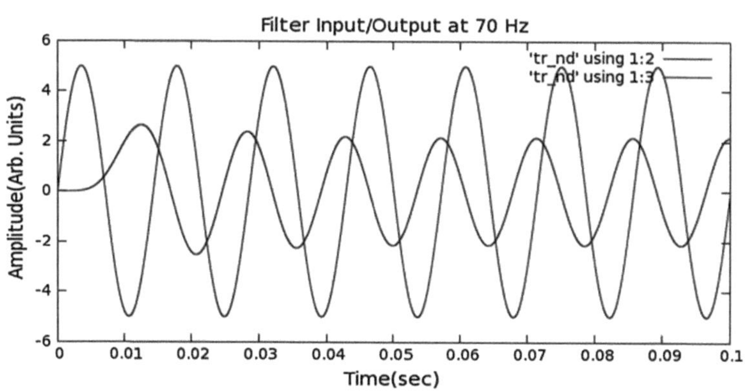


**Fig. 13.10** Fifth-order low-pass unity-gain Butterworths filter output for 70 Hz input, cut-off frequency of 60 Hz

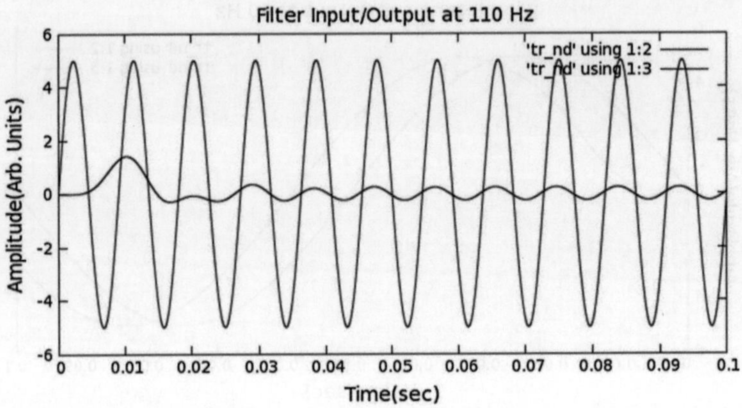


**Fig. 13.11** Fifth-order low-pass unity-gain Butterworths filter output for 110 Hz input, cut-off frequency of 60 Hz

decreases sharply when the input signal frequency is greater than or equal to the cut-off frequency of 60 Hz.

## 13.2 Simple Single Slope Analog-to-Digital Converter

An analog-to-digital converter [3] (ADC) provides the crucial interface between the analog and digital signal domains. There are many types of ADCs (flash, successive approximation, single/dual slope, etc.), with varying levels of accuracy/resolution, quantization noise, processing speed, etc; we present a very simple one here, to illustrate the basic concepts. The design of an accurate, high-resolution, and low noise ADC is in a later chapter.

Figure 13.12 shows a basic single slope ADC. The integration of a reference voltage generates a ramp that is compared to the input voltage. As soon as these two voltages are the same, the comparator output goes high, the integrator's capacitor is discharged via the MOSFET, the counter is stopped, and its latest count value is transferred to the output register, followed by a counter reset. The count value in the register is the number corresponding to the input voltage value read in. There are drawbacks to the design, e.g.,

- Integrator components (capacitor, resistor) must be very accurate (low thermal drift, etc.)
- Negative input voltages cannot be digitized.

To analyze the behavior of this ADC, we model each of the components of Fig. 13.12, as TDF modules. The source code for the ADC and its test harness is in Figs. 13.13, 13.14, 13.16, and 13.17. The input signal source is a sine wave with a DC offset, as the single slope ADC cannot handle any negative valued input (Fig. 13.15).

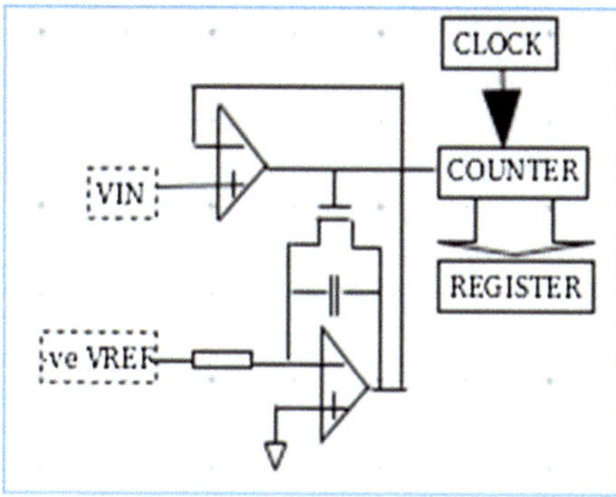


**Fig. 13.12** Single slope ADC design. A positive reference voltage cannot be applied to an inverting integrator. So, the input voltage can only be negative, restricting the utility of the ADC

```
#ifndef ADC_H
#define ADC_H
    using namespace std;

    #include <systemc-ams>
    #include "commonsrcs.h"

    const double TOL = 0.5;
    const double MULTFACTOR = 8.5E+6;

    SCA_TDF_MODULE(integrator)
    {
      /*Declare/define input/output channels and module members */
      sca_tdf::sca_in<bool> cntrlin;
      sca_tdf::sca_in<double> valuein;
      sca_tdf::sca_out<double> valueout;
      bool cntrlvalue;
      double currtime;
      double invalue;
      double lasttime;
      double lastinvalue;
      double lastoutvalue;
      double outvalue;

      /* Set input/output port parameters */
      void set_attributes()
      {
```

**Fig. 13.13** Single slope ADC integrator—uses popular trapezoidal rule for numerical computation

```
        valueout.set_timestep(1.0, sc_core::SC_US);
        cntrlin.set_rate(1);
        valuein.set_rate(1);
        valueout.set_rate(1);
    }
  /* Implement trapezoidal integration rule */
  void processing()
  {
    cntrlvalue = cntrlin.read();
    invalue = valuein.read();
    currtime = valueout.get_time().to_seconds();
    if(cntrlvalue == false)
    {
        outvalue = (currtime - lasttime)*(invalue + lastinvalue)*0.5;
    }
    else if(cntrlvalue == true)
    {
      outvalue = 0.0;
      lastoutvalue = 0.0;
      lastinvalue = 0.0;
      lasttime = currtime;
    }

    lasttime = currtime;
    lastinvalue = invalue;
    lastoutvalue = outvalue;
    lastoutvalue *= MULTFACTOR;
    valueout.write(fabs(lastoutvalue));
  }


  /* Constructor with member initialization */
  SCA_CTOR(integrator):cntrlvalue(false),
                    currtime(0.0),
                    lastinvalue(0.0),
                    lasttime(0.0),
                    outvalue(0.0){ }
  ~integrator(){ } /*Destructor */
};
```

**Fig. 13.13** (continued)

```
SCA_TDF_MODULE(comparator)
{
  /*Declare/define input/output channels and module members */
  sca_tdf::sca_in<double> valuein1;
  sca_tdf::sca_in<double> valuein2;
  sca_tdf::sca_out<bool> valueout;
  double din1;
  double din2;
  bool cntrlout;

  /* Set input/output port parameters */
  void set_attributes()
  {
    valueout.set_timestep(1.0, sc_core::SC_US);
    valuein1.set_rate(1);
    valuein2.set_rate(1);
    valueout.set_rate(1);
    valueout.set_delay(1);
  }

  /* Comparator operation  */
  void processing()
  {
    din1 = valuein1.read();
    din2 = valuein2.read();
    cntrlout = din1 < din2 ? true : false;
    valueout.write(cntrlout);
  }

  /* Constructor with member initialization */
  SCA_CTOR(comparator) : din1(0.0),
                         din2(0.0),
                         cntrlout(false){ }
  ~comparator(){ } /*Destructor */
};
```

**Fig. 13.14** Single slope ADC comparator

The multiplicative constant MULTFACTOR has been used for convenience of simulation only, and there is no real-world counterpart in a real-world single slope ADC.

```
SCA_TDF_MODULE(counter)
{
  /*Declare/define input/output channels and module members */
  sca_tdf::sca_in<bool> cntrlsig;
  sca_tdf::sca_out<unsigned int> countout;
  unsigned int count;
  bool cntrlvalue;
  /* Set input/output port parameters */
  void set_attributes()
  {
    countout.set_timestep(1.0, sc_core::SC_US);
    countout.set_rate(1);
    cntrlsig.set_rate(1);
  }

  /* Counter operation */
  void processing()
  {
    cntrlvalue = cntrlsig.read();
    if(cntrlvalue == true) { count = 0; }
    else { count += 1; }
    countout.write(count);
  }

  /* Constructor with member initialization */
  SCA_CTOR(counter) : count(0){ }
  ~counter() { } /* Destructor */
};
#endif
```

**Fig. 13.15** Single slope ADC counter

```
#include "adc.h"
#include "gendatatrace.h"

int sc_main(int argc, char **argv)
{
  /* Declare/define inter-module channels */
  sca_tdf::sca_signal<bool> cntrl;
  sca_tdf::sca_signal<double> achnl;
  sca_tdf::sca_signal<double> rchnl;
```

**Fig. 13.16** Test harness for single slope ADC

```
sca_tdf::sca_signal<double> ichnl;
sca_tdf::sca_signal<unsigned int> divalue;

/* Declare/define ADC modules */
sinsrc_offset ss_offset("ss_offset");
refsrc refsrc1("refsrc1");
integrator integrator1("integrator1");
comparator comparator1("comparator1");
counter counter1("counter1");

/* Declare/define trace files/objects and assign
   module parameters */
tracedouble tranaloginput("tr_analog_input");
tracedouble trdblintgout("tr_integ_out");
tracebool trcntrlsig("tr_cntrl_sig");
traceint trint("tr_dig_value");
```

**Fig. 13.16** (continued)

```
ss_offset.sigout(achnl);
ss_offset.amplitude = 5.0;
ss_offset.frequency = 250;
ss_offset.offset = 5.5;

/* Connect modules and channels */
refsrc1.refout(rchnl);
refsrc1.refvalue = 0.25;

integrator1.cntrlin(cntrl);
integrator1.valuein(rchnl);
integrator1.valueout(ichnl);

comparator1.valuein1(achnl);
comparator1.valuein2(ichnl);
comparator1.valueout(cntrl);

counter1.cntrlsig(cntrl);
counter1.countout(divalue);

tranaloginput.in(achnl);
trint.in(divalue);
```

**Fig. 13.16** (continued)

```
    trdblintgout.in(ichnl);
    trcntrlsig.in(cntrl);

    /* Run simulation for pre-defined time period and then stop */
    sc_start(50.0, sc_core::SC_MS);
    sc_core::sc_stop();
    return 0;
}
```

**Fig. 13.16** (continued)

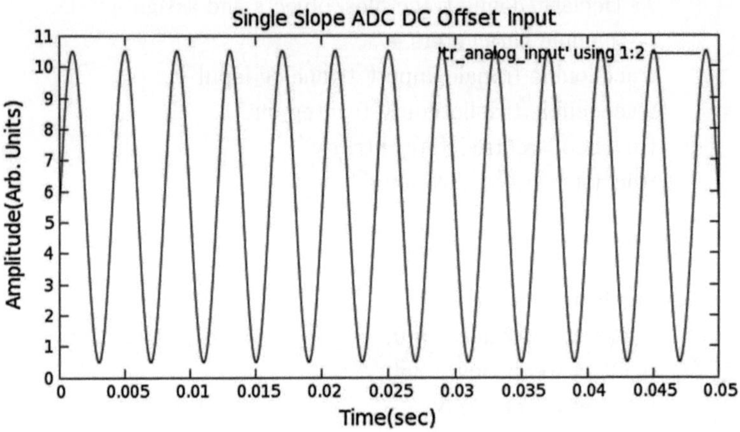


**Fig. 13.17** DC offset sine wave input for single slope ADC

The single slope ADC could also have been implemented as a mixed ELN-TDF design, using SystemC-AMS's idealized operational amplifier sca_eln::sca_nullor for both the comparator and integrator. In that case, the counter/register would be pure TDF modules, and converter ports would interface between ELN and TDF modules.

A single slope ADC is not used very widely for the reasons mentioned earlier In addition, selecting the reference voltage VREF demand all the skills of the designer, to optimize all the desired properties of the ADC. VREF value must fit within the available power supply rails, be large enough with respect to the comparators offset, noise voltage and still be small enough so that nonlinear effects remain controlled. By experimentation, we found that the optimum VREF value is = 0.25 V. The TDF module implementing the numerical integrator uses the trapezoidal rule, which states that the area under curve, *f*(*x*) between two values of the independent variable (*a*, *b*) is $Area = \frac{f(b)-f(a)}{2(b-a)}$. The accuracy/resolution of a single slope ADC is low, and it cannot handle negative input voltages. Consequently, our input to the ADC is a sine wave with a DC offset. *A very interesting fact is the counter module, which is*

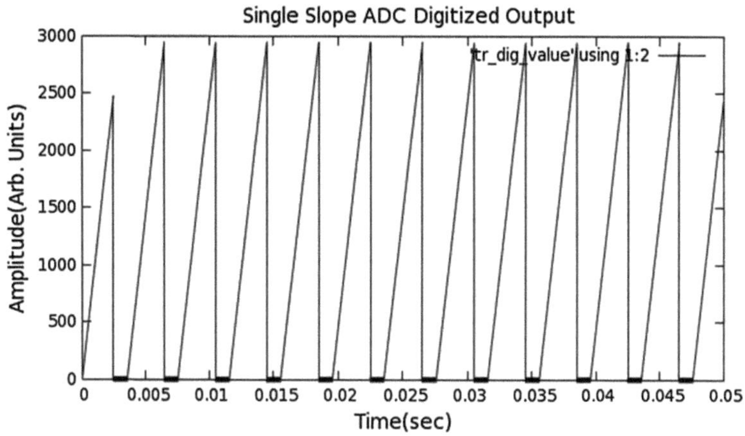


**Fig. 13.18** Single slope ADC output with DC offset sine wave input

*purely digital, but is implemented as SCA_TDF_MODULE, with no external clocks. In this case, the 'set_timestep' method of the TDF module establishes the required clock, by enforcing the time interval between each output operation.* Figures 13.17 and 13.18 show the ADC input and scaled output.

## 13.3 Quadrature Phase-Shift Key Modulation

Quadrature phase-shift keying [4] (QPSK) is reliable and popular(used in Code division multiple access—CDMA) digital modulation technique. Digital modulation techniques arise from the fundamental bandwidth constraints of channels. Any digital signal consists of 1s and 0s, with voltage levels indicating the logic levels 1 and 0. For example, a digital signal consisting of alternate 1s and 0s may be visualized as a square wave of duty cycle 50 %. From fundamental Fourier analysis, a square wave is a linear superposition of odd harmonics of a sine wave, of a pre-defined frequency; if the pre-defined frequency (or fundamental frequency) is very high (e.g., several GHz), then the bandwidth requirements to transmit such a signal are huge.

The problem is circumvented by digital modulation, where a very high frequency (e.g., several GHz) analog carrier sinusoidal signal is 'mixed' with a digital signal, whose fundamental frequency is orders of magnitude lower (e.g., audio signals are in the range 3–8 kHz). This is digital modulation, and QPSK is one such popular digital modulation scheme. The modulated signal is transmitted over a standard channel and the bandwidth requirements are much relaxed. The algorithmic/mathematical details of this scheme may be found in a number of excellent texts and online tutorials. Here, we focus on a simple real-world QPSK modulator (Fig. 13.19).

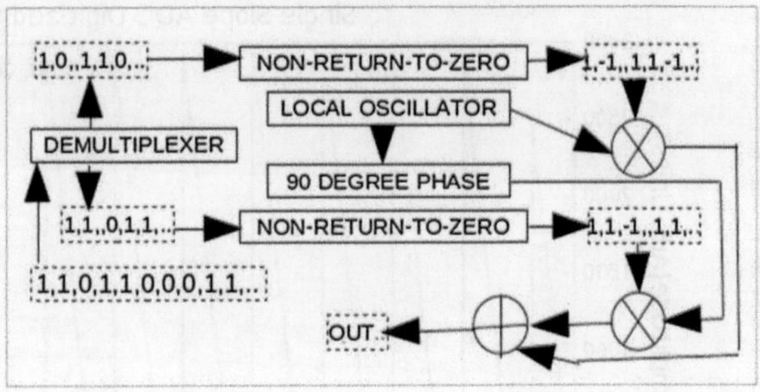


**Fig. 13.19** Simplest QPSK modulation scheme. Example input bitstream is shown

A digital input bitstream is fed into a demultiplexer, which generates two output bitstreams. A local high-frequency oscillator generates the high-frequency carrier signal cos wt (I component) and its 90° phase-shifted counterpartsin wt (Q or quadrature component). Each of the I and Q components is fed into their own analog mixer. The demultiplexed bitstreams are each fed into its own 1-bit digital-to-analog converter (DAC), which in this case is a *non-return-to-zero* converter. A non-return-to-zero converter allows a logic 1 to pass unchanged, but emits a $-1$ for each logic 0 input. Each of the non-return-to-zero converter outputs is fed into the corresponding analog mixers. The outputs from each of the two mixers are fed into an adder, and its output is the QPSK modulated signal. Instead of addition, one of the mixer outputs might be subtracted from the other.

If the demultiplexer outputs are $A_1(t)$ and $A_2(t)$, respectively, then the outputs of the two mixers are $A_1(t)\cos \mathrm{wt}$ and $A_2(t)\sin \mathrm{wt}$, with the final QPSK modulated output being $A_1(t)\cos \mathrm{wt} + A_2(t)\sin \mathrm{wt}$.

The demodulation scheme is slightly complicated. In all real-world applications of QPSK, the transmitter and receiver are not at the same physical location, so that the receiver must first detect the high-frequency carrier frequency accurately. After that, the receiver's local oscillator generates the I and Q components ($\cos wt$ and $\sin \mathrm{wt}$). The incoming signal is then mixed with both the I and Q components, to get (after some mathematical manipulation) $\frac{A_1(t)}{2} + \frac{A_1(t)}{2}\cos 2\mathrm{wt} + \frac{A_2(t)}{2}\sin 2\mathrm{wt}$ and $\frac{A_2(t)}{2} - \frac{A_2(t)}{2}\cos 2\mathrm{wt} + \frac{A_1(t)}{2}\sin 2\mathrm{wt}$ . Low-pass filtering removes the high-frequency sinusoidal components ($\sin 2\mathrm{wt}$, $\cos 2\mathrm{wt}$). The remaining part ($\frac{A_1(t)}{2}$, $\frac{A_2(t)}{2}$) is fed into special decision-making circuitry, followed by multiplexing, to recover the original bitstream sent into the modulator (on the transmitter). A key fact about the QPSK scheme is that the carrier frequency must be very high compared to the data (*baseband*) frequency. From Fig. 5.18, the QPSK module contains a random bit source (to mimic a real-world data source), a demultiplexer, *two not-return-to-zero* (1-bit digital-to-analog converter—DAC) modules, a local oscillator (producing a high-frequency carrier sinusoidal signal and its 90° phase lagged—*quadrature*—version), two mixers and an adder, each of which are implemented as TDF modules, Figs. 13.20, 13.21, 13.22, 13.23, 13.24, and 13.25.

Input/output traces to/from the QPSK modulator are in Fig. 13.26a, b, and c.

```
#ifndef QPSKMODDEMOD_H
#define QPSKMODDEMOD_H

#include <systemc-ams>

const unsigned int MAX_COUNT = 512;
const unsigned int MAX_NUM = 1500;

SCA_TDF_MODULE(bitsrc)
{
  /*Declare/define input/output channels and module members */
  sca_tdf::sca_out<bool> bitout;
  bool b0;
  unsigned int count;

   /* Set input/putput portl parameters */
  void set_attributes()
  {
    bitout.set_timestep(1.0, sc_core::SC_NS);
    bitout.set_rate(1);
  }

   /* Generate random bit stream */
   void processing()
   {
    if(count == MAX_NUM)
    {
      count = 0;
      b0 = std::rand() % 2;
    }
     else
    {
      count += 1;
    }
    bitout.write(b0);
  }

  /* Constructor with member initialization */
  SCA_CTOR(bitsrc):b0(false),count(0) { }
  ~bitsrc() { }  /* Destructor */
};
```

**Fig. 13.20** QPSK modulator random input bit source

```
SCA_TDF_MODULE(demultiplexer)
{
  /*Declare/define input/output channels and module members */
  sca_tdf::sca_in<bool> bitin;
  sca_tdf::sca_out<bool> bitout1;
  sca_tdf::sca_out<bool> bitout2;
  bool togglevalue;
  bool tmp;
   /* Set input/putput port parameters */
  void set_attributes()
  {
    bitout1.set_timestep(1.0, sc_core::SC_NS);
    bitout2.set_timestep(1.0, sc_core::SC_NS);
    bitin.set_rate(1);
    bitout1.set_rate(1);
    bitout2.set_rate(1);
  }

  /* Demultiplexer operation */
  void processing()
  {
    tmp = bitin.read();
    if(togglevalue == false)
    {
      bitout1.write(tmp);
    }
    else if(togglevalue == true)
    {
      bitout2.write(tmp);
    }
    togglevalue = togglevalue == true ? false:true;
  }

  /* Constructor with member initialization */
  SCA_CTOR(demultiplexer) : togglevalue(false) { }

  ~demultiplexer() { }
 /* Destructor */
};
```

**Fig. 13.21** QPSK modulator demultiplexer

```
SCA_TDF_MODULE(nrz)
{
  /*Declare/define input/output channels and module members */
 sca_tdf::sca_in<bool> bitin;
 sca_tdf::sca_out<int> nrzout;
 bool tmpbool;
 int valueout;

  /* Set input/putput port parameters */
 void set_attributes()
 {
   nrzout.set_timestep(1.0, sc_core::SC_NS);
   bitin.set_rate(1);
   nrzout.set_rate(1);
 }

 /* One -bit DAC operation - for each '0' input, emit a -1 */
 void processing()
 {
   tmpbool = bitin.read();
   valueout = tmpbool == true ? 1 : -1;
   nrzout.write(valueout);
 }

 SCA_CTOR(nrz){ } /* Constructor */
 ~nrz() { }  /* Destructor */
};
```

**Fig. 13.22** QPSK modulator 1-bit digital-to-analog converter—DAC—non-return-to-zero

```
SCA_TDF_MODULE(carrier)
{
  /*Declare/define input/output channels and module members */
  sca_tdf::sca_out<double> carrout;
  sca_tdf::sca_out<double> carroutps;
  double sampleperiod;
  double freq;
  double carrvalue;
  double carrvalueps;
  double timevalue;

  /* Set input/putput port parameters */
  void set_attributes()
  {
    /*module time step = output port time step x output port rate */
    carrout.set_timestep(1.0, sc_core::SC_NS);
    carrout.set_rate(1);
  }

  /* Modulation and quadrature signal generation */
  void processing()
  {
      timevalue = carrout.get_time().to_seconds();
    carrvalue = (2.0/sampleperiod)*sin(2.0*3.14*freq*timevalue);
    carrvalueps = (2.0/sampleperiod)*cos(2.0*3.14*freq*timevalue);
    carrout.write(carrvalue);
    carroutps.write(carrvalueps);
  }

  SCA_CTOR(carrier) { } /* Constructor */
  ~carrier() { }  /*Destructor */
};
```

**Fig. 13.23** QPSK modulator high-frequency carrier and quadrature signal generator

```
SCA_TDF_MODULE(qpsk)
{
  /*Declare/define input/output channels and module members */
  sca_tdf::sca_in<int> iin1;
  sca_tdf::sca_in<int> iin2;
  sca_tdf::sca_in<double> carrier;
  sca_tdf::sca_in<double> carrierps;
  sca_tdf::sca_out<double> modsig;

  double c1;
  double c2;
  double outval;
  int i1;
  int i2;

    /* Set input/putput port parameters */
   void set_attributes()
  {
  /*module time step = output port time step x output port rate */
    iin1.set_rate(1);
    iin2.set_rate(1);
    carrier.set_rate(1);
    carrierps.set_rate(1);
    modsig.set_timestep(1.0, sc_core::SC_NS);
    modsig.set_rate(1);
  }

  /* Mixing operation */
  void processing()
  {
    c1 = carrier.read();
    c2 = carrierps.read();
    i1 = iin1.read();
    i2 = iin2.read();
    outval = i1*c1 - i2*c2;
    modsig.write(outval);
  }

  SCA_CTOR(qpsk) { } /* Constructor */
  ~qpsk() { } /* Destructor */
};
```

**Fig. 13.24** QPSK modulator mixer and adder

```
#include "qpskmoddemod.h"
#include <cstring>
#include "gendatatrace.h"
using namespace std;

int sc_main(int argc, char **argv)
{
 double carrierfreq;
 double carriertimesample;
 /* Check and read in input parameters */
 if(argc < 3)
 {
  std::cout<<"incomplete parameter list"<<std::endl;
  std::cout<<"usage"<<std::endl;
  std::cout<<"./sim <carrier freq> <sample rate> "<<std::endl;
  exit(0);
 }
 else
 {
  carrierfreq = strtod(argv[1], NULL);
  carriertimesample = strtod(argv[2], NULL);
 }

 /* Declare/define inter-module signal channels */
 sca_tdf::sca_signal<bool> bitchnl;
 sca_tdf::sca_signal<bool> bitchnl1;
 sca_tdf::sca_signal<bool> bitchnl2;
 sca_tdf::sca_signal<bool> bitrz1;
 sca_tdf::sca_signal<bool> bitrz2;
 sca_tdf::sca_signal<bool> bitchnlout;
 sca_tdf::sca_signal<double> sine1;
 sca_tdf::sca_signal<double> sine2;
 sca_tdf::sca_signal<double> finval;
 sca_tdf::sca_signal<int> intchnl1;
 sca_tdf::sca_signal<int> intchnl2;
 sca_tdf::sca_signal<double> dmout1;
 sca_tdf::sca_signal<double> dmout2;
 sca_tdf::sca_signal<double> filt1out;
 sca_tdf::sca_signal<double> filt2out;
```

**Fig. 13.25** QPSK test harness

```
    /* Declare/define modulator modules */
    bitsrc bitsr_c("bitsr_c");
    carrier car("carrier");
    demultiplexer demux("demux");
    nrz nrz_2("nrz_2");
    qpsk qps_k("qps_k");


  /* Declare/define trace files/objects */
  traceboolcombo tr_bool_dm("tr_bool_dm");
  tracedouble tr_qpsk_mod("tr_qpsk_mod");
  tracedoublecombo tr_dbl_carr("tr_dbl_carr");
  traceintcombo tr_int_nrz("tr_int_nrz");

  /* Connect channels with module ports */
  bitsr_c.bitout(bitchnl);
  demux.bitin(bitchnl);
  demux.bitout1(bitchnl1);
  demux.bitout2(bitchnl2);
  nrz_1.bitin(bitchnl1);
  nrz_1.nrzout(intchnl1);
  nrz_2.bitin(bitchnl2);
  nrz_2.nrzout(intchnl2);
```

**Fig. 13.25** (continued)

```
  car.sampleperiod = carriertimesample;
  car.freq = carrierfreq;
  car.carrout(sine1);
  car.carroutps(sine2);

  qps_k.iin1(intchnl1);
  qps_k.iin2(intchnl2);
  qps_k.carrier(sine1);
  qps_k.carrierps(sine2);
  qps_k.modsig(finval);

  /* Connect trace files with ports */
  tr_dbl_carr.in1(sine1);
  tr_dbl_carr.in2(sine2);
  tr_bool_dm.inbool0(bitchnl);
  tr_bool_dm.inbool1(bitchnl1);
  tr_bool_dm.inbool2(bitchnl2);
  tr_int_nrz.int1(intchnl1);
  tr_int_nrz.int2(intchnl2);
  tr_qpsk_mod.in(finval);

  /* Run simulation for pre-defined time period and then stop */
  sc_core::sc_start(15, sc_core::SC_US);
  sc_core::sc_stop();
  return 0;
}
```

**Fig. 13.25** (continued)

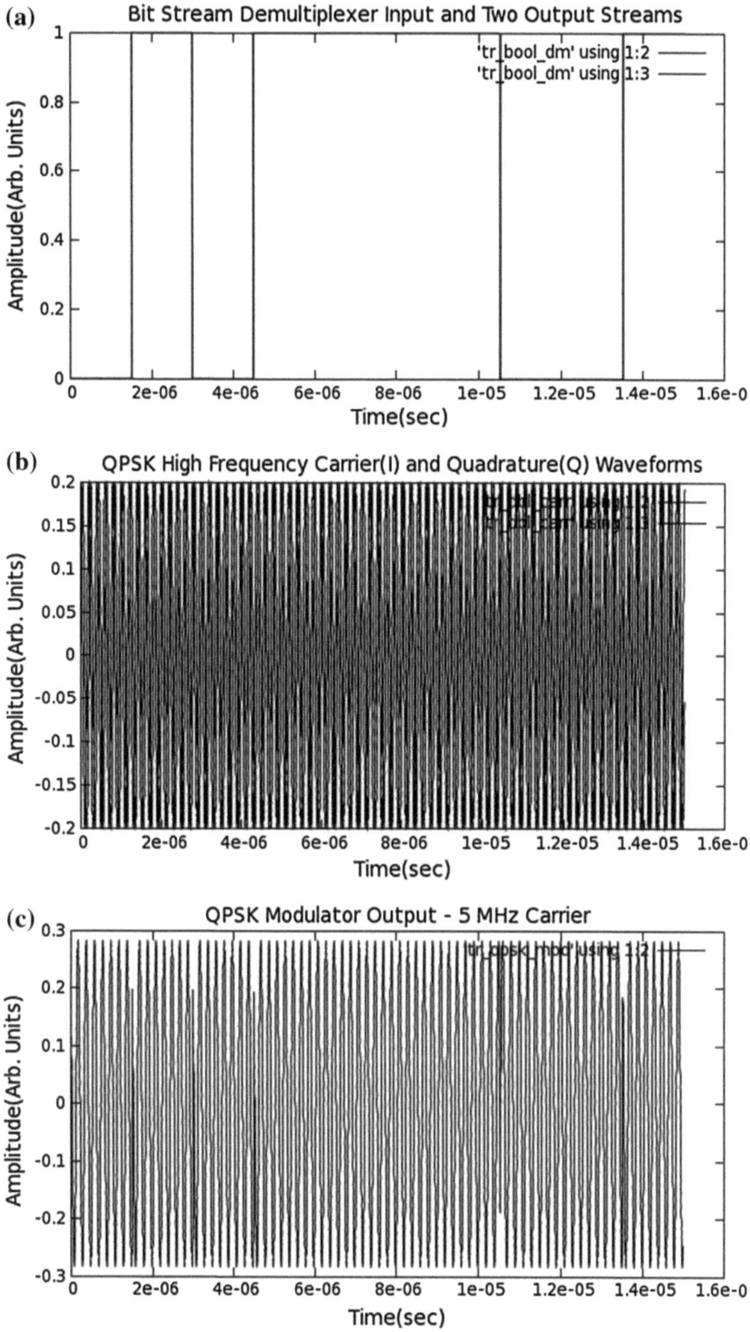


**Fig. 13.26** **a** Random bitstream input to QPSK demultiplexer, and its two output streams to the 1-bit analog-to-digital converters. **b** QPSK carrier ($I$) and quadrature ($Q$) signal, one of which ($I$ or $Q$) is input for one mixer, and the other ($I$ or $Q$) is input to the other mixer. **c** QPSK modulator final output waveform

## References


1. SystemC-AMS Working Group. SystemC-AMS (Analog Mixed Signal)—Accellera Systems Initiative. http://www.accellera.org/downloads/standards/systemc/ams and http://www.accellera.org/activities/committees/systemc-ams/
2. Texas Instruments. Analog Embedded Processing, Semiconductor Company Texas Instruments. *Active Filter Design Techniques, Literature Number SLOA088 Texas Instruments, excerpted from OP-Amps for Everyone, Literature Number SLOD006A* http://www.ti.com/lit/ml/sloa088/sloa088.pdf
3. Maxim. Analog,linear and mixed-signal devices from Maxim *Understanding integrating ADCs Maxim Application Note AN1041* http://pdfserv.maximintegrated.com/en/an/AN1041.pdf
4. Birdsall, T. G. and Saha, D.(1989) Quadrature Phase Shift Keying IEEE Transactions on Communications, 37 (5), 437 - 448

# Chapter 14
# Linear Signal Flow in Practice and Theory

**Abstract** Linear signal flow [1] (LSF) formalism enables us to model, analyze, and explore the design space of any non-conservative system with continuous time and directed real-valued signals. When represented as a graph, LSF modules are the nodes and the unidirectional arcs between the nodes are the channels, and it fits perfectly with the TLM paradigm of processing elements communicating via channels. Specifically the LSF formalism enables modeling/analysis of an AMS system defined in terms of relations between variables of a system of linear algebraic equations. Unlike TDF, the formalism is restrictive—the user cannot write customized code for the model, but only use members of a pre-defined set of LSF models (addition, subtraction, multiplication, derivative, etc.), as required.

## 14.1 Delta-Sigma Modulator

Earlier, we examined a simple single slope analog-to-digital converter (ADC). Here, we examine, model, and analyze the crucial modulator component of a very sophisticated and popular ADC—the *sigma-delta* ADC [2, 3]. A sigma-delta ADC consists of a pure analog block (sigma-delta modulator) and a pure digital block (decimation filter/digital low-pass filter). We focus on the modulator. Our model is more complete and differs significantly from the sigma-delta modulator model presented in the official SystemC-AMS [1] Users' Guide. First, we examine in detail the sigma-delta architecture—Fig. 14.1.

Consider a sine wave input applied to a multi-bit output ADC. The sampling frequency is $F_s$ which the Nyquist theorem must be at least 2× the input bandwidth. The fast fourier transform (FFT) of the digital output shows a single prominent peak and lots of random quantization noise. Quantization noise results from the analog input having an infinite number of states, but the ADC digital output has a fixed resolution/number of states. This constraint introduces some output distortion and information loss, which manifests itself as quantization noise. The sigma-delta ADC cleverly tackles the noise issue with oversampling.

A. Banerjee and B. Sur, *SystemC and SystemC-AMS in Practice*,
DOI: 10.1007/978-3-319-01147-9_14,

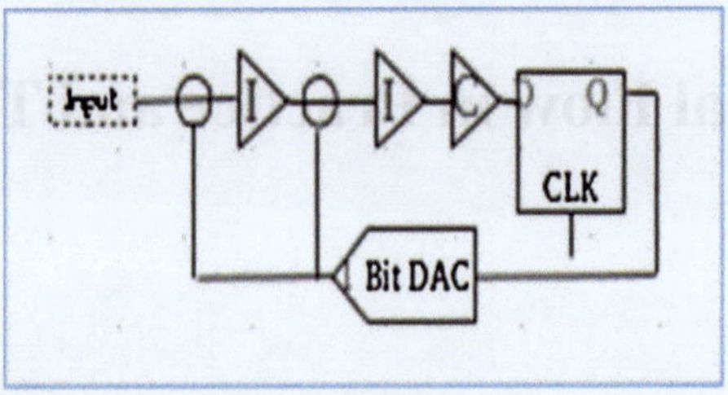


**Fig. 14.1** Popular second-order sigma-delta modulator. *I* indicates integrator, and *C* indicates comparator. A first-order sigma-delta modulator consists of a single integrator, a single 1-bit DAC, a single summing node, and a single comparator. The digital decimation/low-pass filter pair has not been shown

The sampling rate is now increased to $k\,F_s$ where k is a large positive number (typically a power of 2—e.g., 64, 128, 256). An FFT of the output shows that the noise floor has dropped, the signal-to-noise ratio (SNR) is unchanged, and the noise spectrum has been pushed out to a higher-frequency range. The high-frequency noise band can be easily filtered with a low-pass filter, thereby allowing low noise floor, wide dynamic range main signal to pass unhindered. This technique is referred to as *noise shaping*. The term 'sigma' arises from the integrator and the term 'delta' from the 1-bit digital-to-analog converter (DAC) of Fig. 14.1. The comparator acts as a 1-bit ADC, and the 1-bit DAC provides the difference or 'delta' between the 1-ADC output values. Our example is a second-order sigma-delta modulator, but fifth-order delta-sigma modulators for audio applications are commonly available—higher-order modulators cannot be created by connecting a series of first-order sigma-delta modulators as phase noise become prominent. A 1-bit DAC is really a *not-return-to-zero* (NRZ) circuit—its output is $-1$ for logic 0 input and $+1$ otherwise.

Oversampling is achieved with a high-frequency oversampling clock and the *D* flip-flop. *The sigma-delta modulator trades off accuracy in amplitude with accuracy in time.* The modulator consists of two integrators, a comparator, two sum/subtraction nodes, a 1-bit DAC, and a *D* flip-flop. While the comparator and the DAC are TDF modules, the sum/subtraction, gain and integrators are library supplied LSF modules, and the *D* flip-flop is a pure digital (SystemC—*not* System-AMS) module. This is a truly mixed signal design that draws upon pure analog and pure digital modules. Figures 14.2, 14.3, 14.4, 14.5, and 14.6 contain the source code for all the modulator components and the test harness. *This design example also illustrates how TDF and LSF modules may be connected to achieve a specific design goal* (Fig. 14.6).

The structure of the pure LSF module differs markedly from a TDF module. As the LSF module most likely communicates with external TDF modules, converter ports (TDF-> LSF and LSF-> TDF) are essential. The module does not contain any user defined code, but rather a set of LSF library supplied modules interconnected via LSF channels. It consists of two LSF subtraction modules:

sca_lsf::sca_sub sub1, sca_lsf::sca_sub sub2

```
#ifndef DELSIG_H
#define DELSIG_H

using namespace std;

#include <systemc-ams>

SCA_TDF_MODULE(comparator)
{
  /*Declare/define input/output ports and module members */
  sca_tdf::sca_in<double> sigin;
  sca_tdf::sca_in<double> refsig;
  sca_tdf::sca_out<double> sigout;
   /* Converter port */
  sca_tdf::sca_de::sca_out<bool> signature;
  double d1;
  double d2;
  double d3;
  bool deout;

  /* Set input/output port parameters */
  void set_attributes()
  {
    sigoutde.set_timestep(1.0, sc_core::SC_US);
    sigout.set_timestep(1.0, sc_core::SC_US);
    sigin.set_rate(1);
    sigout.set_rate(1);
    sigoutde.set_rate(1);
  }

  /* Comparator operation */
  void processing()
  {
    d1 = sigin.read();
    d2 = refsig.read();
    d3 = (d1 > d2) ? 1.0 : 0.0;
    deout = (d3 == 1.0) ? true : false;
    sigout.write(d3);
    sigoutde.write(deout);
  }
```

**Fig. 14.2** Delta-sigma modulator comparator—TDF module

```
/* Constructor with member initialization */
SCA_CTOR(comparator) : d1(0.0),
                       d2(0.0),
                       d3(0.0),
                       deout(false)
                       {}
~comparator(){} /* Destructor &/
};
```

**Fig. 14.2** (continued)

```
SC_MODULE(lssd2)
{
  private:
   /* Internal signal channels */
  sca_lsf::sca_signal sig1;
  sca_lsf::sca_signal sig2;
  sca_lsf::sca_signal sig3;
  sca_lsf::sca_signal sig4;
  sca_lsf::sca_signal sig5;
  sca_lsf::sca_signal sig6;
  sca_lsf::sca_signal sig7;
  sca_lsf::sca_signal sig8;

  public:
  /* Input/output ports and members */
  sca_tdf::sca_in<double> sigin;
  sca_tdf::sca_in<double> fdbksigin;
  sca_tdf::sca_out<double> sigout;
  /* LSF-TDF converter ports */
  sca_lsf::sca_tdf::sca_source tdf2lsf0;
  sca_lsf::sca_tdf::sca_source tdf2lsf1;
  sca_lsf::sca_tdf::sca_sink lsf2tdf;
  /* SystemC-AMS supplied library LSF modules */
  sca_lsf::sca_gain gain1;
  sca_lsf::sca_gain gain2;
  sca_lsf::sca_integ integ1;
  sca_lsf::sca_integ integ2;
```

**Fig. 14.3** Delta-sigma modulator pure LSF module consisting of library supplied LSF modules and channels

```
sca_lsf::sca_sub sub1;
sca_lsf::sca_sub sub2;

/* Constructor with member initialization */
lssd2(sc_core::sc_module_name,
          double d0,
          double d1):sigin("sigin"),
           fdbksigin("fdbksigin"),
           sigout("sigout"),
           tdf2lsf0("tdf2lsf0"),
           tdf2lsf1("tdf2lsf1"),
           lsf2tdf("lsf2tdf"),
           gain1("gain1", d0),
           gain2("gain2", d1),
           integ1("integ1"), integ2("integ2"),
             sub1("sub1"), sub2("sub2"),
             sig1("sig1"), sig2("sig2"),
             sig3("sig3"), sig4("sig4"),
             sig5("sig5"), sig6("sig6"),
             sig7("sig7"), sig8("sig8")
  {
   /* Connect input/output ports, and library LSF
      modules with internal signal channels */
   tdf2lsf0.inp(sigin);
   tdf2lsf0.y(sig1);
   tdf2lsf1.inp(fdbksigin);
   tdf2lsf1.y(sig4);
   lsf2tdf.x(sig8);
   lsf2tdf.outp(sigout);
   sub1.x1(sig1);
   sub1.x2(sig4);
   sub1.y(sig2);

   sub2.x1(sig5);
   sub2.x2(sig4);
   sub2.y(sig6);

   gain1.x(sig3);
   gain1.y(sig5);
```

**Fig. 14.3** (continued)

```
    gain2.x(sig7);
    gain2.y(sig8);

    integ1.x(sig2);
    integ1.y(sig3);
    integ2.x(sig6);
    integ2.y(sig7);
  }

  ~lssd2() { } /*Destructor */
};
```

**Fig. 14.3** (continued)

```
SCA_TDF_MODULE(onebitdac)
{
  /*Declare/define input/output ports and members */
  sca_tdf::sca_in<double> sigin;
  sca_tdf::sca_out<double> sigout;
  double d0;
  double d1;

  /* Set input/output port parameters */
  void set_attributes()
  {
    sigout.set_timestep(1.0, sc_core::SC_US);
    sigout.set_rate(1);
    sigout.set_delay(1);
    sigin.set_rate(1);
  }

  /* One bit DAC operation */
  void processing()
  {
    d0 = sigin.read();
    d1 = d0 == 1.0 ? 1.0 : -1.0;
    sigout.write(d1);
  }
```

**Fig. 14.4** Delta-sigma modulator 1-bit digital-to-analog converter, or not-return-to-zero converter

```
    SCA_CTOR(onebitdac){ } /* Constructor */
   ~onebitdac(){ } /* Destructor */
};
#endif
```

**Fig. 14.4** (continued)

```
#ifndef DFF1_H
#define DFF1_H

#include <systemc>

SC_MODULE(dff1)
{
 /* Clock input port, and input/output ports */
 sc_core::sc_in<bool> clock;
 sc_core::sc_in<bool> din;
 sc_core::sc_out<bool> dout;

 /* Main clocked thread */
 void ffoperation()
 {
   while(true)
   {
     wait();
     dout.write(din.read());
   }
 }

 /* Constructor */
 SC_CTOR(dff1)
 {
   /* Sensitize main tread to clock */
   SC_CTHREAD(ffoperation, clock.pos());
  }

 ~dff1(){} /*Destructor */
};
```

**Fig. 14.5** Delta-sigma modulator *D* flip-flop pure SystemC module

```
#include "commonsrcs.h"
#include "delsig.h"
#include "dff1.h"
#include "gendatatrace.h"
#include "sysctrace.h"
#include <cstdlib>
#include <cstring>

int sc_main(int argc, char **argv)
{
  /* Variables to hold input data values */
  double amplitude;
  double frequency;
  double refvalue;

   /* Check input */
   if(argc < 4)
  {
    std::cout<<"Missing arguments ..."<<std::endl;
    std::cout<<"usage ./sim <amplitude> <frequency> <refval>"<<std::endl;
    exit(0);
  }

  /* Extract  values of input   parameters */  ?
  amplitude = strtod(argv[1], NULL);
  frequency = strtod(argv[2], NULL);
  refvalue  = strtod(argv[3], NULL);

 /* Declare/define inter-module signal channels */
  sc_core::sc_signal<bool> dffin;
 sc_core::sc_signal<bool> dffout;
 sca_tdf::sca_signal<double> sigin;
 sca_tdf::sca_signal<double> refsig;
 sca_tdf::sca_signal<double> sigcmpin;
 sca_tdf::sca_signal<double> sigdacin;
 sca_tdf::sca_signal<double> sigdacout;
```

**Fig. 14.6** Delta-sigma modulator test harness

```
/* Declare/define clock for  D flip-flop  */
sc_core::sc_clock clock("clock", 1.6, sc_core::SC_US, 0.5);
/* Declare/define internal modules */
comparator c0("c0");
refsrc ref_src("ref_src");
lssd2 delsig_core("delsig_core", 60.0, 60.0);
dff1 dff_1("dff_1");
onebitdac one_bit_dac("one_bit_dac");
sinsrc ssrc("sin_src");
/* Declare/define trace files/objects */
/* First the trace file/object for SystemC D flip-flop */
ssctracesglbool trbool("tr_dff_out");
/* Declare/define trace file/object for delta-sigma modulator output */
tracedoublecombo trdbl("tr_delsigcore_dac");

/* Assign parameters, connect modules and channels */
ssrc.sigout(sigin);
ssrc.amplitude = amplitude;
ssrc.frequency = frequency;

ref_src.refout(refsig);
ref_src.refvalue = refvalue;

delsig_core.sigin(sigin);
delsig_core.fdbksigin(sigdacout);
delsig_core.sigout(sigcmpin);

c0.sigin(sigcmpin);
c0.refsig(refsig);
c0.sigout(sigdacin);
c0.sigoutde(dffin);

dff_1.clock(clock);
dff_1.din(dffin);
dff_1.dout(dffout);

one_bit_dac.sigin(sigdacin);
one_bit_dac.sigout(sigdacout);
```

**Fig. 14.6** (continued)

```
    trdbl.in1(sigin);
    trdbl.in2(sigdacin);
    trbool.clk(clock);
    trbool.din0(dffout);

    /* Run simulation for pre-defined time period and then stop */
    sc_core::sc_start(100.0, sc_core::SC_MS);
    sc_core::sc_stop();
    return 0;
}
```

**Fig. 14.6** (continued)

Two LSF integration modules:
sca_lsf::sca_integ integ1, sca_lsf::sca_integ integ2
Two LSF gain modules:
sca_lsf::sca_gain gain1, sca_lsf::sca_gain gain2
The ports are pure TDF, as well TDF-> LSF and LSF-> TDF *converter ports*:
sca_tdf::sca_in < double > sigin
sca_tdf::sca_in < double > sigout and
sca_tdf::sca_in < double > fdbksigin
sca_lsf::sca_tdf::sca_source tdf2lsf0
sca_lsf::sca_tdf::sca_source tdf2lsf1
sca_lsf::sca_tdf::sca_sink lsf2tdf.
The sca_source ports convert TDF format data to LSF format data, and the sca_sink ports convert LSF format data to TDF format data.

All internal data transfer between the subtraction and integration modules are via a set of sca_lsf::sca_signal channels:

Figures 14.7 and 14.8 show sample input to and output from the delta-sigma modulator. The sigma-delta modulator circumvents issues of the single slope ADC. The modulator indicates the *average* value of the input (sigma-delta modulator trades accuracy in amplitude with accuracy in time) so that no precision integrator components are required, and negative value inputs are easily digitized. The biggest advantage is that the noise spectrum is pushed out to the higher-frequency bands, thereby reducing the noise floor of the actual signal. The final low-pass filter stage easily removes the noise spectrum. It is assumed that the signal bandwidth is 1.0 kHz, so that the sampling frequency is chosen as $f_s = 10000.0\,\text{Hz}$, that satisfies the Nyquist condition $f_s \geq 2x$ Bandwidth and the oversampling frequency is $Kf_s = 64xf_s = 640000.0\,\text{Hz}$. The output (periodic sequence of 1s and 0s) clearly shows that accuracy in amplitude has been traded with accuracy in time. A metric often used with sigma-delta modulators is effective number of bits (ENOB) and is defined as $\text{ENOB} = \frac{\text{SNR}-1.76\,\text{dB}}{6.02\,\text{dB}}$. As the analog input signal attains

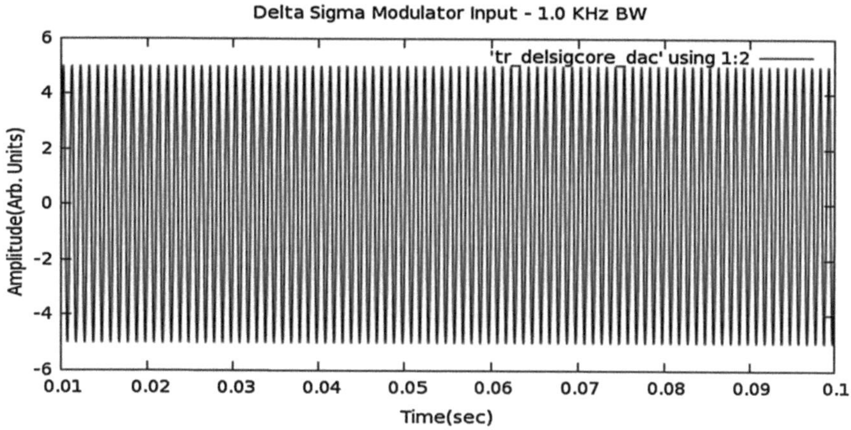


**Fig. 14.7** Analog signal input to sigma-delta modulator

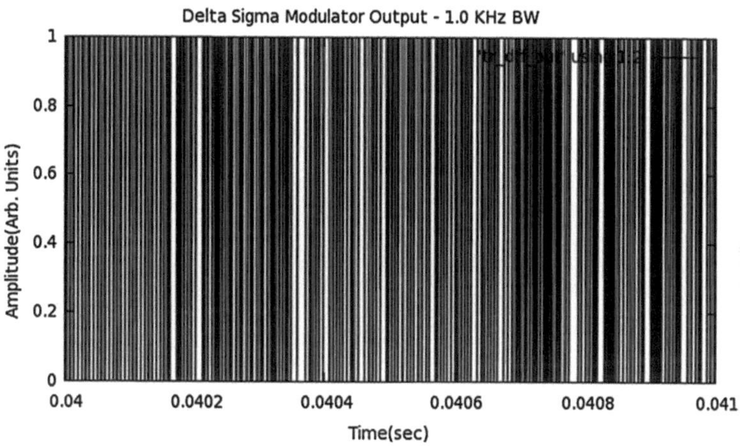


**Fig. 14.8** Sigma-delta modulator output corresponding to audio signal input of Fig. 2

values higher than a pre-defined mean value, the number of 'ones' in the output bit stream increases, and conversely, as the analog input signal attains values below the same pre-defined mean, the number of 'zeros' in the output bit stream increases. So, from a very simplistic standpoint, the bit stream output represents the *average value* of the analog input signal. For a sine wave input (Fig. 14.7), the analog input has value greater than zero half of each cycle and has value less than zero the other half of the cycle.

## References

1. SystemC-AMS Working Group. SystemC-AMS (Analog Mixed Signal)—Accellera Systems Initiative. http://www.accellera.org/downloads/standards/systemc/ams and http://www.accellera.org/activities/committees/systemc-ams/ Additional information [WWW] URL http://www.accellera.org/activities/committees/systemc-ams/
2. Boser, B. E. and Wooley, B. A.(1988) The Design of Sigma-Delta Modulation Analog-to-Digital Converters *IEEE Journal of Solid-State Circuits, Vol: 23 (6), 1298 – 1308*
3. Baker, Bonnie. Texas Instruments. Analog Embedded Processing, Semiconductor Company. *How Delta-Sigma ADCs Work Part 1* http://www.ti.com/lit/an/slyt423/slyt423.pdf

# Chapter 15
# Electrical Linear Networks in Practice and Theory

**Abstract** The ELN [1] formalism allows one to analyze/design electrical networks consisting of *linear elements only* (capacitor, inductor, resistor, current/voltage source, gyrator, nullor, controlled current/voltage sources, etc.). No nonlinear device, e.g., the simplest p-n junction diode, is supported, thereby limiting the utility of this framework. In a very rudimentary way, a bi-junction transistor (BJT) or metal-oxide-semiconductor field-effect transistor (MOSFET) may be modeled as a switch. The ELN module structure is very similar to that of LSF—SystemC-AMS [1] built-in modules connected appropriately via channels. Moreover, like LSF, ELN modules need to interact with either TDF or pure SystemC modules to read in/out data. In the ELN examples examined subsequently, SystemC-AMS library module *sca_eln::sca_node* is used as the electrical terminal for each electrical device, rather than the *sca_eln::sca_terminal*. This is because the latter has to be bound to the sca_eln::sca_node in the final circuit, for correct simulation.

## 15.1 Fifth-Order Unity-Gain Low-Pass Butterworths Filter

In the TDF chapter, the fifth-order unity-gain low-pass filter [2] was treated as a set of mathematical equations (pole–zero form). Here, the same filter is examined as a pure electrical circuit, with the aim of comparing its behavior from the pole–zero case, with respect to that obtained from the current analysis. This analysis is based on the ELN built-in class *sca_eln::sca_nullor* (idealized operational amplifier). Using identical formalism as in Chap. 5, we determine the capacitor and resistor values for each of the three sub-components.

$a_1 = w_c R_1 C_1$ $R_1 C_1 = \frac{a_1}{w_c} = 0.0026$ from look-up tables for the unity-gain first-order low-pass stage of the fifth-order unity-gain low-pass Butterworths filter, $a_1 = 1$ so that we choose $C_1 = 1.0\,\mathrm{uF}$ and $\mathrm{R}_1 = 2650.0\,\Omega$.

For the unity-gain second-order low-pass filter (Sallen-Key) second stage of the fifth-order unity-gain low-pass Butterworths filter, $a_2 = 1.618$ and $b_2 = 1$ and $C_1 (R_1 + R_2) = \frac{1.618}{w_c} = 0.0043$ $C_1 C_2 R_1 R_2 = \frac{1}{w_c^2} = 0.000007$ and

A. Banerjee and B. Sur, *SystemC and SystemC-AMS in Practice*,
DOI: 10.1007/978-3-319-01147-9_15,

$C_2 \geq \frac{4b_2}{a_2^2} C_1$. Selecting $C_1 = 1.0E - 6F$ and using the relationship between $C_2$ and $C_1$ gives $C_2 \geq 1.42E - 6F$, so $C_2$ is chosen to be $C_2 = 2.5E - 6F$. Substituting chosen capacitor values in the equations for $a_2$ and $b_2$ provides the quadratic equation—$R_1^2 - 4.46 \times 10^3 R_1 + 2.8 \times 10^6 = 0$ after some mathematical manipulation. Solving to get $R_1$ and using the result to get $R_2$ gives the following pair of values $R_1 = 3.697 \times 10^3\ \Omega$, $R_2 = 0.763 \times 10^3\ \Omega$, and $R_1 = 0.763 \times 10^3\ \Omega$, $R_2 = 3.697 \times 10^3\ \Omega$.

Identical treatment of the third stage (second second-order unity-gain low-pass Sallen-Key filter) of the fifth-order unity-gain low-pass Butterworths filter, using standard coefficient values ($a_2 = 0.618$, $b_3 = 1.0$), provides, for chosen capacitor value of $C_1 = 1.0E - 6F$, $C_2 = 15.0E - 6F$. Using these capacitor values, the corresponding resistor value pairs are $R_1 = 1.21 \times 10^3\ \Omega$, $R_2 = 0.39 \times 10^3\ \Omega$ and $R_1 = 0.39 \times 10^3\ \Omega$, $R_2 = 1.21 \times 10^3\ \Omega$.

The SystemC–AMS source file and its test bench are Figs. 15.1 and 15.2. The frequency and phase responses are in Figs. 15.3 and 15.4, respectively.

Closer examination of the ELN module shows the similarity of its structure with the LSF module—TDF ports to communicate with external TDF modules,

```
#ifndef FILTFORDBTWELN_H
#define FILTFORDBTWELN_H

#include <systemc-ams>

SC_MODULE(filtfordbtweln)
{
  /* TDF input/output ports to allow data to be
     read in/out from/to TDF modules */
  sca_tdf::sca_in<double> inp;
  sca_tdf::sca_out<double> outp;

  /* TDF-ELN/ELN-TDF converter voltage sources */
  sca_eln::sca_tdf::sca_vsource v_in;
  sca_eln::sca_tdf::sca_vsink   v_out;

  /* Idealized operational amplifiers */
  sca_eln::sca_nullor nullr0;
  sca_eln::sca_nullor nullr1;
  sca_eln::sca_nullor nullr2;

  /* Capacitors */
  sca_eln::sca_c c0;
```

**Fig. 15.1** ELN module representation of fifth-order low-pass Butterworths filter

```
    sca_eln::sca_c c1;
    sca_eln::sca_c c2;
    sca_eln::sca_c c3;
    sca_eln::sca_c c4;

    /* Resistors */
    sca_eln::sca_r r0;
    sca_eln::sca_r r1;
    sca_eln::sca_r r2;
    sca_eln::sca_r r3;
    sca_eln::sca_r r4;

    /* Electrical netlist nodes */
    sca_eln::sca_node n1;
    sca_eln::sca_node n2;
    sca_eln::sca_node n3;
    sca_eln::sca_node n4;
    sca_eln::sca_node n5;
    sca_eln::sca_node n6;
    sca_eln::sca_node n7;

    sca_eln::sca_node n8;
    sca_eln::sca_node n9;

  /* Special ground node */
  sca_eln::sca_node_ref gnd;

  /* Set input/output port parameters */
  void set_attributes()
  {
    inp.set_rate(1);
    outp.set_timestep(1.0, sc_core::SC_US);
    outp.set_rate(1);
  }

  /* Constructor - member initialization */
  SC_CTOR(filtfordbtweln): v_in("v_in", 1.0),
                           v_out("v_out", 1.0),
                           nullr0("nullr0"),
                           nullr1("nullr1"),
```

**Fig. 15.1** (continued)

```
                    nullr2("nullr2"),
                    c0("c0", 1.0E-6),
                    c1("c1", 1.0E-6),
                    c2("c2", 2.5E-6),
                    c3("c3", 1.0E-6),
                    c4("c4", 15.0E-6),
                    r0("r0", 2650.0),
                    r1("r1", 3697.0),
                    r2("r2", 763.0),
                    r3("r3", 1210.0),
                    r4("r4", 390.0)
{
  /* Connect all electrical components to node:
     to form netlist */
  v_in.inp(inp);
  v_in.p(n1);
  v_in.n(gnd);

  v_out.outp(outp);
  v_out.p(n9);
  v_out.n(gnd);

  nullr0.nip(n2);
  nullr0.nin(n3);
  nullr0.nop(n3);
  nullr0.non(gnd);

  r0.p(n1);
  r0.n(n2);
  c0.p(n2);
  c0.n(gnd);

  r1.p(n3);
  r1.n(n4);
  r2.p(n4);
  r2.n(n5);

  nullr1.nip(n5);
  nullr1.nin(n6);
  nullr1.nop(n6);
```

**Fig. 15.1** (continued)

```
        nullr1.nop(n6);
        nullr1.non(gnd);

        c1.p(n4);
        c1.n(n6);

        c2.p(n5);
        c2.n(gnd);

        c2.p(n5);
        c2.n(gnd);

        r3.p(n6);
        r3.n(n7);

        r4.p(n7);
        r4.n(n8);

        nullr2.nip(n8);
        nullr2.nin(n9);
        nullr2.nop(n9);
        nullr2.non(gnd);

        c3.p(n7);
        c3.n(n9);
        c4.p(n8);
        c4.n(gnd);
      }

    ~filtfordbtweln(){ }
    /* Destructor */
  };

#endif
```

**Fig. 15.1** (continued)

```
#include "commonsrcs.h"
#include "filtfordbtweln.h"
#include "gendatatrace.h"
#include <cstdlib>
```

**Fig. 15.2** Test harness for ELN fifth-order low-pass Butterworths filter

```
#include <cstring>

int sc_main(int argc, char **argv)
{
  /* Inter-module signal channels */
  sca_tdf::sca_signal<double> sig0;
  sca_tdf::sca_signal<double> sig1;
  /* Input signal amplitude and frequency */
  double amplitude;
  double frequency;

  /* Check input parameters */
  if(argc < 3)
  {
    std::cout<<"missing input ..."<<std::endl;
    std::cout<<"usage ./sim <amplitude> <frequency>"<<std::
    exit(0);
  }

   /* Extract amplitude and frequency from input parameters
   amplitude = strtod(argv[1], NULL);
   frequency = strtod(argv[2], NULL);

  /* Declare/define internal modules and trace file/object */

   sinsrc sin_src("sin_src");
   filtfordbtweln filt("filt");
   tracedoublecombo trdbl("tr_filt_5_ord_btw");

  /* Connect module ports and channels and assign
     parameters */
  sin_src.sigout(sig0);
  sin_src.amplitude = amplitude;
  sin_src.frequency = frequency;

  filt.inp(sig0);
  filt.outp(sig1);

  trdbl.in1(sig0);
  trdbl.in2(sig1);
  /* Run simulation for pre-defined time period and then sto
```

**Fig. 15.2** (continued)

```
    sc_core::sc_start(200.0, sc_core::SC_MS);
    sc_core::sc_stop();
    return 0;
  }
```

**Fig. 15.2** (continued)

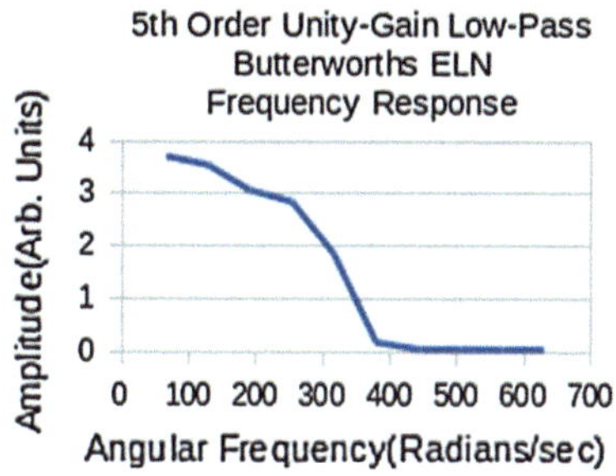


**Fig. 15.3** Fifth-order low-pass unity-gain Butterworths filter frequency response

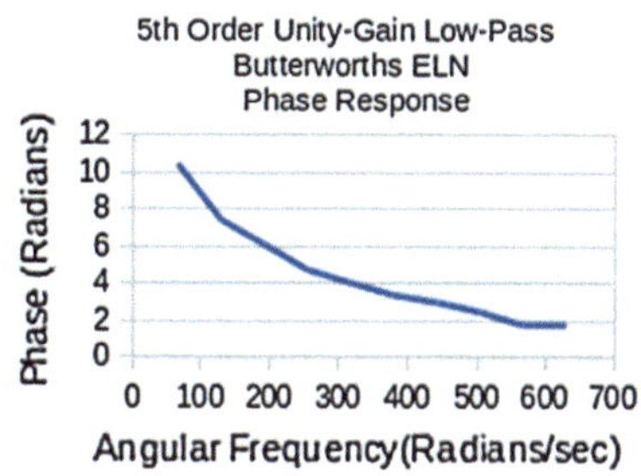


**Fig. 15.4** Fifth-order low-pass unity-gain Butterworths filter phase response

converter ports to convert to/from TDF to LSF and vice versa, use of ELN library modules and primitives and complete absence of user code.

Typical console output on executing this model looks like:

```
./sim 5 500
        SystemC 2.3.0-ASI—Jul 19 2012 18:53:11
    Copyright (c) 1996–2012 by all Contributors,
    ALL RIGHTS RESERVED

  SystemC AMS extensions 1.0 Version: 1.0—BuildRevision: 1541 2013_02_18
    Copyright (c) 2010–2013 by Fraunhofer-Gesellschaft
        Institut Integrated Circuits/EAS
    Licensed under the Apache License, Version 2.0

  Info: SystemC-AMS:
    17 SystemC-AMS modules instantiated
    2 SystemC-AMS views created
    3 SystemC-AMS synchronization objects/solvers instantiated
```

```
Info: SystemC-AMS:
   1 dataflow clusters instantiated
     cluster 0:
       3 dataflow modules/solver, contains e.g. module: sin_src
       3 elements in schedule list,
       1 us cluster period,
       ratio to lowest:        1 e.g. module: sin_src
       ratio to highest:       1 sample time e.g. module: sin_src
       0 connections to SystemC de, 0 connections from SystemC de
Info: SystemC-AMS:
   ELN solver instance: sca_linear_solver_0 (cluster 0)
      has 18 equations for 15 modules (e.g. filt.v_in),
      1 inputs and 1 outputs to other (TDF) SystemC-AMS domains,
      0 inputs and 0 outputs to SystemC de.
      1 us initial time step.
```

Figures 15.3 and 15.4 show the frequency and phase response of the fifth-order low-pass Butterworths filter with a cut-off frequency of 60 Hz, just like in the pole–zero analysis. These plots look similar to those obtained from the pole–zero analysis.

## 15.2 5.0 kHz Mid-frequency Bandpass Filter

An inductor connected in series with a capacitor forms a bandpass filter [3]. A parallel connection of the same two components forms a band stop or notch filter. The mid-frequency of pass/cut-off band may be obtained from the values of the two

```
#ifndef ELNBNDPASS_H
#define ELNBNDPASS_H

#include <systemc-ams>

SC_MODULE(elnbndpass)
{
   /* TDF input/output ports to allow data to be
      read in/out from/to TDF modules */
  sca_tdf::sca_in<double> inp;
  sca_tdf::sca_out<double> outp;

  /* TDF-ELN/ELN-TDF converter voltage sources */
  sca_eln::sca_tdf::sca_vsource v_in;
  sca_eln::sca_tdf::sca_vsink   v_out;
```

**Fig. 15.5** ELN bandpass filter with mid-frequency of 5 kHz

```
    sca_eln::sca_c cap0;
    sca_eln::sca_l indt0;

    /* Electrical netlist nodes */
    sca_eln::sca_node n1;
    sca_eln::sca_node n2;
    sca_eln::sca_node_ref gnd;

    /* Constructor - member initialization */
    SC_CTOR(elnbndpass):v_in("v_in", 1.0),
                        v_out("v_out", 1.0),
                        cap0("cap0", 1.0E-6),
                        indt0("indt0", 1.0E-3)
    {
      /* Connect all electrical components to nodes
         to form netlist */
      v_in.inp(inp);
      v_in.p(n1);
      v_in.n(gnd);

      v_out.outp(outp);
      v_out.p(n2);
      v_out.n(gnd);

      cap0.p(n1);
      cap0.n(n2);

      indt0.p(n2);
      indt0.n(gnd);
    }

    ~elnbndpass(){ } /*Destructor */
};
```

**Fig. 15.5** (continued)

```
#include "commonsrcs.h"
#include "elnbndpass.h"

#include "gendatatrace.h"
```

**Fig. 15.6** ELN bandpass filter test harness

```
#include <cstdlib>
#include <cstring>

int sc_main(int argc, char **argv)
{
  /* Inter-module signal channels */
  sca_tdf::sca_signal<double> sigin;
  sca_tdf::sca_signal<double> sigout;
  /* Input signal amplitude and frequency */
  double amplitude;
  double frequency;

  /* Check input parameters */
  if(argc < 3)
  {
    std::cout<<"insufficient parameters ..."<<std::endl;
    std::cout<<"usage : ./sim <amplitude> <frequency>"<<std::endl;
    exit(0);
  }

  /* Extract amplitude and frequency from input parameters */
  amplitude = strtod(argv[1], NULL);
  frequency = strtod(argv[2], NULL);

  /* Declare/define internal modules and trace file/object */
  elnbndpass eln_bp("eln_bp");
  sinsrc sin_src("sin_src");
  tracedoublecombo trdbl("tr_bndpass");

  /* Connect module ports and channels and assign
     parameters */
  sin_src.sigout(sigin);
  sin_src.amplitude = amplitude;
  sin_src.frequency = frequency;

  eln_bp.inp(sigin);
  eln_bp.outp(sigout);
  trdbl.in1(sigin);
  trdbl.in2(sigout);
```

**Fig. 15.6** (continued)

```
    /* Run simulation for pre-defined time period and then stop */
    sc_core::sc_start(100.0, sc_core::SC_MS);
    sc_core::sc_stop();
    return 0;
}
```

Fig. 15.6 (continued)

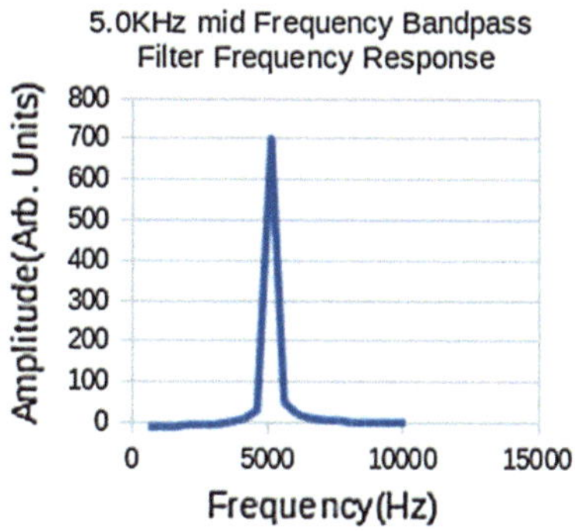


Fig. 15.7 5.0 kHz mid-band frequency bandpass filter frequency response

components chosen. While operational amplifiers may be used as a circuit component in low-frequency range applications, they are inappropriate at RF frequency ranges. Our design is applicable to both the frequency domains. A typical LC bandpass filter has a transfer function $\frac{V_0}{V_i} = \frac{1}{1-\frac{1}{w^2LC}}$, where $w$ is the angular frequency. The mid (or center)-frequency is $w_c^2 = \frac{1}{LC}$. The source code for the filter and test bench is in Figs. 15.5 and 15.6, while the frequency response is in Fig. 15.7.

## 15.3 Simple CMOS Inverter

The ELN framework provides linearized implementations of some typical nonlinear electrical/electronic devices, e.g., MOSFET—*rswitch*. A versatile module, it can switch on/off current or voltage between two electrical terminals, controlled by a Boolean control signal which may be generated by a pure SystemC

```
#ifndef ELNTDFINV_H
#define ELNTDFINV_H

#include <systemc-ams>
```

Fig. 15.8 CMOS inverter using two sca_eln::sca_tdf::sca_rswitch modules

```
SC_MODULE(elntdfinv)
{
 /* TDF input/output ports to allow data to be
     read in/out from/to TDF modules, First
     switch control ports */
 sca_tdf::sca_in<bool> in0;
 sca_tdf::sca_in<bool> in1;

 /* Inverter data input/output ports */
 sca_tdf::sca_in<double> vconst;
 sca_tdf::sca_out<double> out0;

 /* ELN switch modules */
 sca_eln::sca_r r0;
 sca_eln::sca_r r1;
 sca_eln::sca_tdf::sca_rswitch swtch0;
 sca_eln::sca_tdf::sca_rswitch swtch1;

 /* TDF-ELN/ELN-TDF converter voltage source/sink */
 sca_eln::sca_tdf::sca_vsource vsrc;
 sca_eln::sca_tdf::sca_vsink vout;

 /* Electrical netlist nodes */
 sca_eln::sca_node n1;
 sca_eln::sca_node n2;
 sca_eln::sca_node n3;
 sca_eln::sca_node n4;
 sca_eln::sca_node_ref gnd;

 /* Constructor - member initialization */
 SC_CTOR(elntdfinv):r0("r0", 500.0),
              r1("r1", 5000.0),
              swtch0("swtch0"),
              swtch1("swtch1"),
              vout("vout",1.0),
              vsrc("vsrc",1.0)
 {
    /* Connect all electrical components to nodes
      to form netlist */
   r0.p(n4);
   r0.n(gnd);
```

**Fig. 15.8** (continued)

```
    r1.p(n2);
    r1.n(n3);

    swtch0.p(n1);
    swtch0.n(n2);
    swtch0.ctrl(in0);

    swtch1.p(n3);
    swtch1.n(n4);
    swtch1.ctrl(in1);

    vsrc.inp(vconst);
    vsrc.p(n1);
    vsrc.n(gnd);

    vout.outp(out0);
    vout.p(n3);
    vout.n(gnd);
  }
  ~elntdfinv(){ }
 /* Destructor */
};

#endif
```

**Fig. 15.8** (continued)

```
SCA_TDF_MODULE(trgsrc)
{
  /* Declare/define input/output ports and module members */
  sca_tdf::sca_out<bool> out0;
  sca_tdf::sca_out<bool> out1;
  bool b0;
  bool b1;

  unsigned int rval;

  /* Set port parameters &/
  void set_attributes()
```

**Fig. 15.9** TDF module for generating Boolean control signals for sca_eln::sca_tdf::sca_rswitch

```
{
  out0.set_timestep(1.0, sc_core::SC_US);
  out1.set_timestep(1.0, sc_core::SC_US);
  out0.set_rate(1);
  out1.set_rate(1);
}


/* Module p[eration */
void processing()
{
  rval = ((unsigned int)(10.0*drand48()) % 2);
  b0 = rval == 1 ? true : false;
  b1 = !b0;
  out0.write(b0);
  out1.write(b1);
  std::cout<<name()<<" "<<b0<<" "<<b1<<std::endl;
}

/* Constructor - initilaize members */
SCA_CTOR(trgsrc):b0(false),b1(false),rval(0){ }

~trgsrc(){ }
/* Destructor */
};
```

**Fig. 15.9** (continued)

```
#include "commonsrcs.h"
#include "elntdfinv.h"
#include "gendatatrace.h"

int sc_main(int argc, char **argv)
{
  /* Declare/define signal channels */
  sca_tdf::sca_signal<bool> sig0;
  sca_tdf::sca_signal<bool> sig1;
  sca_tdf::sca_signal<double> sig2;
  sca_tdf::sca_signal<double> sig3;

  /* Declare/define modules - inverter, reference signal source
```

**Fig. 15.10** Test harness for ELN-TDF inverter

```
    trigger and trace object (/
  elntdfinv eln_tdf_inv("eln_tdf_inv");
  refsrc ref_src("ref_src");
  trgsrc trg_src("trg_src");
  tracedouble trdbl("tr_eln_tdf_inv");

  /( Connect module ports and signal channels */
  eln_tdf_inv.in0(sig0);
  eln_tdf_inv.in1(sig1);
  eln_tdf_inv.vconst(sig3);
  eln_tdf_inv.out0(sig2);
  ref_src.refout(sig3);
  ref_src.refvalue = 5.0;

  trg_src.out0(sig0);
  trg_src.out1(sig1);
  trdbl.in(sig2);

  /* Run simulation for pre-defined time period and then stop */
  sc_core::sc_start(100.0, sc_core::SC_US);
  sc_core::sc_stop();
  return 0;
}
```

**Fig. 15.10** (continued)

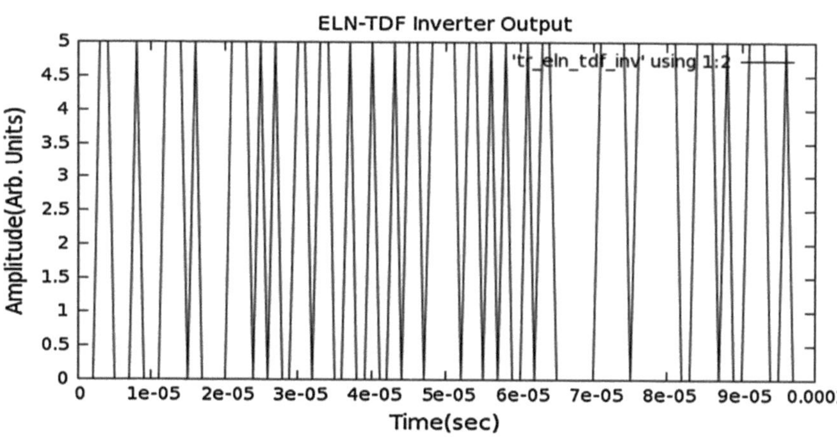


**Fig. 15.11** Sample inverter output

module(SC_MODULE) or a SystemC-AMS module(SCA_TDF_MODULE). This approach is fully justified, since SystemC-AMS is designed to help analyze/ examine the behavior of a circuit, rather than performance characteristics. Here, we examine the generation of the Boolean control signal from a SystemC-AMS module.

A CMOS inverter consists of a NMOS and a PMOS, which switch on/off on the two phases of the same clock. SystemC-AMS provides only the rswitch module. To mimic the behavior of a NMOS and a PMOS, two rswitch modules are used, that are switched on using the opposite phases of the same clock. The source code is in Figs. 15.8 and 15.9, and the test harness and sample output are in Figs. 15.10 and 15.11.

## References

1. SystemC-AMS Working Group. SystemC-AMS (Analog Mixed Signal)—Accellera Systems Initiative. http://www.accellera.org/downloads/standards/systemc/ams and http://www.accellera.org/activities/committees/systemc-ams/
2. Texas Instruments. Analog Embedded Processing, Semiconductor Company Texas Instruments. *Active Filter Design Techniques, Literature Number SLOA088 Texas Instruments, excerpted from OP-Amps for Everyone, Literature Number SLOD006A* http://www.ti.com/lit/ml/sloa088/sloa088.pdf
3. K, Puglia V. *A General Design Procedure for Bandpass Filters Derived from Low Pass Prototype Elements: Part I.* Microwave Journal, 1 Dec. 2000. http://www.microwavejournal.com/articles/3102-a-general-design-procedure-for-bandpass-filters-derived-from-low-pass-prototype-elements-part-i

# Chapter 16
# Real-World Electrical Linear Networks, Linear Signal Flow, and Timed Data Flow Combinations

**Abstract** In the previous chapters and examples, we have presented either pure timed data flow (TDF) or electrical linear networks (ELN) and TDF or linear signal flow (LSF) and TDF combinations. Such combinations are facilitated by SystemC-AMS's [1] built-in converter ports that allow data to flow back and forth between ELN, LSF, and TDF modules, without the designer having to create any custom SystemC-AMS code to achieve the same. A few real-world examples that combine all three formalisms are examined here.

## 16.1 Band-Pass Filter Second-Order Sigma-Delta Modulator

A good example of a combination of the combination of all three formalisms is the second-order sigma-delta modulator [2, 3]. The second-order sigma-delta modulator examined previously had integrators (basically low-pass filters) in it, resulting in the noise spectrum being pushed out to high-frequency bands. This noise is easily removed by a digital decimation and low-pass filter combination.

There are some specialized applications (telecommunications) in which more noise filtering is necessary, i.e., noise spectrum may be pushed to frequency bands above and below frequencies of interest. Then, the integrator of the sigma-delta modulator is replaced by a band-pass filter. The band-pass filter is modeled as electrical linear networks (ELN), while the rest of the modulator is modeled as a linear signal flow (LSF)–timed data flow (TDF) combination. A band-pass filter may also be modeled in the pole-zero form by replacing the $s$ of a low-pass filter with $\frac{1}{w_d\left(s+\frac{1}{s}\right)}$, where $w_d$ is the frequency band of the band-pass filter. Figure 16.1 shows the modified sigma-delta modulator, Figs. 16.2, 16.3, 16.4, 16.5, and 16.6 show the source code and Fig. 16.7 shows test harness. For illustration purposes, the center frequency for the band-pass filter in this example has been set to 60.0 Hz.

A. Banerjee and B. Sur, *SystemC and SystemC-AMS in Practice*,
DOI: 10.1007/978-3-319-01147-9_16,

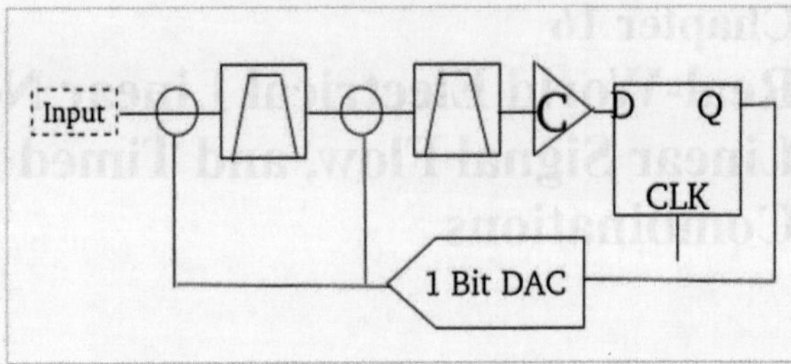


**Fig. 16.1** Sigma-delta modulator with band-pass filter replacing the integrator of a traditional modulator. A "C" represents a comparator

```
#ifndef DELSIGBP_H
#define DELSIGBP_H

#include <systemc-ams>

SC_MODULE(elnbndpass60)
{
  /* TDF input/output ports to allow data to be
      read in/out from/to TDF modules */
  sca_tdf::sca_in<double> inp;
  sca_tdf::sca_out<double> outp;

  /* Converter TDF-ELN/ELN-TDF voltage source/sink */
  sca_eln::sca_tdf::sca_vsource v_in;
  sca_eln::sca_tdf::sca_vsink   v_out;

  /* ELN capacitor - inductor for filter */
  sca_eln::sca_c cap0;
  sca_eln::sca_l indt0;

  /* Electrical netlist nodes */
  sca_eln::sca_node n1;
  sca_eln::sca_node n2;
  sca_eln::sca_node_ref gnd;

  /* Constructor - member initialization */
  SC_CTOR(elnbndpass60):v_in("v_in", 1.0),
                      v_out("v_out", 1.0),
                      cap0("cap0", 2.645E-3),
                      indt0("indt0", 2.645E-3)
```

**Fig. 16.2** 60 Hz mid-band band-pass filter delta-sigma modulator ELN band-pass filter

```
    {
      /* Connect all electrical components to nodes
         to form netlist */
      v_in.inp(inp);
      v_in.p(n1);
      v_in.n(gnd);

      v_out.outp(outp);
      v_out.p(n2);
      v_out.n(gnd);

      cap0.p(n1);
      cap0.n(n2);

      indt0.p(n2);
      indt0.n(gnd);
    }

    ~elnbndpass60(){ } /* Destructor */
  };
```

**Fig. 16.2** (continued)

```
SCA_TDF_MODULE(comparator)
{
  /* Declare/define input/output ports and module members */
  sca_tdf::sca_in<double> sigin;
  sca_tdf::sca_in<double> refsig;
  sca_tdf::sca_out<double> sigout;

  /* Output port to pure SystemC nodule */
  sca_tdf::sca_de::sca_out<bool> sigoutde;
  double d1;
  double d2;
  double d3;
  bool deout;

  /* Set input/output port attributes */
  void set_attributes()
  {
    sigoutde.set_timestep(1.0, sc_core::SC_US);
    sigout.set_timestep(1.0, sc_core::SC_US);
```

**Fig. 16.3** 60 Hz mid-band band-pass filter delta-sigma modulator TDF comparator

```
    sigin.set_rate(1);
    sigout.set_rate(1);
    sigoutde.set_rate(1);
  }

  /* Comparator operation */
  void processing()
  {
    d1 = sigin.read();
    d2 = refsig.read();
    d3 = (d1 > d2) ? 1.0 : 0.0;
    deout = (d3 == 1.0) ? true : false;
    sigout.write(d3);
    sigoutde.write(deout);
  }

  /* Constructor - member initialization */
  SCA_CTOR(comparator) : d1(0.0),
                        d2(0.0),
                        d3(0.0),
                        deout(false)
                        {}
  ~comparator(){} /*Destructor */
};
```

**Fig. 16.3** (continued)

```
SCA_TDF_MODULE(onebitdac)
{
  /* Declare/define input/output ports
  and module members */
  sca_tdf::sca_in<double> sigin;
  sca_tdf::sca_out<double> sigout;
  double d0;
  double d1;

  /* Set input/output port parameters */
  void set_attributes()
  {
  sigout.set_timestep(1.0, sc_core::SC_US);
```

**Fig. 16.4** 60 Hz mid-band band-pass filter delta-sigma modulator TDF 1-bit digital-to-analog converter

```
sigout.set_rate(1);
sigout.set_delay(1);
sigin.set_rate(1);
}

/* Digital to analog conversion */
void processing()
{
d0 = sigin.read();
d1 = d0 == 1.0 ? 1.0 :-1.0;
sigout.write(d1);
}

/* Constructor */
SCA_CTOR(onebitdac){ }
~onebitdac(){ } /*Destructor */
};
```

**Fig. 16.4** (continued)

```
SC_MODULE(sumwrap)
{
  private:
   /* Internal inter LSF module signal channels */
   sca_lsf::sca_signal sig1;
   sca_lsf::sca_signal sig2;
   sca_lsf::sca_signal sig3;

  public:
   /* TDF input/output ports */
   sca_tdf::sca_in<double> sigin;
   sca_tdf::sca_in<double> fdbksigin;
   sca_tdf::sca_out<double> sigout;
   /* LSF-TDF/TDF-LSF converter ports */
   sca_lsf::sca_tdf::sca_source tdf2lsf0;
   sca_lsf::sca_tdf::sca_source tdf2lsf1;
   sca_lsf::sca_tdf::sca_sink lsf2tdf;
```

**Fig. 16.5** 60 Hz mid-band band-pass filter delta-sigma modulator wrapper around primitive LSF sum/subtraction module

```
    /* LSF summing node */
    sca_lsf::sca_sub sub_b;

    /* Constructor - initialize converter ports, members */
    sumwrap(sc_core::sc_module_name):sigin("sigin"),
                                     fdbksigin("fdbksigin"),
                                     sigout("sigout"),
                                     sub_b("sub_b"),
                                     tdf2lsf0("tdf2lsf0"),
                                     tdf2lsf1("tdf2lsf1"),
                                     lsf2tdf("lsf2tdf")
    {
      /* Connect ports, modules and internal signal channels */
      tdf2lsf0.inp(sigin);
      tdf2lsf0.y(sig1);
      tdf2lsf1.inp(fdbksigin);
      tdf2lsf1.y(sig2);
      sub_b.x1(sig1);
      sub_b.x2(sig2);
      sub_b.y(sig3);

      lsf2tdf.x(sig3);
      lsf2tdf.outp(sigout);
    }
    ~sumwrap(){ } /*Destructor */
};
```

**Fig. 16.5** (continued)

```
#ifndef DFF1_H
#define DFF1_H

#include <systemc>

SC_MODULE(dff1)
{
 /* Declare/define clock and input/output ports *
 sc_core::sc_in<bool> din;
 sc_core::sc_in<bool> clock;
```

**Fig. 16.6** 60 Hz mid-band band-pass filter delta-sigma modulator pure systemC D flip-flop

```
  sc_core::sc_out<bool> dout;

  /* Flip-flop operation thread */
  void ffoperation()
  {
    while(true)
    {
      wait();
      dout.write(din.read());
    }
  }

  /* Constructor */
  SC_CTOR(dff1)
  {
    /* Sensitize thread to clock */
    SC_CTHREAD(ffoperation, clock.pos());
  }

  ~dff1(){}  /* Destructor */
};
#endif
```

**Fig. 16.6** (continued)

```
#include "delsigbp.h"
#include "dff1.h"
#include "gendatatrace.h"
#include "sysctrace.h"
#include <cstdlib>
#include <cstring>

int sc_main(int argc, char **argv)
{
  /* SystemC compliant signal channels for Boolean signals */
  sc_core::sc_signal<bool> blsig0;
  sc_core::sc_signal<bool> blsig1;

  /* Inter-module  signal channels */
```

**Fig. 16.7** 60 Hz mid-band band-pass filter delta-sigma modulator test harness

```
sca_tdf::sca_signal<double> sig0;
sca_tdf::sca_signal<double> sig1;
sca_tdf::sca_signal<double> sig2;
sca_tdf::sca_signal<double> sig3;
sca_tdf::sca_signal<double> sig4;
sca_tdf::sca_signal<double> sig5;
sca_tdf::sca_signal<double> sig6;
sca_tdf::sca_signal<double> sig7;

/* Variables to hold input parameters */
double amplitude;
double frequency;
double refval;

/* Check input */
if(argc < 4)
{
  std::cout<<"insufficient parameters ..."<<std::endl;
  std::cout<<"usage : ./sim <amplitude> <frequency> <reference
value>"<<std::endl;
  exit(0);
}

/* Extract parameters from input */
amplitude = strtod(argv[1], NULL);
frequency = strtod(argv[2], NULL);
refval = strtod(argv[3], NULL);

/* Declare/define clock for  SystemC D flip-flop */ */
sc_core::sc_clock clock("clock", 2.0, sc_core::SC_US, 0.5);
/* Declare/define modules */
comparator comp("c0");
constsrc cnstsrc("constsrc");
dff1 df_f_1("df_f_1");
elnbndpass60 eln_bp_0("eln_bp_0");
elnbndpass60 eln_bp_1("eln_bp_1");
onebitdac one_bit_dac("one_bit_dac");
sinsrc sin_src("sin_src");
```

**Fig. 16.7** (continued)

```
sumwrap sum_wrap_0("sum_wrap_0");
sumwrap sum_wrap_1("sum_wrap_1");
/* Declare/define trace files/objects */
ssctracesglbool trbool("tr_dff_out");
tracedoublecombo trdbl("tr_bndpass_bp");

/* Connect modules and signal channels */
sin_src.sigout(sig0);
sin_src.amplitude = amplitude;
sin_src.frequency = frequency;
cnstsrc.sigout(sig1);
cnstsrc.amplitude = refval;
comp.sigin(sig6);
comp.refsig(sig1);
comp.sigout(sig7);
comp.sigoutde(blsig0);
df_f_1.din(blsig0);
df_f_1.clock(clock);
df_f_1.dout(blsig1);
eln_bp_0.inp(sig3);
eln_bp_0.outp(sig4);
eln_bp_1.inp(sig5);
eln_bp_1.outp(sig6);
one_bit_dac.sigin(sig7);
one_bit_dac.sigout(sig2);
sum_wrap_0.sigin(sig0);
sum_wrap_0.fdbksigin(sig2);
sum_wrap_0.sigout(sig3);
sum_wrap_1.sigin(sig4);
sum_wrap_1.fdbksigin(sig2);
sum_wrap_1.sigout(sig5);
trdbl.in1(sig0);
trdbl.in2(sig7);
trbool.clk(clock);
trbool.din0(blsig1);
```

**Fig. 16.7** (continued)

```
/* Run simulator;ation for pre-defined time period and stop */
sc_core::sc_start(100.0, sc_core::SC_MS);
sc_core::sc_stop();
return 0;
}
```

**Fig. 16.7** (continued)

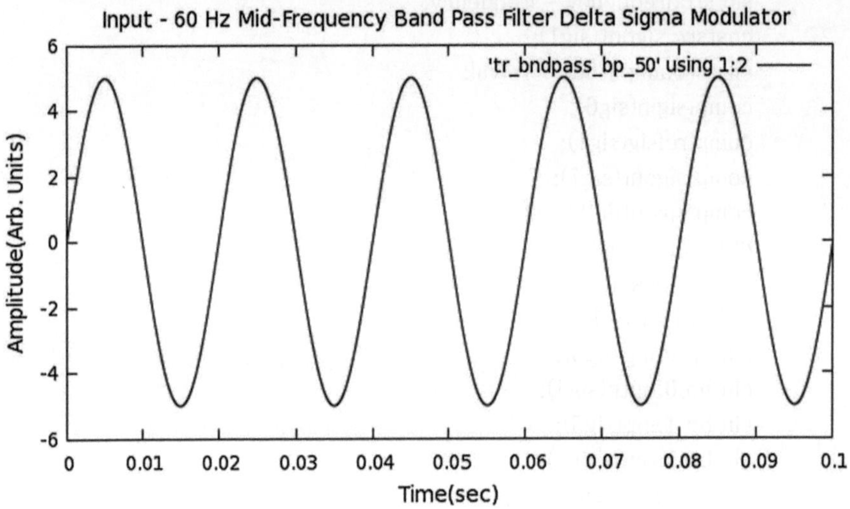


**Fig. 16.8** 60 Hz sine wave input to 60 Hz mid-frequency band-pass filter sigma-delta modulator

When the analog input signal frequency is set at 60 Hz (mid-frequency of band-pass filter, the output waveform consists of 1s and 0s, equal number of each, as expected. When the analog input signal frequency is increased to 1 kHz, then at first the band-pass filter's filtering action is not apparent, but soon after, the output corresponds to an input of 60 Hz, as expected. This band-pass-filter-based delta-sigma modulator contains two sum/subtraction modules, two band-pass filters, a comparator, a D flip-flop, and a 1-bit digital-to-analog converter. While the sum/subtraction modules are wrapper classes around primitive LSF modules, the band-pass filter is a pure ELN module, the 1-bit digital-to-analog converter a TDF module, and the D flip-flop a pure digital module.

Figures 16.8, 16.9, 16.10, and 16.11 show sample input to and output from the 60 Hz mid-band band-pass filter delta-sigma modulator.

The input signal frequency is now increased, and Figs. 16.10 and 16.11 show the input and modulator output, respectively.

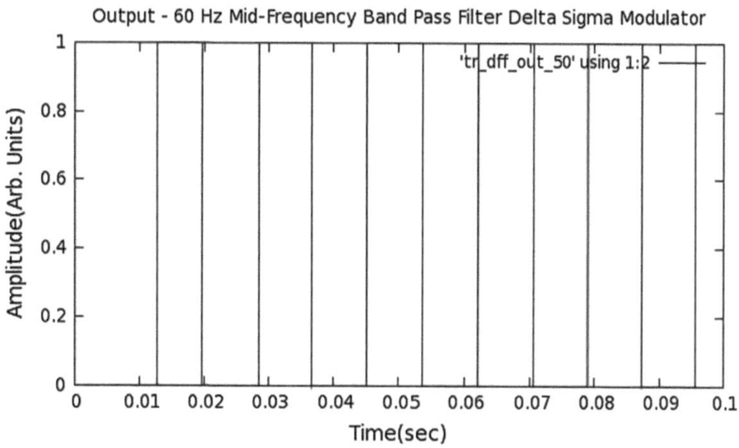


**Fig. 16.9** Output from 60 Hz mid-frequency band-pass filter sigma-delta modulator, for input of Fig. 16.8

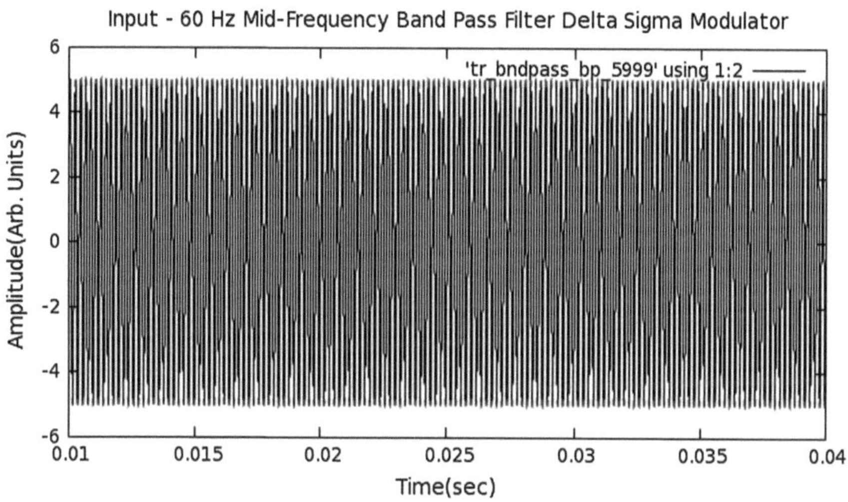


**Fig. 16.10** High-frequency sine wave input to 60 Hz mid-frequency band-pass filter sigma-delta modulator

## 16.2 Position-Sensitive Detector and CD-ROM Reader

A CD-ROM reader/writer or its more recent variation, the DVD reader/writer, is a common component of any laptop/personal computer. In the reader, the CD-ROM or DVD disc rotates at a very high angular velocity, while an electronic optical detection circuit detects/reads out the 1s and 0s imprinted on the CD-ROM/DVD disc's surface.

The electronic optical circuit consists of a semiconductor laser and a light-sensitive semiconductor device (PIN diode structure) whose electrical output signal varies

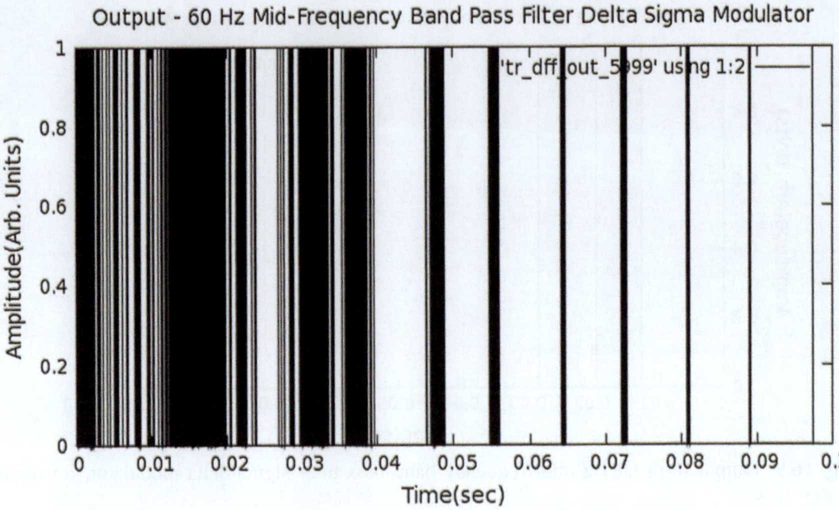


**Fig. 16.11** Output from 60 Hz mid-frequency band-pass filter sigma-delta modulator, for input of Fig. 16.10

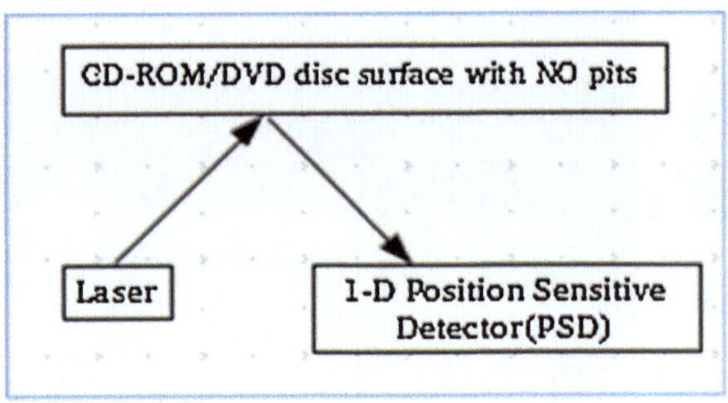


**Fig. 16.12** Laser light reflected off a CD-ROM/DVD surface with no pits—light lands at position A on PSD

with where on its surface the reflected laser light (from CD-ROM/DVD disc surface) falls. This semiconductor device is called a position-sensitive detector [3] (PSD) and is available in one-dimensional/two-dimensional forms. A one-dimensional PSD has two electrodes at its extremities, and its *working principle is that the current output at each electrode is directly proportional to the distance of the spot where the reflected laser light spot hits the light-sensitive surface, from that electrode.* The surface of any CD-ROM/DVD is pitted (corresponding to a stored 1 or 0, or vice versa). As the light from the laser hits a pit (on CD-ROM/DVD surface), the reflected light falls at a different spot on the light-sensitive surface of the PSD than the case when there is no pit (Figs. 16.12, 16.13). If a PSD is placed such that the light (reflected from CD-ROM/DVD surface) falls on a different spot when the light gets reflected from inside a pit as compared to when no pit is present, then the PSD output can be processed to detect 1s and 0s stored on the CD-ROM/DVD surface.

A one-dimensional PSD works as follows—Fig. 16.14.

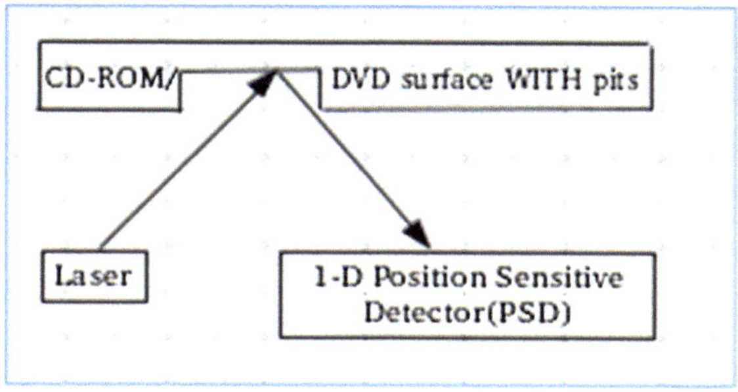


**Fig. 16.13** Laser light reflected off a CD-ROM/DVD surface with pits—light lands at position B on PSD (different from position A of Fig. 16.12)

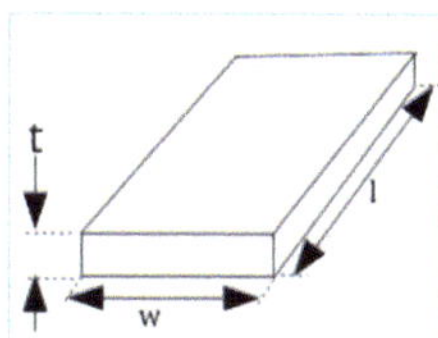


**Fig. 16.14** A one-dimensional position-sensitive detector (PSD). Current flows parallel to the direction of *L*. The semiconducting material has some resistivity rho

Let the resistivity of the semiconducting material of the PSD be rho, where $L$ is the length, $W$ is the width, and $t$ is the thickness of the rectangular parallelepiped. For purposes of this analysis, current always flows parallel to $L$. The resistance of any resistor is: $R = \frac{\text{rho}L}{\text{Area}}$, where Area $=$ tW. Some manipulation gives $R = \frac{\text{rho}L}{\text{tW}}$ and $R = \left(\frac{\text{rho}}{\text{Area}}\right)L$. The quantity $\frac{\text{rho}}{\text{Area}}$ is called *sheet resistance*, with units of Ohms/Square. The corresponding resistance of a small length $x$ of the semiconducting material, parallel to $L$, is $R(x) = \text{sheet} - \text{resistance}\, x$. When reflected light (from the CD-ROM/DVD surface) falls on the PSD surface at some position $p$, where $0 \leq p$ and $p < L$, the PSD is a variable voltage divider, as the distance to the electrodes at the extremities differ, and the resistance. Current flowing through the two extremity electrodes varies as the spot of light moves on the light-sensitive surface. With appropriate encoding, this current variation (equivalently voltage drop) can be used to denote 1s and 0s.

The PSD output is fed into an instrumentation amplifier, and then, its output is digitized for processing by digital signal processing algorithms (DSP). So, a simple CD-ROM/DVD reader would consist of the position-sensitive detector (PSD), followed by instrumentation amplifier and then the analog-to-digital converter (ADC) and then DSP components. We focus on the PSD (TDF module) and the instrumentation amplifier (ELN module). Figures 16.15, 16.16, and 16.17 contain the PSD, instrumentation amplifier, and test harness source code.

```
#ifndef CDREADER_H
#define CDREADER_H

#include <systemc-ams>

SCA_TDF_MODULE(psd)
{
  /* Declare/define input/output ports and module members */
  sca_tdf::sca_in<double> inlightpos;
  sca_tdf::sca_out<double> voltage1;
  sca_tdf::sca_out<double> voltage2;
  double resistivity;
  double L;
  double W;
  double thickness;
  double sheet_resistance;
  double const_current;
  double voltage_val_1;
  double voltage_val_2;
  double pos;

  /* Initialize member */
  void initialize()
  {
    sheet_resistance = resistivity/(W*thickness);
  }

  /* Set input/output port parameters */
  void set_attributes()
  {
    voltage1.set_timestep(1.0, sc_core::SC_US);
    voltage2.set_timestep(1.0, sc_core::SC_US);
    inlightpos.set_rate(1);
    voltage1.set_rate(1);
    voltage2.set_rate(1);
  }
```

**Fig. 16.15** CD-ROM/DVD reader PSD

```
/* Position sensitive detector signal processing */
void processing()
{
  pos = inlightpos.read();
  voltage_val_1 = sheet_resistance*pos*const_current;
  voltage_val_2 = sheet_resistance*(L - pos)*const_current;
  voltage1.write(voltage_val_1);
  voltage2.write(voltage_val_2);
}

/* Constructor - member initialization */
SCA_CTOR(psd):pos(0.0),
              voltage_val_1(0.0),
              voltage_val_2(0.0),
              const_current(0.05)
{

 /* Destructor */
~psd(){ }
};
```

**Fig. 16.15** (continued)

```
SC_MODULE(instamplifier)
{
  /* TDF input/output ports to allow data to be
     read in/out from/to TDF modules */
 sca_tdf::sca_in<double> inp0;
 sca_tdf::sca_in<double> inp1;
 sca_tdf::sca_out<double> outp;
  /* Converter TDF-ELN/ELN-TDF voltage source/sink */
 sca_eln::sca_tdf::sca_vsource v_in0;
 sca_eln::sca_tdf::sca_vsource v_in1;
 sca_eln::sca_tdf::sca_vsink   v_out;

 /* ELN elements - ideal operational amplifier s,
    resistors */
 sca_eln::sca_nullor nullr0;
 sca_eln::sca_nullor nullr1;
```

**Fig. 16.16** CD-ROM/DVD reader instrumentation amplifier

```
sca_eln::sca_nullor nullr2;
sca_eln::sca_r r0;
sca_eln::sca_r r1;
sca_eln::sca_r r2;
sca_eln::sca_r r3;

/* Electrical netlist modes */
sca_eln::sca_node n1;
sca_eln::sca_node n2;
sca_eln::sca_node n3;
sca_eln::sca_node n4;
sca_eln::sca_node n5;
sca_eln::sca_node n6;
sca_eln::sca_node n7;
sca_eln::sca_node_ref gnd;

/* Constructor - initialize members */
SC_CTOR(instamplifier):v_in0("v_in0", 1.0),
              v_in1("v_in1", 1.0),
              v_out("v_out", 1.0),
              nullr0("nullr0"),
              nullr1("nullr1"),
              nullr2("nullr2"),
              r0("r0", 500.0),
              r1("r1", 500.0),
              r2("r2", 75000000.0),
              r3("r3", 500.0)
{
  /* Connect ports, ELN elements and netlist nodes */
  v_in0.inp(inp0);
  v_in0.p(n1);
  v_in0.n(gnd);

  v_in1.inp(inp1);
  v_in1.p(n2);
  v_in1.n(gnd);

  v_out.outp(outp);
  v_out.p(n7);
  v_out.n(gnd);
```

**Fig. 16.16** (continued)

```
    nullr0.nip(n1);
    nullr0.nin(n3);
    nullr0.nop(n3);
    nullr0.non(gnd);

    nullr1.nip(n2);
    nullr1.nin(n4);
    nullr1.nop(n4);
    nullr1.non(gnd);
    nullr2.nip(n5);
    nullr2.nin(n6);
    nullr2.nop(n7);
    nullr2.non(gnd);

    r0.p(n3);
    r0.n(n5);

    r1.p(n4);
    r1.n(n6);

    r2.p(n5);
    r2.n(n7);

    r3.p(n6);
    r3.n(gnd);
  }

  ~instamplifier(){ }
 /*Destructor */
};
#endif
```

**Fig. 16.16** (continued)

```
#include "commonsrcs.h"
#include "cdreader.h"
#include "gendatatrace.h"

#include <cstdlib>
#include <cstring>

int sc_main(int argc, char **argv)
{
  /* Check input parameters */
  if(argc < 5)
  {
    std::cout<<"insufficient parameters ..."<<std::endl;
    std::cout<<"usage ./sim <resistivity> <length> <width> <thickness> in
metric units"<<std::endl;
    exit(0);
  }

   /* Declare/define inter-module signal channels */
  sca_tdf::sca_signal<double> rawdata;
  sca_tdf::sca_signal<double> psdout0;
  sca_tdf::sca_signal<double> psdout1;
  sca_tdf::sca_signal<double> ampout;

  /* Declare/define variables to hold input parameters  */
  double resistivity;
  double length;
  double width;
  double thickness;

  /* Read in and set input parameters */
  resistivity = strtod(argv[1], NULL);
  length = strtod(argv[2], NULL);
  width = strtod(argv[3], NULL);
  thickness = strtod(argv[4], NULL);
```

**Fig. 16.17** CD-ROM/DVD reader test harness

```
/* Declare/define modules, assign module parameters
   trace file(s)/object(s) */
randata ran_data("ran_data");
psd ps_d("ps_d");
ps_d.resistivity = resistivity;
ps_d.L = length;
ps_d.W = width;
ps_d.thickness = thickness;
instamplifier inst_amp("inst_amp");
tracetriplecombo trpll("tr_cd_reader");

/* Connect module ports and signal channels */
ran_data.randatasigout(rawdata);
ps_d.inlightpos(rawdata);
ps_d.voltage1(psdout0);
ps_d.voltage2(psdout1);
inst_amp.inp0(psdout0);
inst_amp.inp1(psdout1);
inst_amp.outp(ampout);
trpll.in1(psdout0);
trpll.in2(psdout1);
trpll.in3(ampout);

/* Run simulation for pre-defined time period and then stop */
sc_core::sc_start(500.0, sc_core::SC_US);
sc_core::sc_stop();
return 0;
}
```

**Fig. 16.17** (continued)

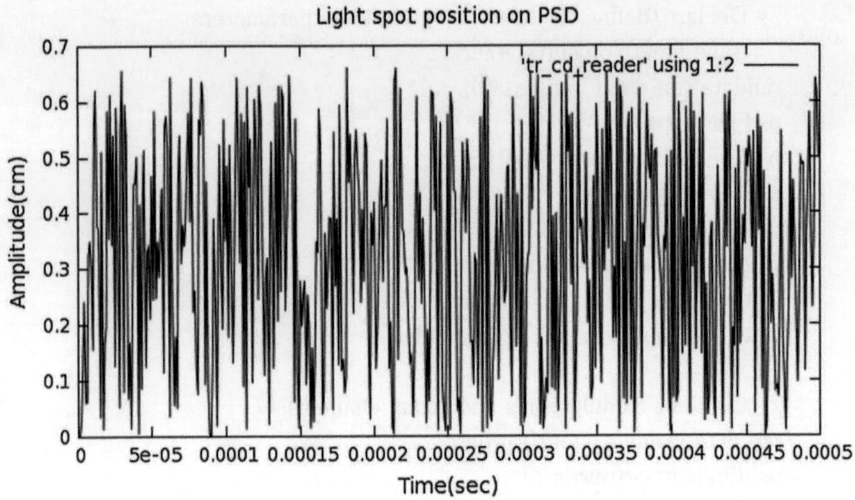


**Fig. 16.18** Position of reflected light (from CD-ROM/DVD disc surface) on light-sensitive surface of PSD

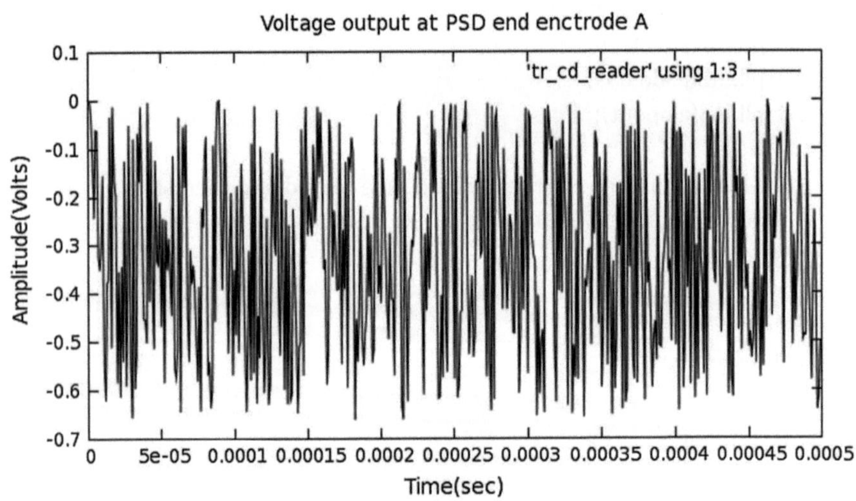


**Fig. 16.19** Voltage output at electrode A of PSD, with light spot as shown in Fig. 16.18

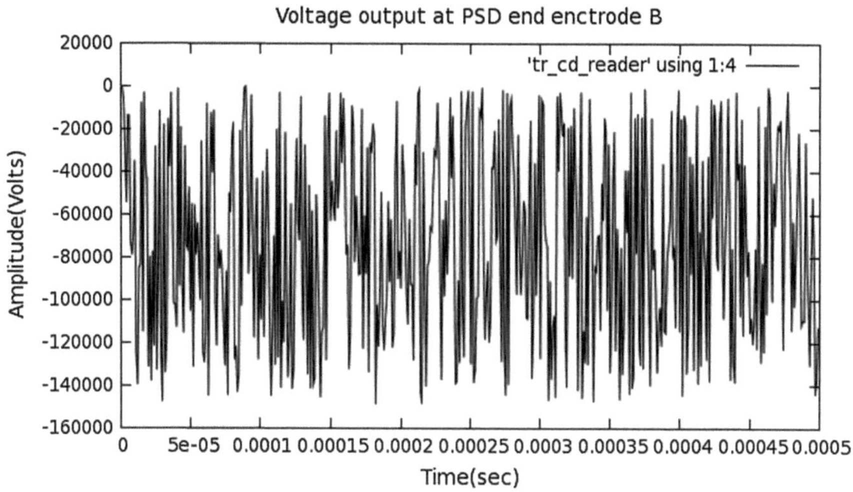


**Fig. 16.20** Voltage output at electrode B of PSD, with light spot as shown in Fig. 16.18

Figures 16.18, 16.19, and 16.20 show, respectively, the position of the laser light spot on the light-sensitive PSD surface and the voltage outputs at the two PSD electrodes. The differential amplifier stage of the instrumentation amplifier is an inverting amplifier.

# References


1. SystemC-AMS Working Group. SystemC-AMS (Analog Mixed Signal) — Accellera Systems Initiative. http://www.accellera.org/downloads/standards/systemc/ams and http://www.acceller a.org/activities/committees/systemc-ams/
2. Schreier, R., Snelgrove, M. (1989) Band Pass Sigma Delta Modulation *Electronics Letters 25 (23), 1560–1561*
3. Chang, T-Y, Bibyk, S. B.(1999) Exact Analysis of Second Order Band-Pass Delta-Sigma Modulator with Sinusoidal Inputs *Proceedings of the IEEE International Symposium on Circuits and Systems, ISCAS '99. 2, 372–375*
4. Makynen A. Position-Sensitive Devices and Sensor Systems for Optical Tracking and Displacement Sensing Application. Dissertation, Faculty of Technology, University of Oulu, 2000 [WWW] URL http://herkules.oulu.fi/isbn9514257804/
5. Andersson H. Position Sensitive Detectors : Device Technology and Applications in Spectroscopy. Dissertation, Department of Information Technology and Media, Mid Sweden University, Sundsvall, Sweden, 2008, ISBN 9789185317912

# Chapter 17
# SystemC-AMS and SystemC Combinations

**Abstract** All examples considered so far consist of interactions between modules implementing one of the SystemC-AMS formalisms (mostly TDF-LSF, TDF-ELN). As SystemC-AMS was designed for mixed signal system simulation, the designer has to analyze a system that involves interactions between pure SystemC modules and pure SystemC-AMS [1] modules. A number of examples are examined here, starting with some simple cases (e.g., SystemC–SystemC-AMS LSF, ELN) and moving to complex real-world systems. Two important details must be remembered to enable these mixed mode simulations to run correctly.

- Correct matching of input/output port types, especially across pure SystemC and pure System-AMS boundary
- The timing (specifically time stepping) of pure SystemC-AMS modules must be explicitly specified. Pure SystemC sequential logic modules always synchronize with a clock. With pure SystemC-AMS modules, especially those that use a SystemC clock input, the time stepping has to be explicitly specified and carefully adjusted to eliminate any mismatch with the SystemC clock input.

## 17.1 SystemC Discrete-Event Clock-Driven SystemC-AMS Demultiplexer

A simple system consisting of a pure SystemC module interacting with a SystemC-AMS LSF demultiplexer is analyzed. The sequential logic pure SystemC module generates random double values synchronized with a master clock. These double values are input to a pure SystemC-AMS LSF demultiplexer, which separates this input into two parallel output streams. The demultiplexer uses the master SystemC clock as its select input, and the two demultiplexed output streams are sent to pure SystemC modules. Figures 17.1, 17.2, and 17.3 show the source code and test harness, while Figs. 17.4, 17.5, and 17.6 show the input and output traces.

A. Banerjee and B. Sur, *SystemC and SystemC-AMS in Practice*,
DOI: 10.1007/978-3-319-01147-9_17,

```
SC_MODULE(desrc)
{
  /*Declare/define SystemC input/output ports */
  sc_core::sc_in<bool> clk;
  sc_core::sc_out<double> outp;
  double d;

  /* Module operation thread */
  void desrc_proc0 ()
  {
   while(1)
   {
     wait();
     d = 10.0*drand48();
     outp.write(d);
   }
  }

  /* Constructor */
 SC_CTOR(desrc):d(0)
 {
   /*Declare/assign thread and its
     sensitivity list */
   SC_THREAD(desrc_proc0);
   sensitive << clk.pos();
 }
 ~desrc(){ } /* Destructor */
};
```

**Fig. 17.1** Pure SystemC double-valued random number generator

```
SC_MODULE(delsfde)
{
  /* Internal LSF signal channels */
  private:
   sca_lsf::sca_signal sig0;
   sca_lsf::sca_signal sig1;
   sca_lsf::sca_signal sig2;

  public:
   /* SystemC input/output ports */
   sc_core::sc_in<bool> clk;
```

**Fig. 17.2** SystemC-AMS LSF demultiplexer wrapper module, with converter ports and SystemC clock input as select

```
sc_core::sc_in<double> inp;
sc_core::sc_out<double> out0;
sc_core::sc_out<double> out1;
/* SystemC discrete-event--SystemC-AMS LSF
converter ports */
sca_lsf::sca_de::sca_source de2lsf;
sca_lsf::sca_de::sca_sink lsf2de0;
sca_lsf::sca_de::sca_sink lsf2de1;
 /* SystemC-AMS LSF demultiplexer with SystemC
discrete-event control signal */
sca_lsf::sca_de::sca_demux lsf2dedemux;

/* Constructor--initialize internal
    modules/channels */
SC_CTOR(delsfde):clk("clk"),
              de2lsf("de2lsf",1.01),
              inp("inp"),
              lsf2de0("lsf2de0", 1.02),
              lsf2de1("lsf2de1", 1.03),
              lsf2dedemux("lsf2dedemux"),
              out0("out0"),
              out1("out1"),
              sig0("sig0"),
              sig1("sig1"),
              sig2("sig2")
 {
   /* Connect all module ports, signal channels */
   de2lsf.inp(inp);
  de2lsf.y(sig0);
   /* Set time step for LSF module */
   de2lsf.set_timestep(1.0, sc_core::SC_NS);
    lsf2dedemux.x(sig0);
    lsf2dedemux.ctrl(clk);
    lsf2dedemux.y1(sig1);
    lsf2dedemux.y2(sig2);
   /* Set time step for LSF module */
   lsf2dedemux.set_timestep(1.0, sc_core::SC_NS);
   lsf2de0.x(sig1);
   lsf2de0.outp(out0);
```

**Fig. 17.2** (continued)

```
    /* Set time step for LSF module */
    lsf2de0.set_timestep(1.0, sc_core::SC_NS);
    lsf2de1.x(sig2);
    lsf2de1.outp(out1);

    /* Set time step for LSF module */
    lsf2de1.set_timestep(1.0, sc_core::SC_NS);
  }

  ~delsfde(){ } /* Destructor */
};
#endif
```

**Fig. 17.2** (continued)

```
#include "delsfde.h"

#include "sysctrace.h"
int sc_main(int argc, char **argv)
{
  /* Pure SystemC signal channels */
  sc_core::sc_signal<double> sig0;
  sc_core::sc_signal<double> sig1;
  sc_core::sc_signal<double> sig2;

  /* Declare/define clock for SystemC clocks */
  sc_core::sc_clock clk("clk", 2.0, sc_core::SC_NS, 0.5);

  /* Declare/define other modules */
  desrc de_src("de_src");
  delsfde de_lsfde("de_lsfde");

  /* Declare/define trace file/object */
  ssctracetrdbl trace_tr_dbl("tr_delsfde");

  /* Connect module ports and signal channels */
  de_src.clk(clk);
  de_src.outp(sig0);
  de_lsfde.clk(clk);
  de_lsfde.inp(sig0);
  de_lsfde.clk(clk);
```

**Fig. 17.3** Test harness SystemC input/output SystemC-AMS LSF demultiplexer with SystemC clock as select

```
    de_lsfde.inp(sig0);
    de_lsfde.out0(sig1);
    de_lsfde.out1(sig2);

    trace_tr_dbl.clk(clk);
    trace_tr_dbl.din0(sig0);
    trace_tr_dbl.din1(sig1);
    trace_tr_dbl.din2(sig2);

    /* run simulation for pre-defined time period and ther
    sc_core::sc_start(500.0, sc_core::SC_NS);
    sc_core::sc_stop();
    return 0;
}
```

**Fig. 17.3** (continued)

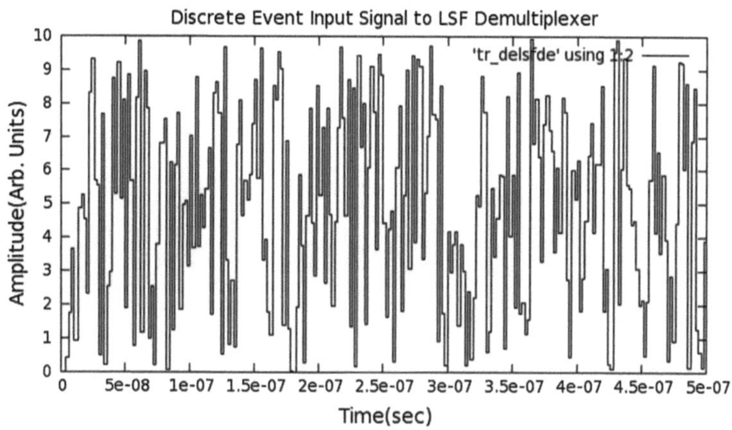


**Fig. 17.4** SystemC module generated random double value input for SystemC-AMS LSF demultiplexer

Note that each SystemC-AMS LSF library module must have its time-related information set explicit with the 'set_timestep (….,….)' method. Although one can set it for one module, and allow that information to propagate through the cluster, using 'set_timestep' with each LSF module does not hurt. This issue does not arise when, e.g., using LSF modules coupled with TDF modules, because in that case, the time information set explicitly in each TDF module propagates through the cluster, which includes the LSF modules. This example uses the built-in System-AMS LSF demultiplexer sca_lsf::sca_de::sca_demux.

All output is sent to a pure SystemC module.

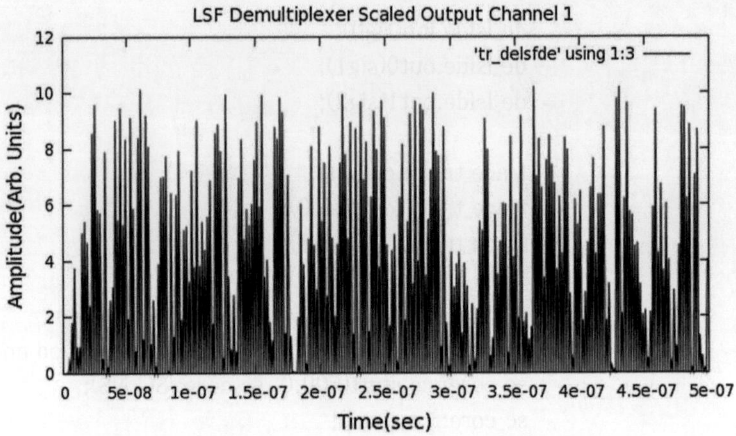


**Fig. 17.5** Scaled channel 1 output from SystemC-AMS LSF demultiplexer

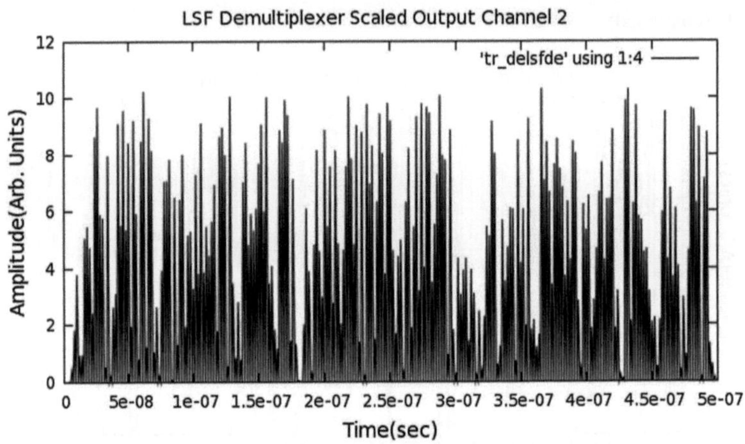


**Fig. 17.6** Scaled channel 2 output from SystemC-AMS LSF demultiplexer

Note that sometimes in these mixed mode simulations, the SystemC-AMS runtime system generates 'Info' messages such as:

Info: SystemC-AMS:
LSF solver instance: sca_linear_solver_0 (cluster 0) has calculated 500 times steps the equation system was 2 times re-initialized.
The following max.10 modules requested the most re-initializations:
de_lsfde.lsf2dedemux 2
de_lsfde.de2lsf 0
de_lsfde.de2lsf 0
de_lsfde.de2lsf 0

de_lsfde.de2lsf 0
de_lsfde.de2lsf 0
de_lsfde.de2lsf 0
de_lsfde.de2lsf 0
de_lsfde.de2lsf 0
de_lsfde.de2lsf 0.

This message indicates that one/more default argument value(s) in the constructor of one/more of the LSF module instances was not overridden. This slows the simulation, but does not affect the simulation process or the quality of the results.

Similar mixed mode designs may be constructed via a judicious combination of one or more pure SystemC module(s) and a SystemC-AMS TDF or a ELN modules, or a general combination of pure SystemC modules and each of SystemC-AMS ELN, LSF, and TDF modules. The programmer's skill is tested when the synchronization between these modules have to be correctly set up.

## Reference

1. SystemC-AMS Working Group. SystemC-AMS (Analog Mixed Signal)—Accellera Systems Initiative. http://www.accellera.org/downloads/standards/systemc/ams and http://www.accellera.org/activities/committees/systemc-ams/

# Index




A. Banerjee and B. Sur, *SystemC and SystemC-AMS in Practice*,
DOI: 10.1007/978-3-319-01147-9,

MIX
Papier aus verantwortungsvollen Quellen
Paper from responsible sources
FSC® C105338

If you have any concerns about our products,
you can contact us on
**ProductSafety@springernature.com**

In case Publisher is established outside the EU,
the EU authorized representative is:
**Springer Nature Customer Service Center GmbH**
**Europaplatz 3, 69115 Heidelberg, Germany**

Printed by Libri Plureos GmbH
in Hamburg, Germany